MULTIHAZARD RISK ATLAS OF MALDIVES

Economy and Demographics —Volume III

MARCH 2020

6 ADB Avenue, Mandaluyong City, 1550 Metro Manila, Philippines
Tel +63 2 8632 4444; Fax +63 2 8636 2444
www.adb.org

ISBN 978-92-9262-048-6 (print); 978-92-9262-049-3 (electronic); 978-92-9262-050-9 (ebook)
Publication Stock No. TCS200051
DOI: http://dx.doi.org/10.22617/TCS200051

Notes:
In this publication, "$" refers to United States dollars.
The maps presented in this atlas reflect airports based on 2017 data from the Civil Aviation Authority of Maldives.

On the cover: An aerial view shows 1 of 26 natural atolls that make up Maldives, which also includes nearly 1,200 small coral islands and some of the world's most beautiful beaches. Recognized as the seventh-largest in the world, the coral reefs and associated ecosystems of Maldives are key foundations for food security and means of livelihood. Yet, they are considered as among the most vulnerable to climate change (photo by Roberta Gerpacio).

Contents

Tables and Maps

Tables

Maps

Foreword

Maldives is among the countries most vulnerable to the impacts of climate change as it is a small island nation with extremely low elevations. Maldives is also very vulnerable to impacts of rising air and sea surface temperatures and changes in rainfall patterns. Climate change impacts will therefore impose significant negative consequences on the Maldivian economy and society. Some of the priority vulnerabilities to climate change are land loss and beach erosion, infrastructure damage, degradation of coral reefs, and adverse impacts on water resources, food security, human health, and the overall economy.

Sustainable coastal resources management is of particular importance to Maldives, such that all regulations involving various development activities have coastal components. Despite the government's continued efforts in improving and sustaining coastal resources management, critical issues remain, such as the need for systematized coastal monitoring, clear definition of coastal boundaries and coastal development, enhanced regulatory and monitoring capacities for coastal resources protection, and sustainable long-term strategies on land reclamation and marine area protection. At a time when climate is rapidly changing and extreme weather events are frequently occurring, the critical roles that marine and coastal environments play in mitigating and adapting to climate change need to be sufficiently documented and properly recognized. It is therefore essential for Maldives to develop and establish a comprehensive digital database of marine and coastal ecosystem features and services that can be regularly monitored.

The *Multihazard Risk Atlas of Maldives* was developed through the project "Establishing a National Geospatial Database for Mainstreaming Climate Change Adaptation into Development Activities and Policies in Maldives" under the Asian Development Bank's regional knowledge and support (capacity development) technical assistance Action on Climate Change in South Asia (2013–2018). This five-volume atlas aims to promote the sustainable development of coastal and marine ecosystems and their various components, by enhancing the awareness of stakeholders on and enjoining them to address climate and disaster risks (including hazards, exposures, and vulnerabilities) to which ecosystems are exposed. The atlas presents spatial information and maps necessary for assessing future development investments in terms of their risks to climate and geophysical hazards.

The target audience of the *Multihazard Risk Atlas of Maldives* are the concerned stakeholders with current or planned development activities in the country, including public and private sectors, nongovernment organizations, research and academic community, development partner agencies, other financial institutions, and the general public. The atlas will also be a useful reference for other developing countries with similar geographical and environmental conditions, particularly small island developing states. It is envisioned that the atlas will significantly contribute to rendering important sector development investments more resilient to hazard-specific risk scenarios in the short, medium, and long terms.

H.E. Dr. Hussain Rasheed Hassan
Minister
Ministry of Environment, Malé

Shixin Chen
Vice-President for Operations 1
Asian Development Bank, Manila

Acknowledgments

Government Ministries, Departments, and Agencies in Maldives
Civil Aviation Authority
Land and Survey Authority
Marine Research Institute
Meteorological Service
Ministry of Economic Development
Ministry of Education
Ministry of Environment
Ministry of Fisheries, Marine Resources and Agriculture
Ministry of Health
Ministry of National Planning and Infrastructure
Ministry of Tourism
National Bureau of Statistics
National Disaster Management Center

International Institutions
Manila Observatory
Marine Spatial Ecology Lab, University of Queensland, Australia
SANDER + PARTNER
United Nations Development Programme

International Institutions in Maldives
International Union for Conservation of Nature, Maldives
United Nations Development Programme, Maldives

National Consultant Team
Ahmed Jameel, Integrated Coastal Zone Management Specialist
Faruhath Jameel, Geographic Information Systems Specialist and Team Leader
Hussain Naeem, Coastal Ecosystems and Biodiversity Specialist
Mahmood Riyaz, Climate Change Risk Assessment Specialist

Abbreviations

BODC	–	British Oceanographic Data Centre
CAA	–	Maldives Civil Aviation Authority
GEBCO	–	General Bathymetric Chart of the Oceans
ha	–	hectare
IHO	–	International Hydrographic Organization
IOC	–	Intergovernmental Oceanographic Commission
IUCN	–	International Union for Conservation of Nature, Maldives
kW	–	kilowatt
ME	–	Ministry of Environment
MED	–	Ministry of Economic Development
MLSA	–	Maldives Land and Survey Authority
MNPI	–	Ministry of National Planning and Infrastructure
MOE	–	Ministry of Education
MOH	–	Ministry of Health
MOT	–	Ministry of Tourism
NBS	–	National Bureau of Statistics
UNDP	–	United Nations Development Programme
UTM	–	Universal Transverse Mercator
WGS	–	World Geodetic System

Demographics and Economy

1

This volume examines Maldivians by focusing on the spatial characteristics of the demographics and economic activities in the country. It concentrates on the people at risk to climate and geologic hazards. This population runs the country—and its economy and politics—and cultivates a unique blend of culture.

Specifically, this volume looks at the distribution of the population across the islands; the allocation of educational and health facilities, tourist spots, transportation and other infrastructure; and the facilities and locations of economic activities, particularly sand mining. All these elements have varying exposure and vulnerability to climate and disaster risks.

Food for the soul. A woman prepares a traditional Maldivian dish for her family (photo by Ibrahim Asad).

2

Humans of Maldives

Life in Maldives. Livelihood is dependent on oceans and seas in the country composed of some 1,200 islands. Majority of Maldivians are Sunni Muslim although there is a cultural mix with South Indian, Sinhalese, and Arab influences (*clockwise from top left*, photos by Ibrahim Asad, Tripadvisor, Mark Fischer, and ADB).

Population

Population fundamentally measures exposure to hazards. The number of people exposed to certain hazards are based on population density.

Maldives has the lowest population density among countries in Asia and the Pacific. More than 400,000 Maldivians are scattered across the roughly 200 inhabited islands in 21 atolls (National Bureau of Statistics 2014). Roughly 38% of the population resides in Malé City, the largest city in Maldives. The rest are distributed across the atolls.

Table III.1: Atoll Population in Maldives

Atoll	Population 2014	%
Malé City	153,904	38.28
Kaafu Atoll	27,360	6.80
Seenu Atoll	21,957	5.46
Haa Dhaalu Atoll	19,649	4.89
Raa Atoll	16,649	4.14
Haa Alifu Atoll	14,666	3.65
Baa Atoll	13,507	3.36
Laamu Atoll	13,498	3.36
Alifu Dhaalu Atoll	13,256	3.30
Gaafu Dhaalu Atoll	13,104	3.26
Shaviyani Atoll	13,095	3.26
Noonu Atoll	12,831	3.19
Gaafu Alifu Atoll	10,862	2.70
Lhaviyani Atoll	10,659	2.65
Thaa Atoll	9,893	2.46
Alifu Alifu Atoll	9,075	2.26
Gnaviyani Atoll	8,510	2.12
Dhaalu Atoll	7,448	1.85
Meemu Atoll	5,478	1.36
Faafu Atoll	4,627	1.15
Vaavu Atoll	2,043	0.51

Note: Total may not add up due to rounding.

Source: National Bureau of Statistics, 2014.

People of Maldives. People enjoy spending time in the beaches of Maldives, which are famous for their natural beauty, clear, turquoise waters, and fine white sand (photo by Adam Azim).

Map III.1: Maldives, Population

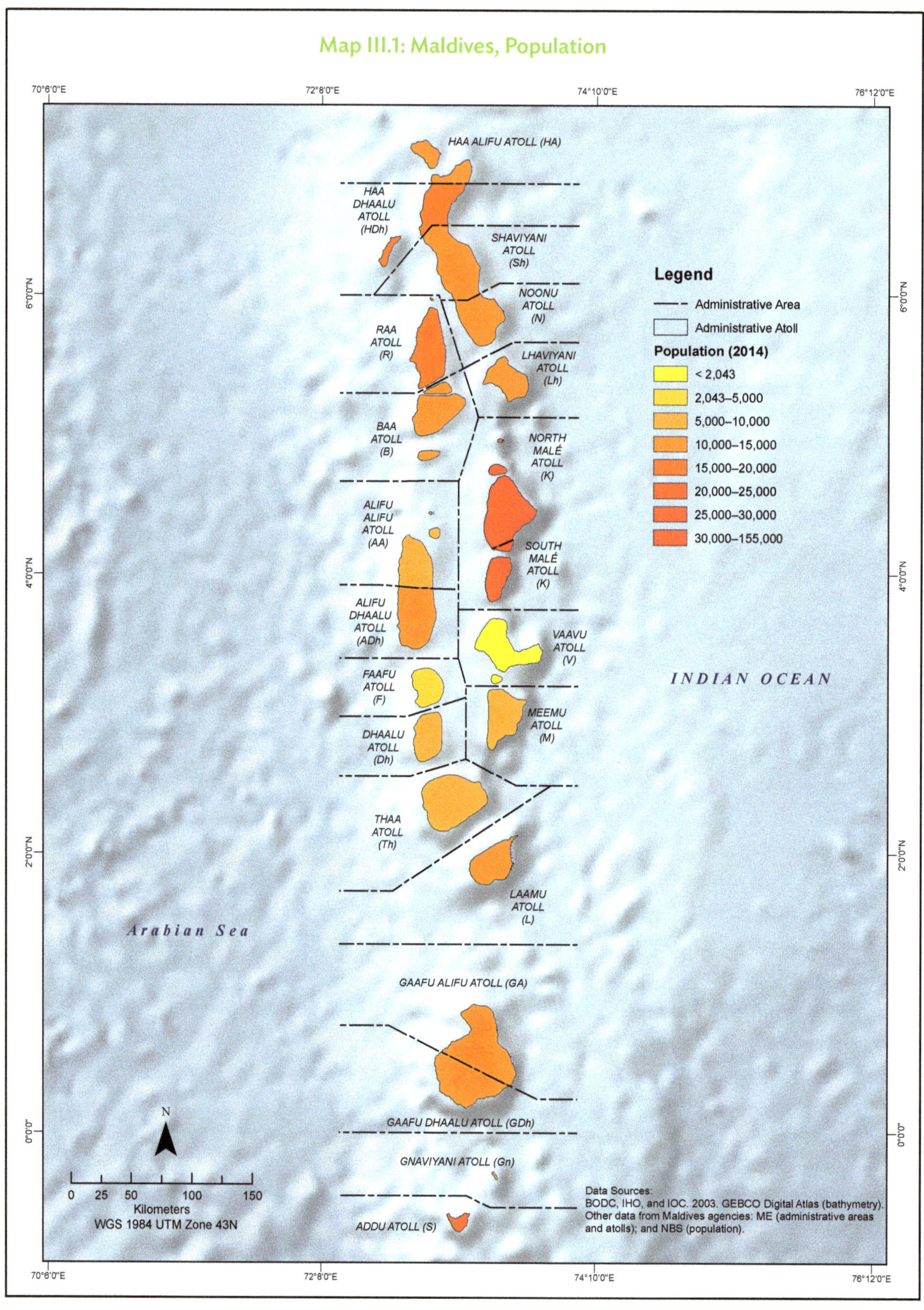

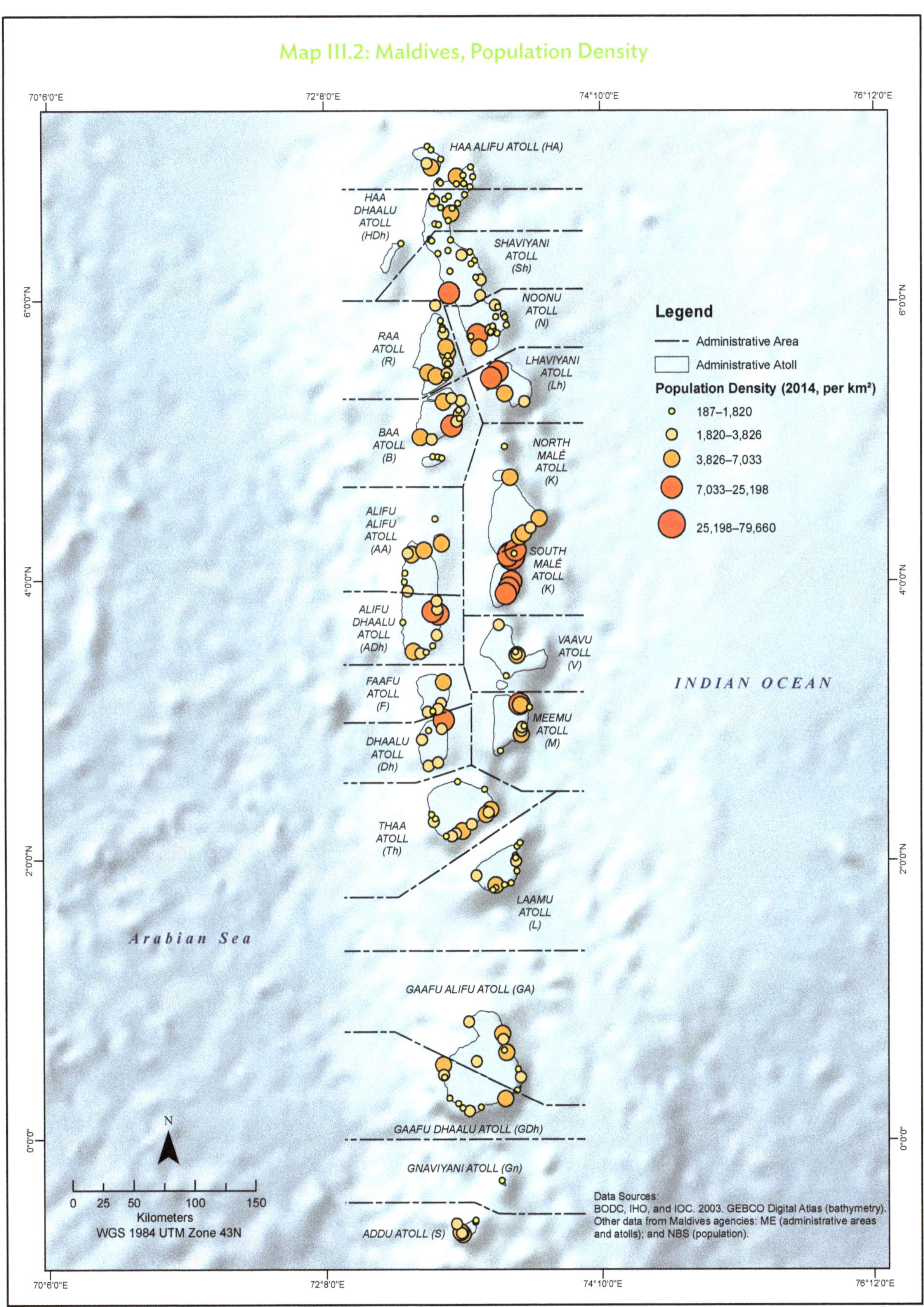
Map III.2: Maldives, Population Density
70°6'0"E
72°8'0"E
74°10'0"E
76°12'0"E
6°0'0"N
4°0'0"N
2°0'0"N
0°0'0"
HAA ALIFU ATOLL (HA)
HAA DHAALU ATOLL (HDh)
SHAVIYANI ATOLL (Sh)
NOONU ATOLL (N)
RAA ATOLL (R)
LHAVIYANI ATOLL (Lh)
BAA ATOLL (B)
NORTH MALÉ ATOLL (K)
ALIFU ALIFU ATOLL (AA)
SOUTH MALÉ ATOLL (K)
ALIFU DHAALU ATOLL (ADh)
VAAVU ATOLL (V)
FAAFU ATOLL (F)
MEEMU ATOLL (M)
DHAALU ATOLL (Dh)
THAA ATOLL (Th)
LAAMU ATOLL (L)
GAAFU ALIFU ATOLL (GA)
GAAFU DHAALU ATOLL (GDh)
GNAVIYANI ATOLL (Gn)
ADDU ATOLL (S)
Arabian Sea
INDIAN OCEAN
Legend
Administrative Area
Administrative Atoll
Population Density (2014, per km²)
187–1,820
1,820–3,826
3,826–7,033
7,033–25,198
25,198–79,660
N
0 25 50 100 150
Kilometers
WGS 1984 UTM Zone 43N
Data Sources:
BODC, IHO, and IOC. 2003. GEBCO Digital Atlas (bathymetry).
Other data from Maldives agencies: ME (administrative areas and atolls); and NBS (population).

Map III.3: Addu City, Population

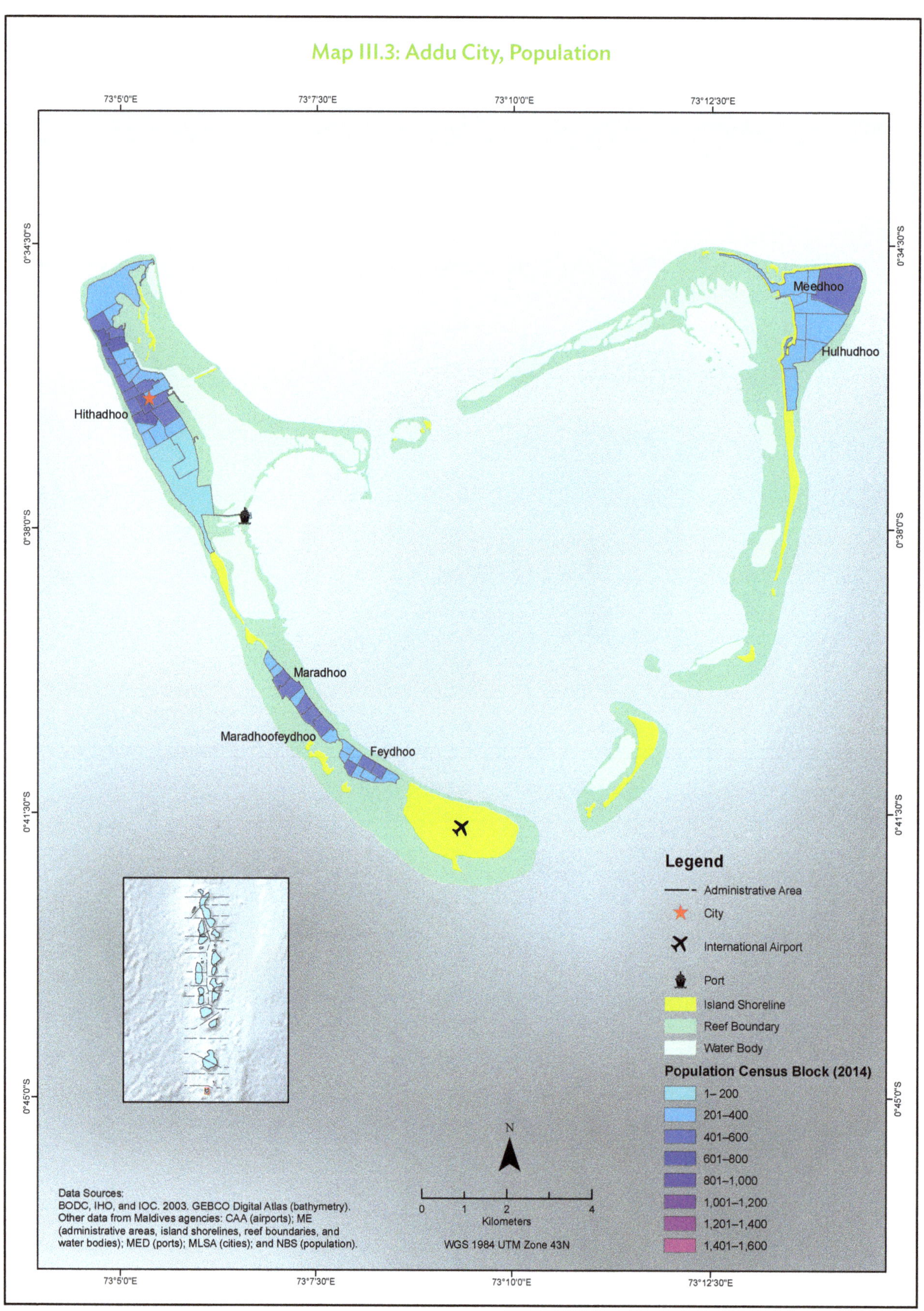

Map III.4: Alifu Alifu Atoll, Population

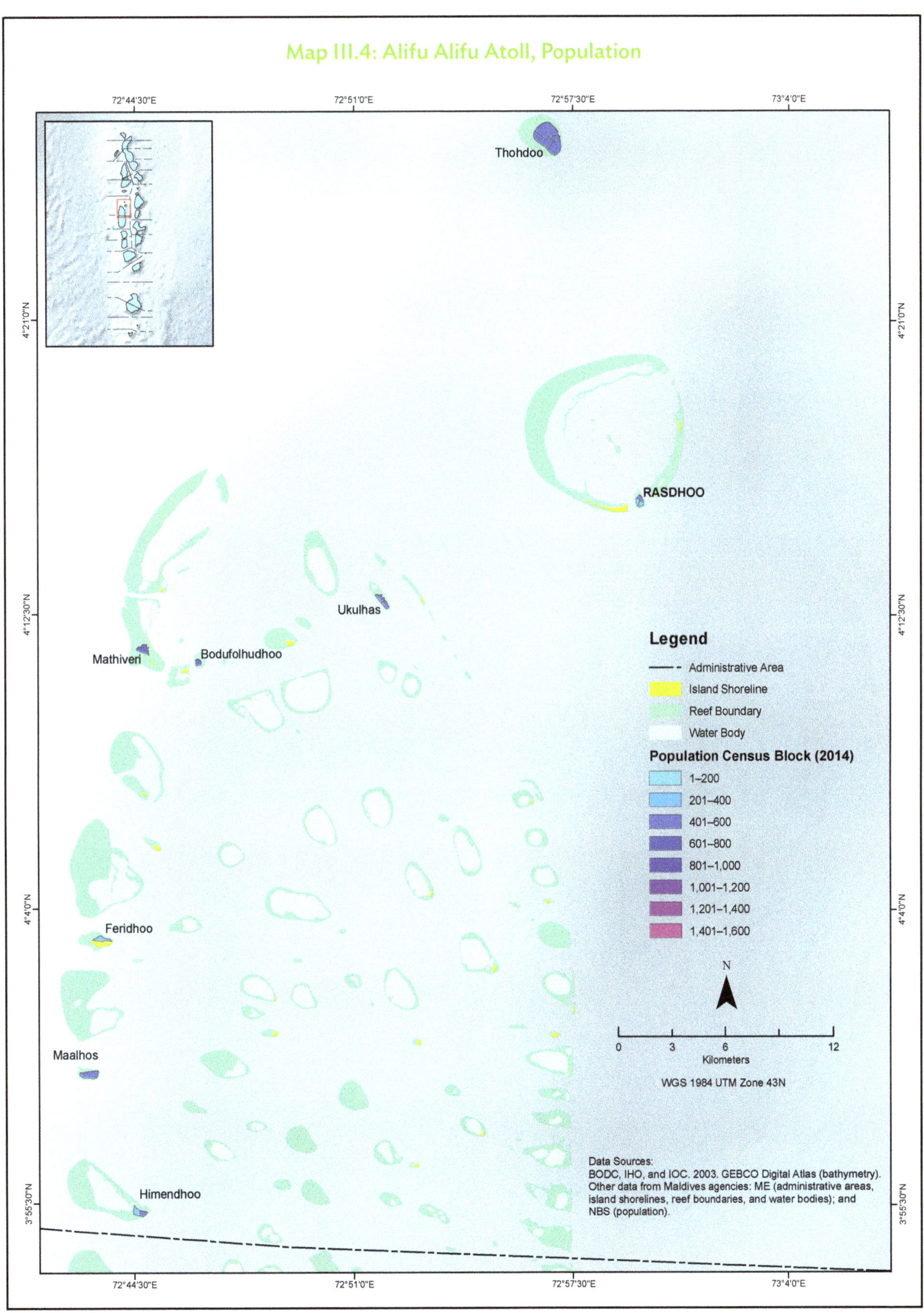

Map III.5: Alifu Dhaalu Atoll, Population

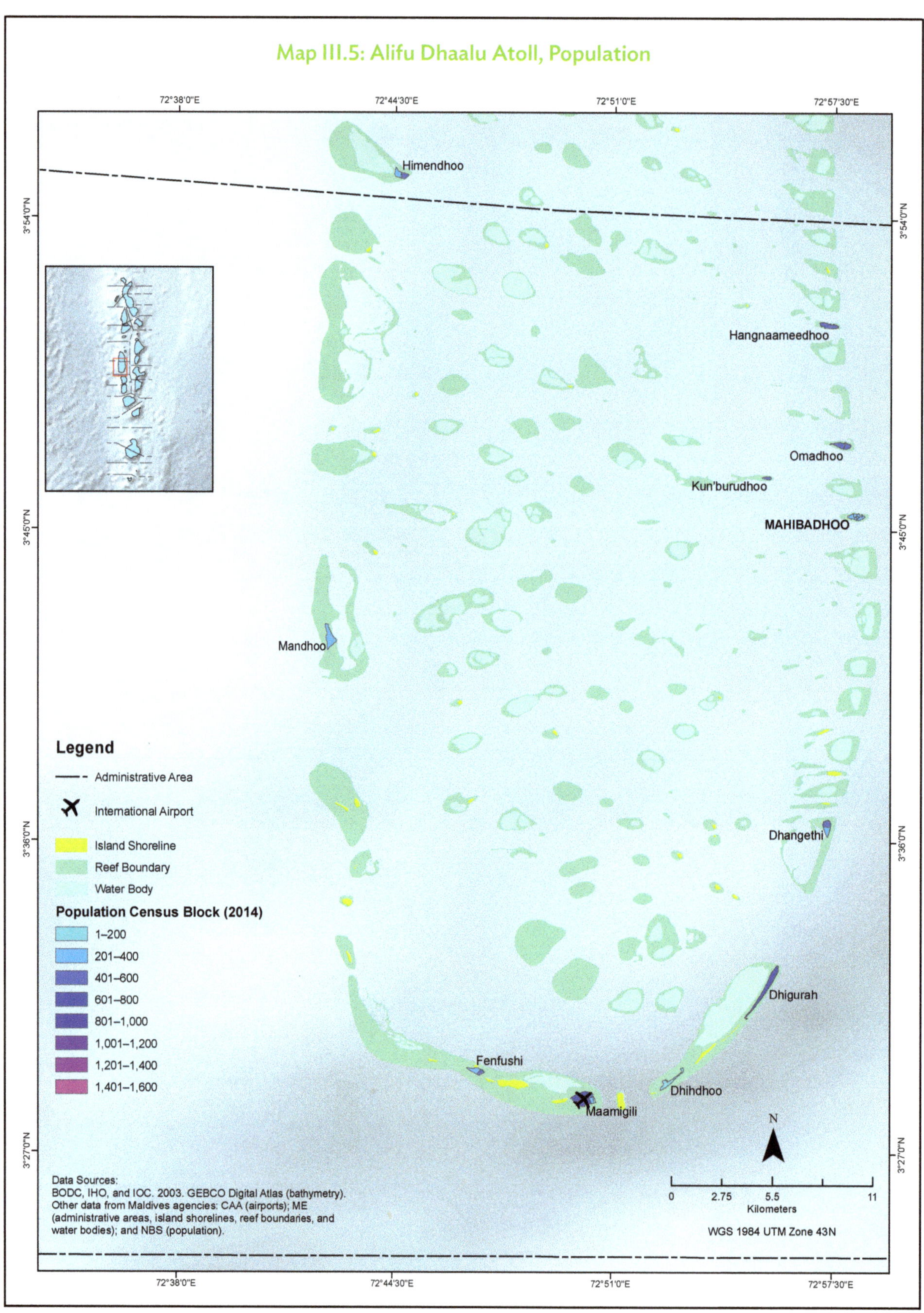

Data Sources:
BODC, IHO, and IOC. 2003. GEBCO Digital Atlas (bathymetry).
Other data from Maldives agencies: CAA (airports); ME (administrative areas, island shorelines, reef boundaries, and water bodies); and NBS (population).

Map III.6: Baa Atoll, Population

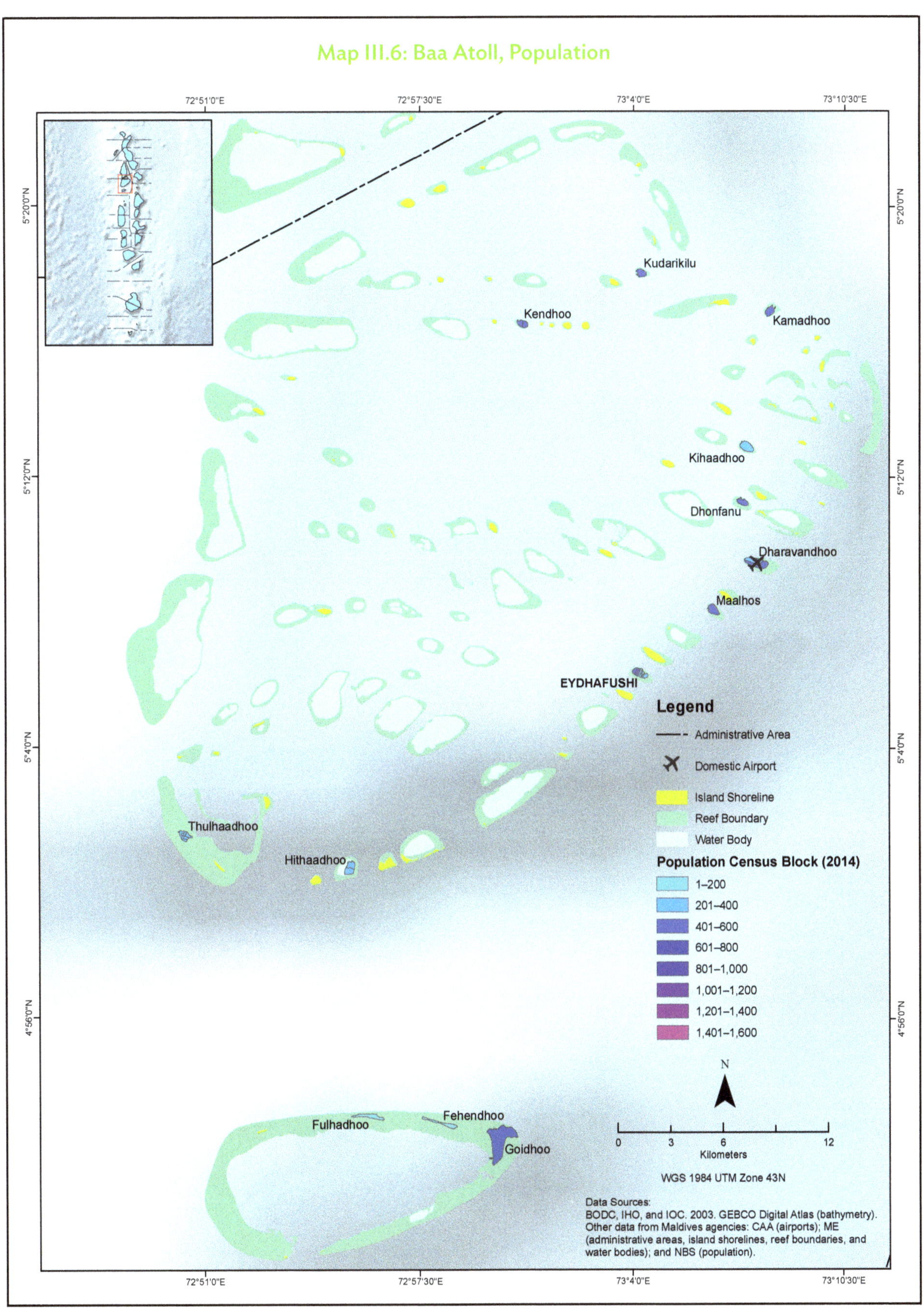

Map III.7: Dhaalu Atoll, Population

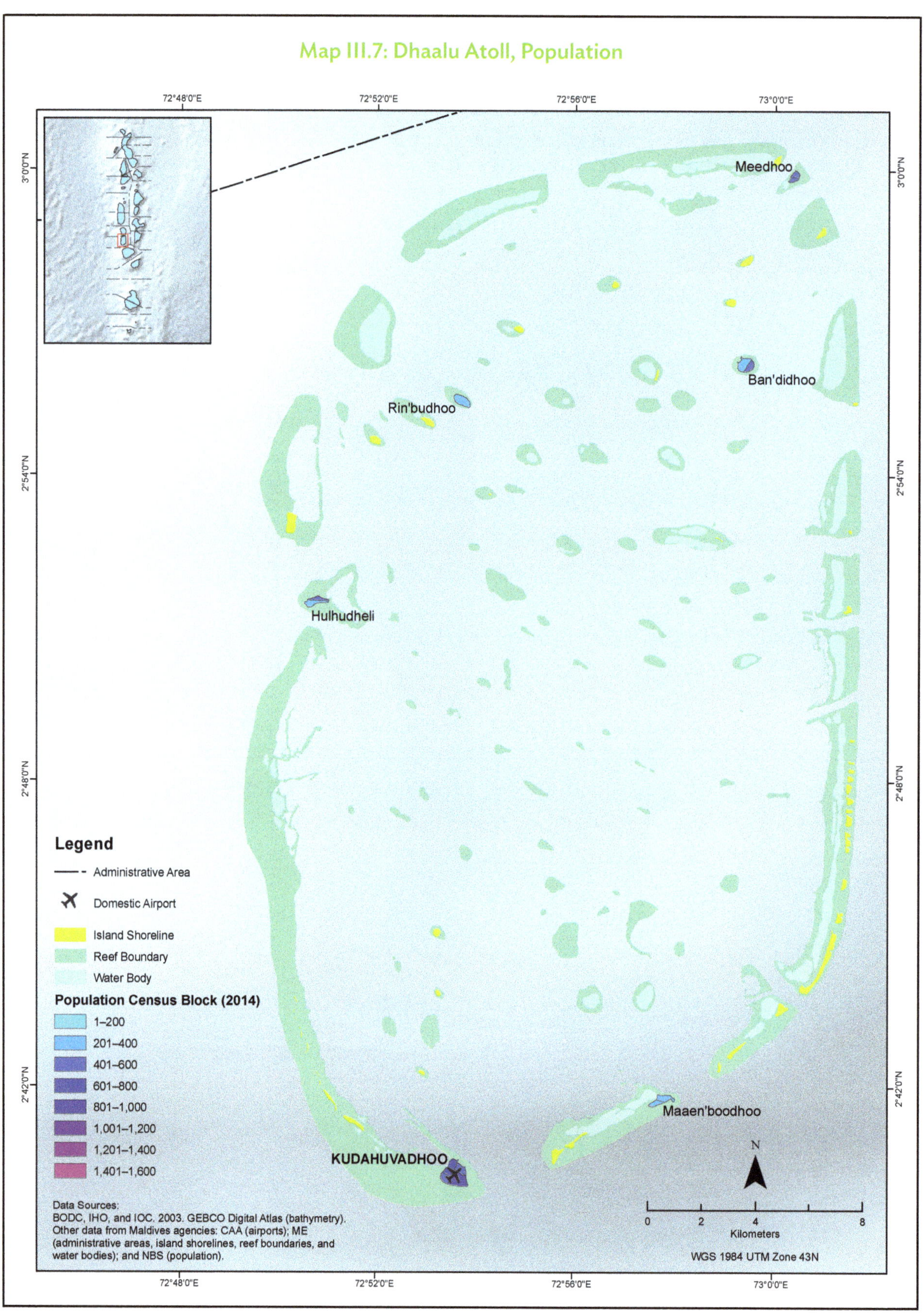

Map III.8: Faafu Atoll, Population

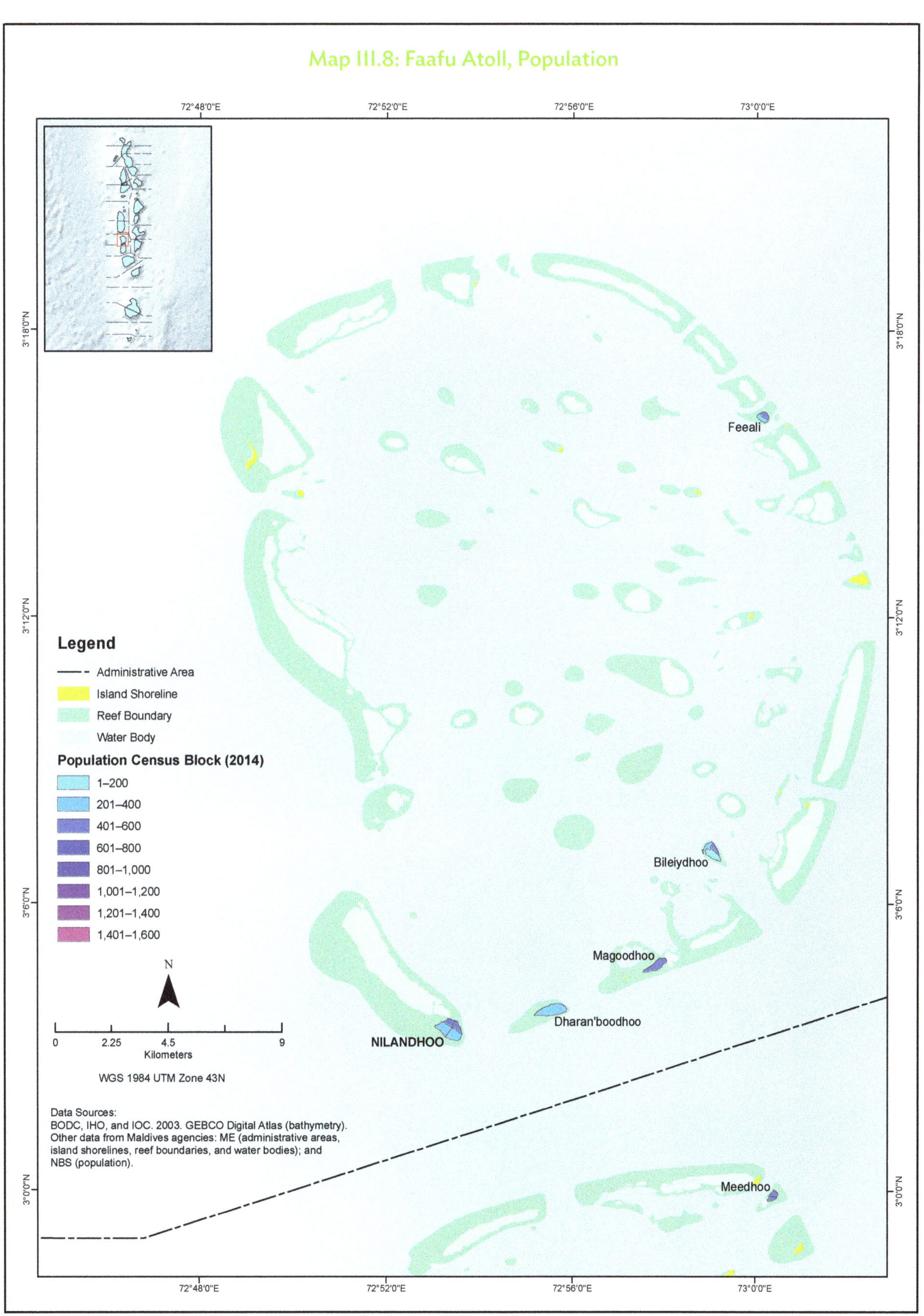

Map III.9: Gaafu Alifu Atoll, Population

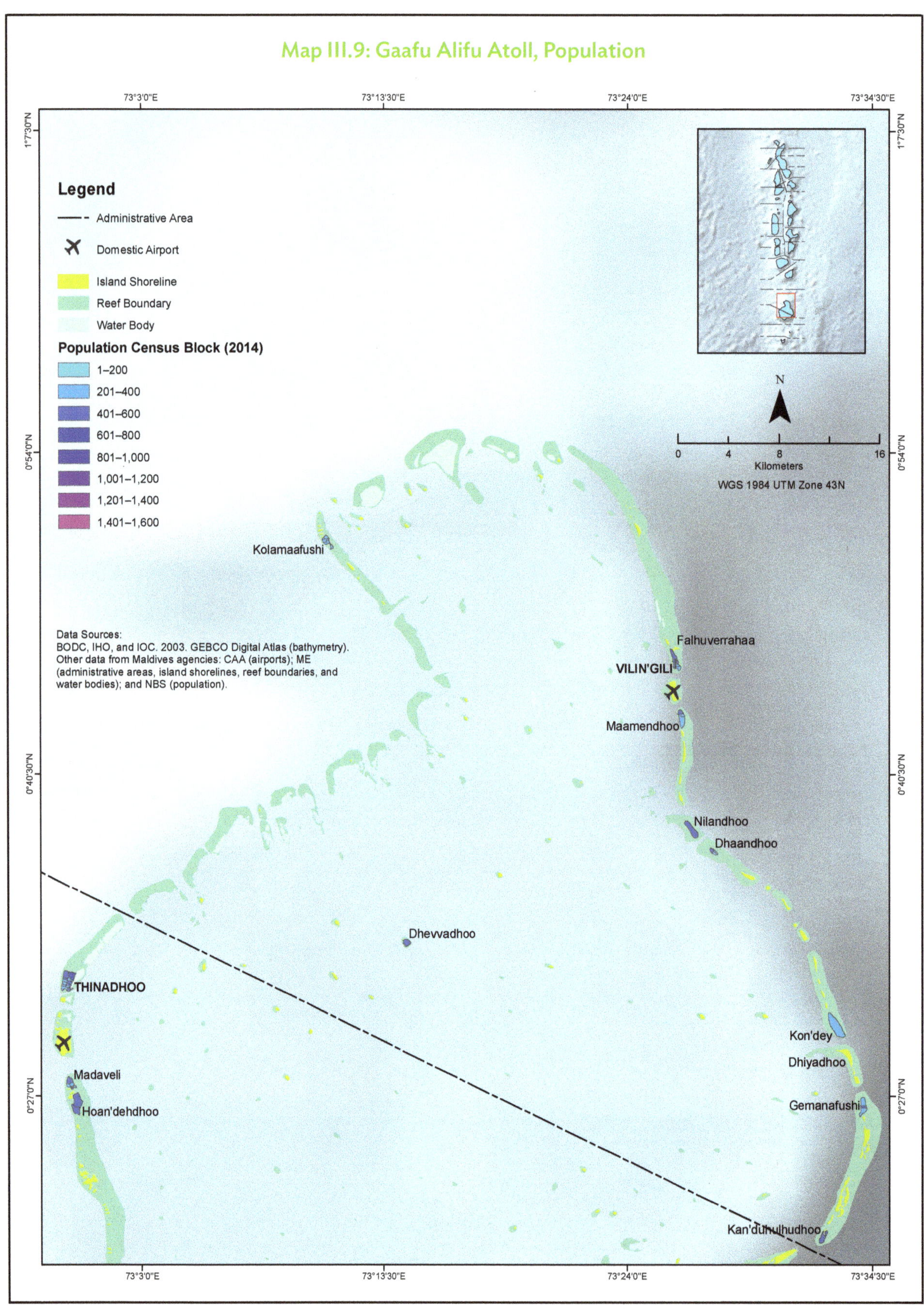

Map III.10: Gaafu Dhaalu Atoll, Population

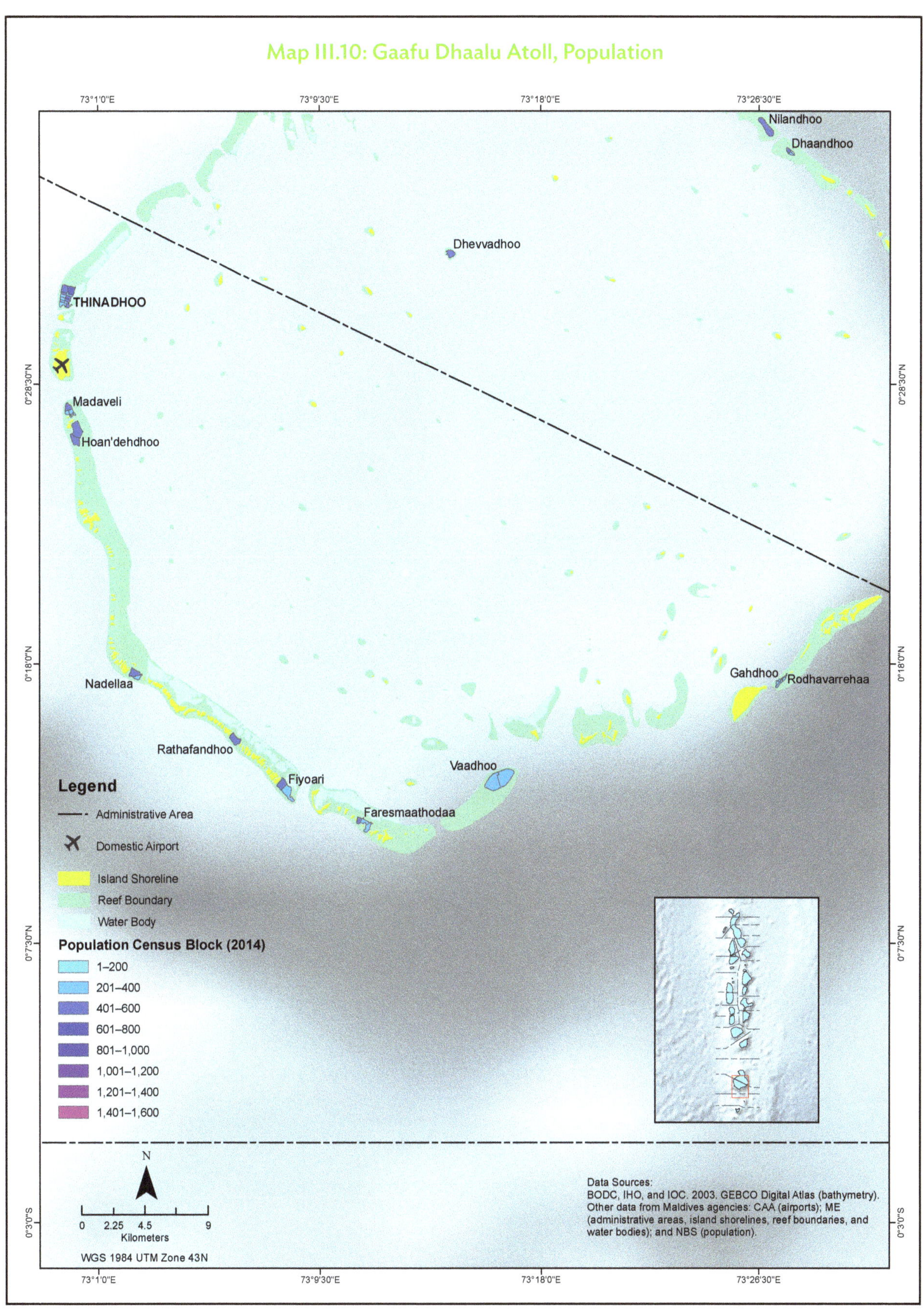

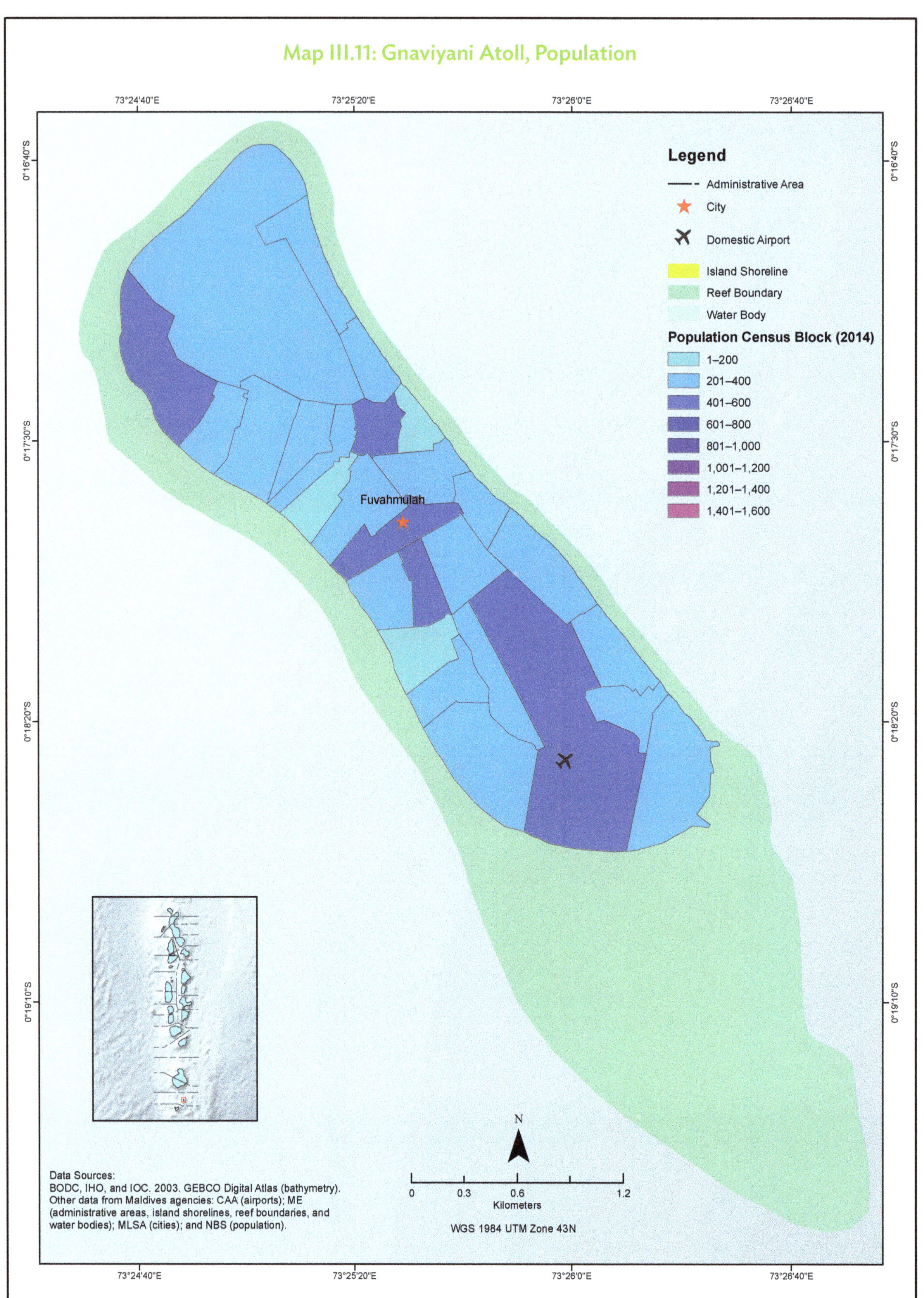
Map III.11: Gnaviyani Atoll, Population
73°24'40"E
73°25'20"E
73°26'0"E
73°26'40"E
0°16'40"S
0°17'30"S
0°18'20"S
0°19'10"S
Legend
Administrative Area
City
Domestic Airport
Island Shoreline
Reef Boundary
Water Body
Population Census Block (2014)
1–200
201–400
401–600
601–800
801–1,000
1,001–1,200
1,201–1,400
1,401–1,600
Fuvahmulah
N
0
0.3
0.6
1.2
Kilometers
WGS 1984 UTM Zone 43N
Data Sources:
BODC, IHO, and IOC. 2003. GEBCO Digital Atlas (bathymetry).
Other data from Maldives agencies: CAA (airports); ME (administrative areas, island shorelines, reef boundaries, and water bodies); MLSA (cities); and NBS (population).

Map III.12: Haa Alifu Atoll, Population

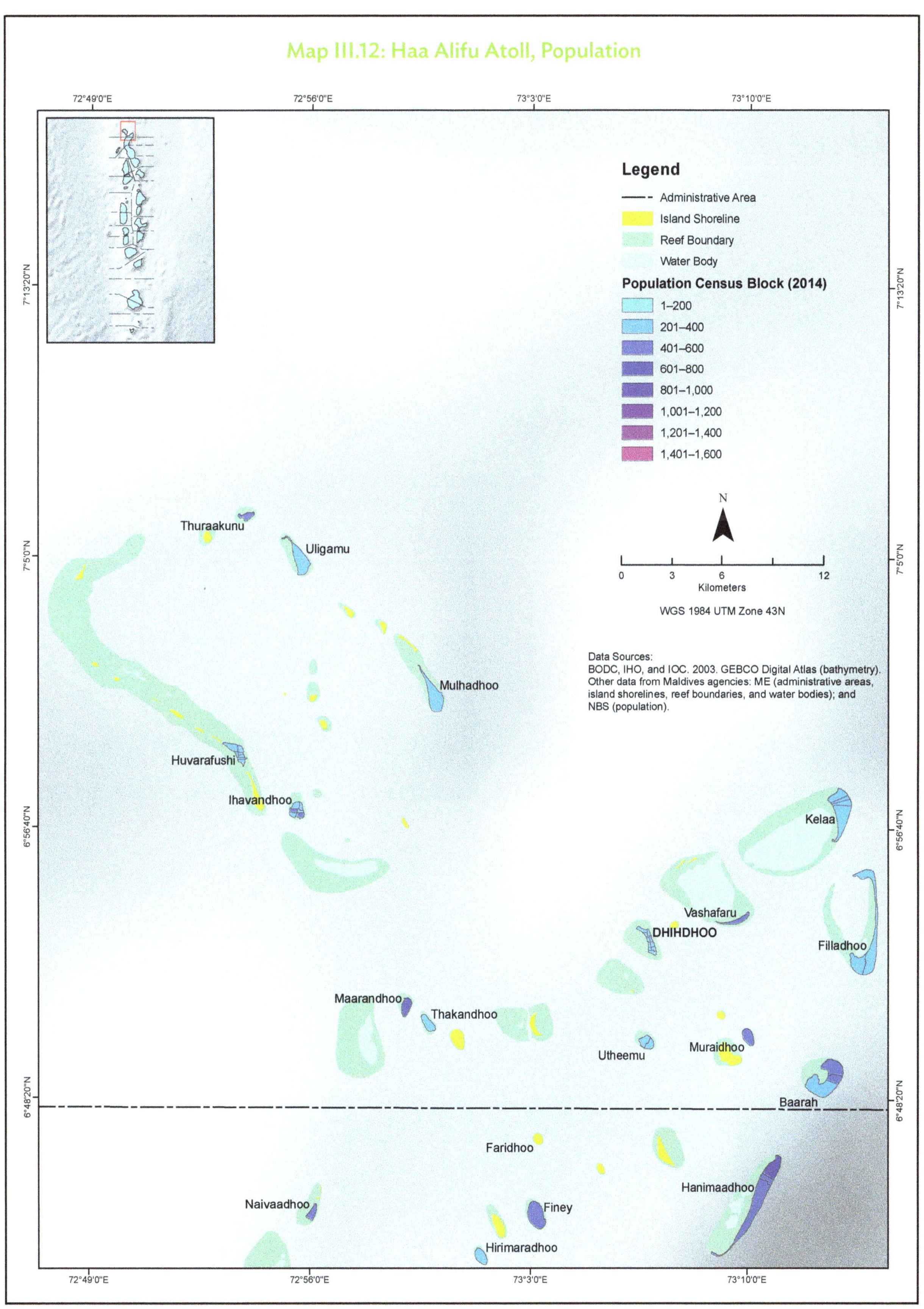

Map III.13: Haa Dhaalu Atoll, Population

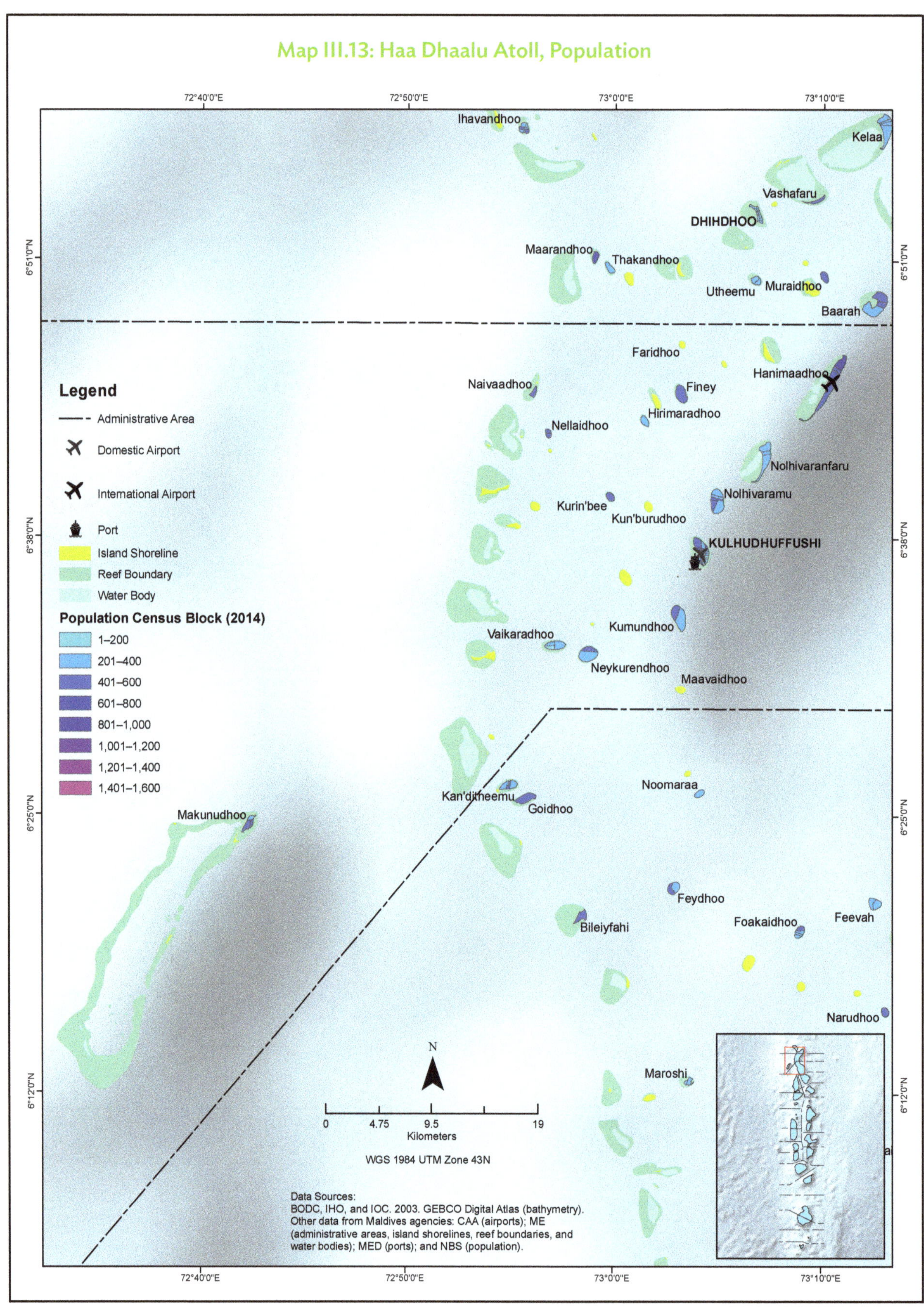

Map III.14: Laamu Atoll, Population

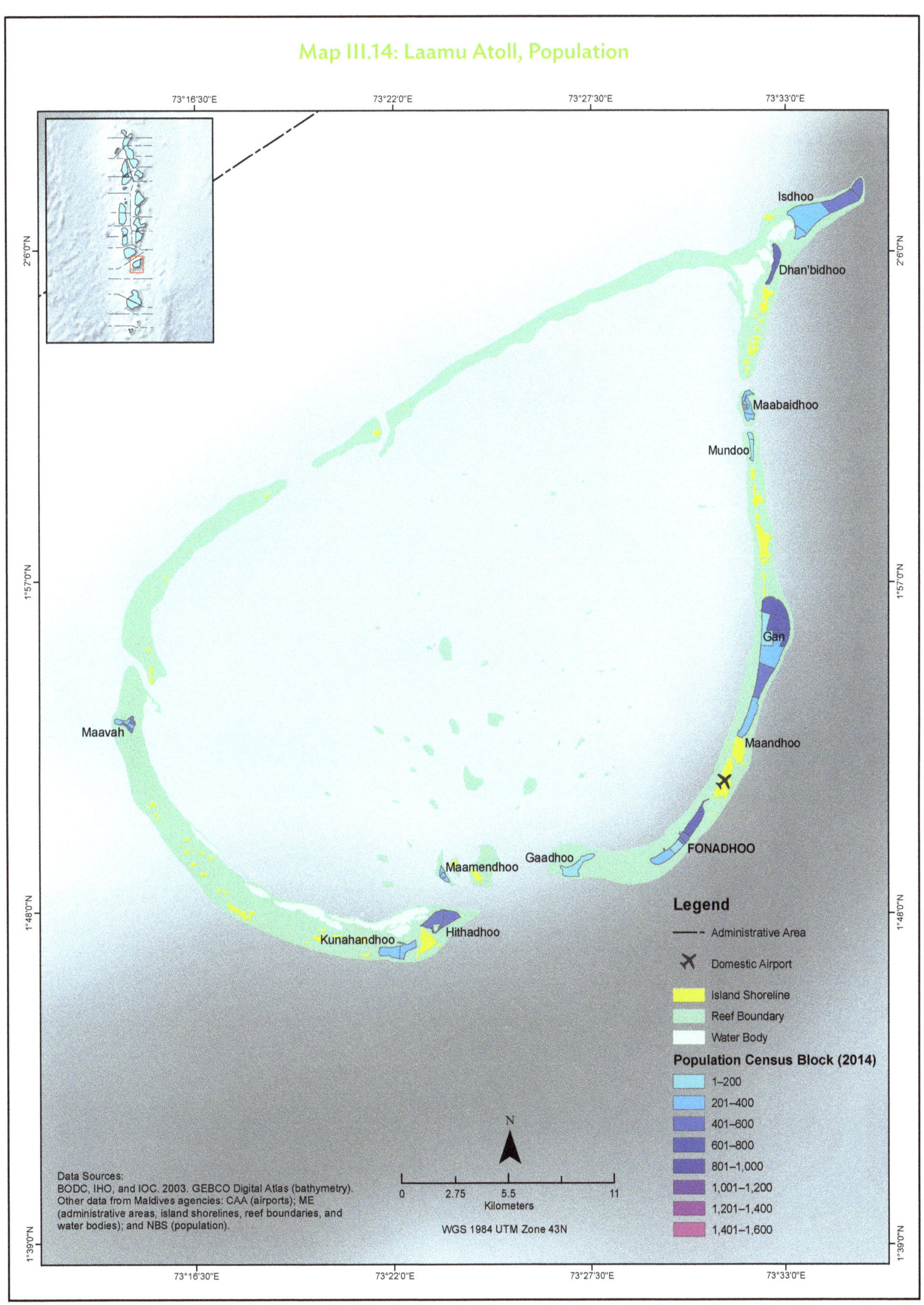

Map III.15: Lhaviyani Atoll, Population

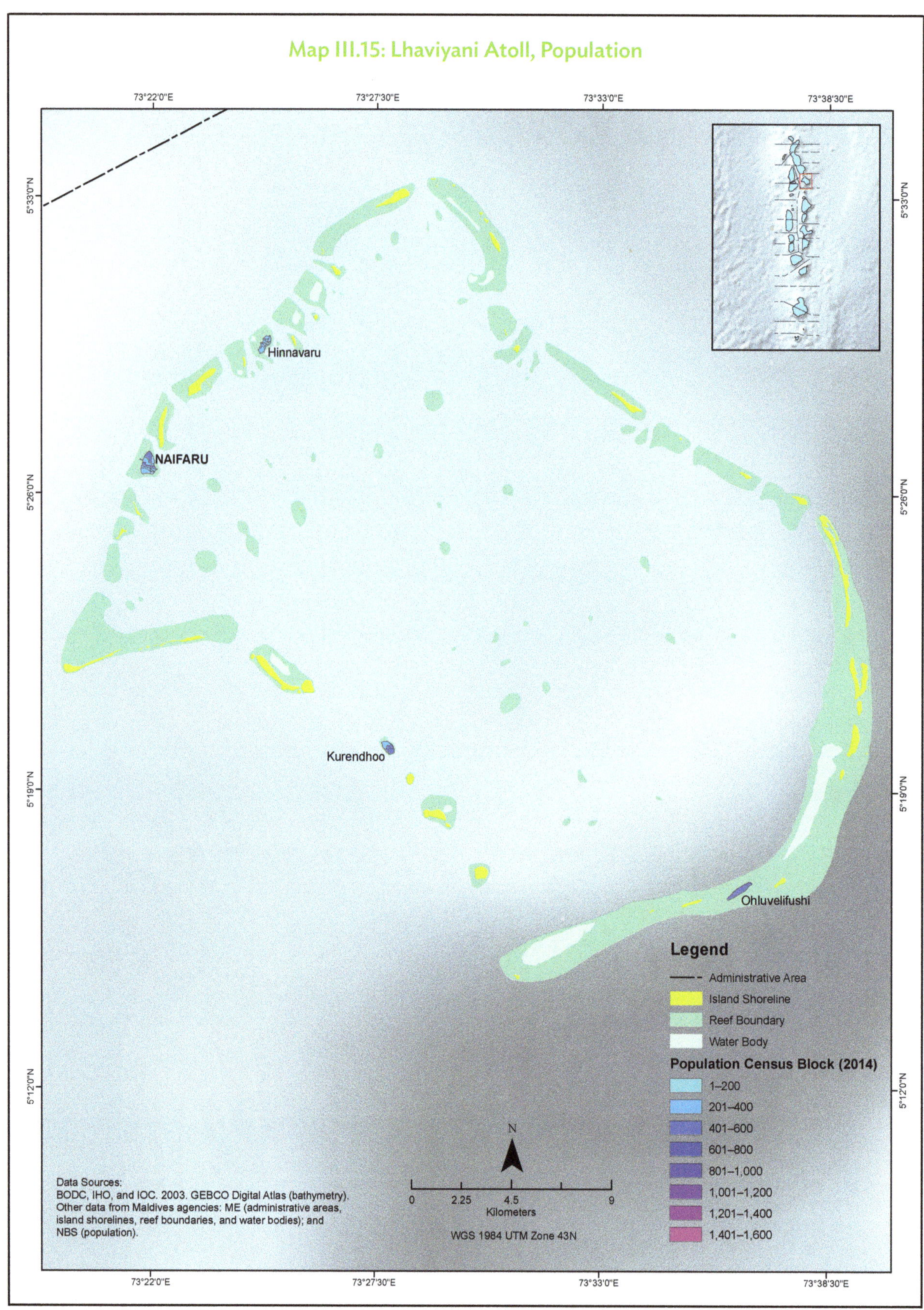

Map III.16: Meemu Atoll, Population

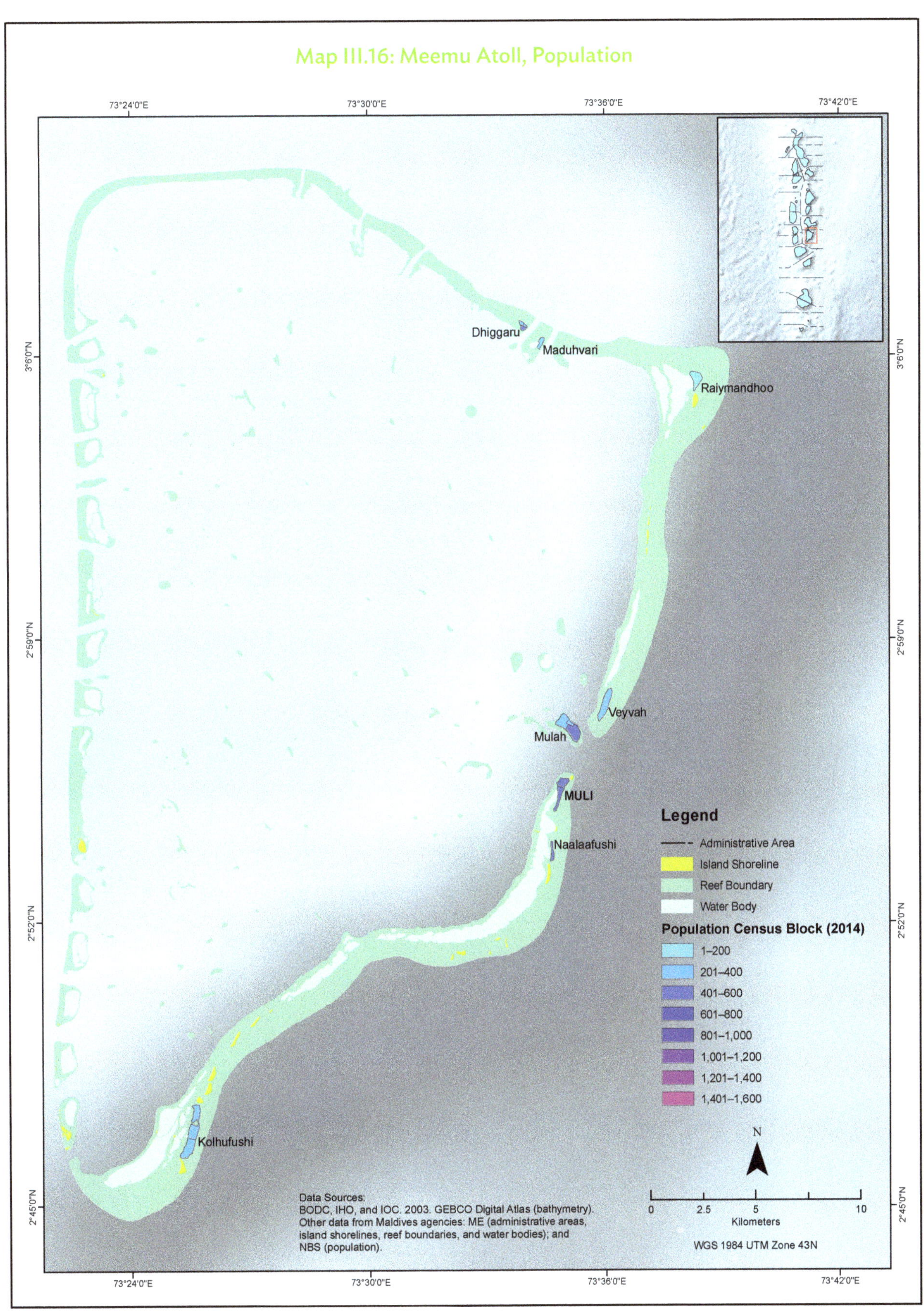

Map III.17: Noonu Atoll, Population

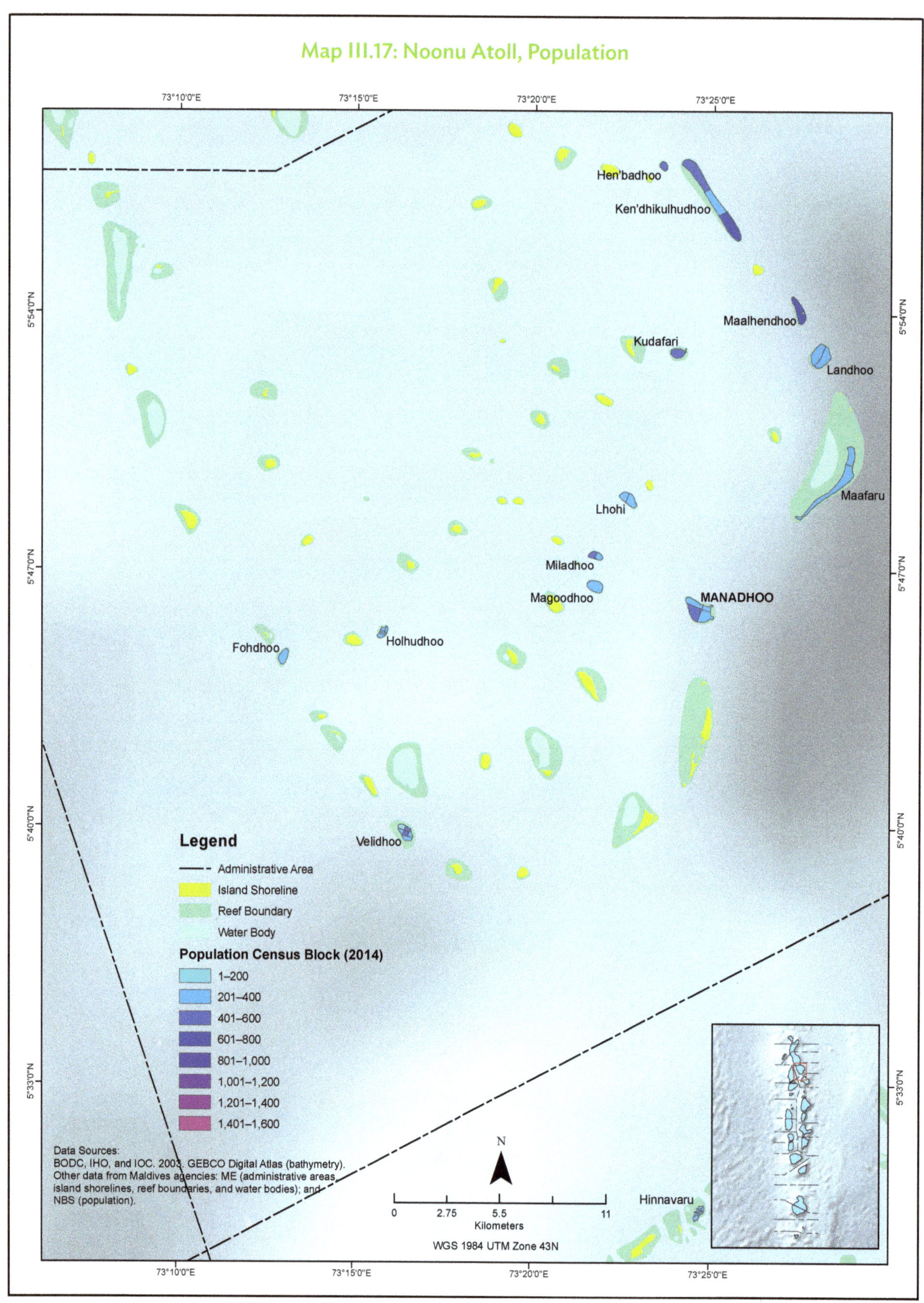

Map III.18: North Malé Atoll, Population

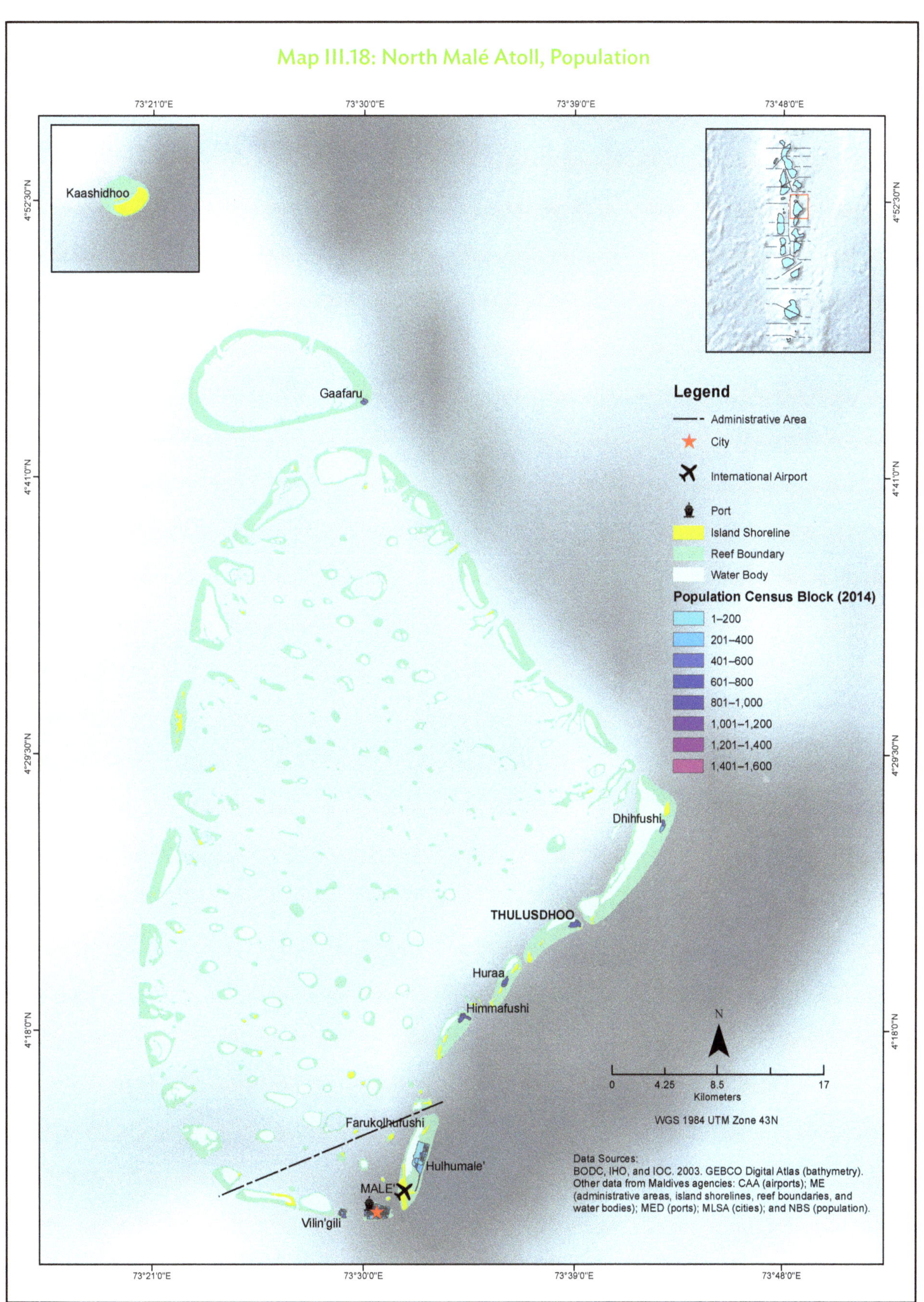

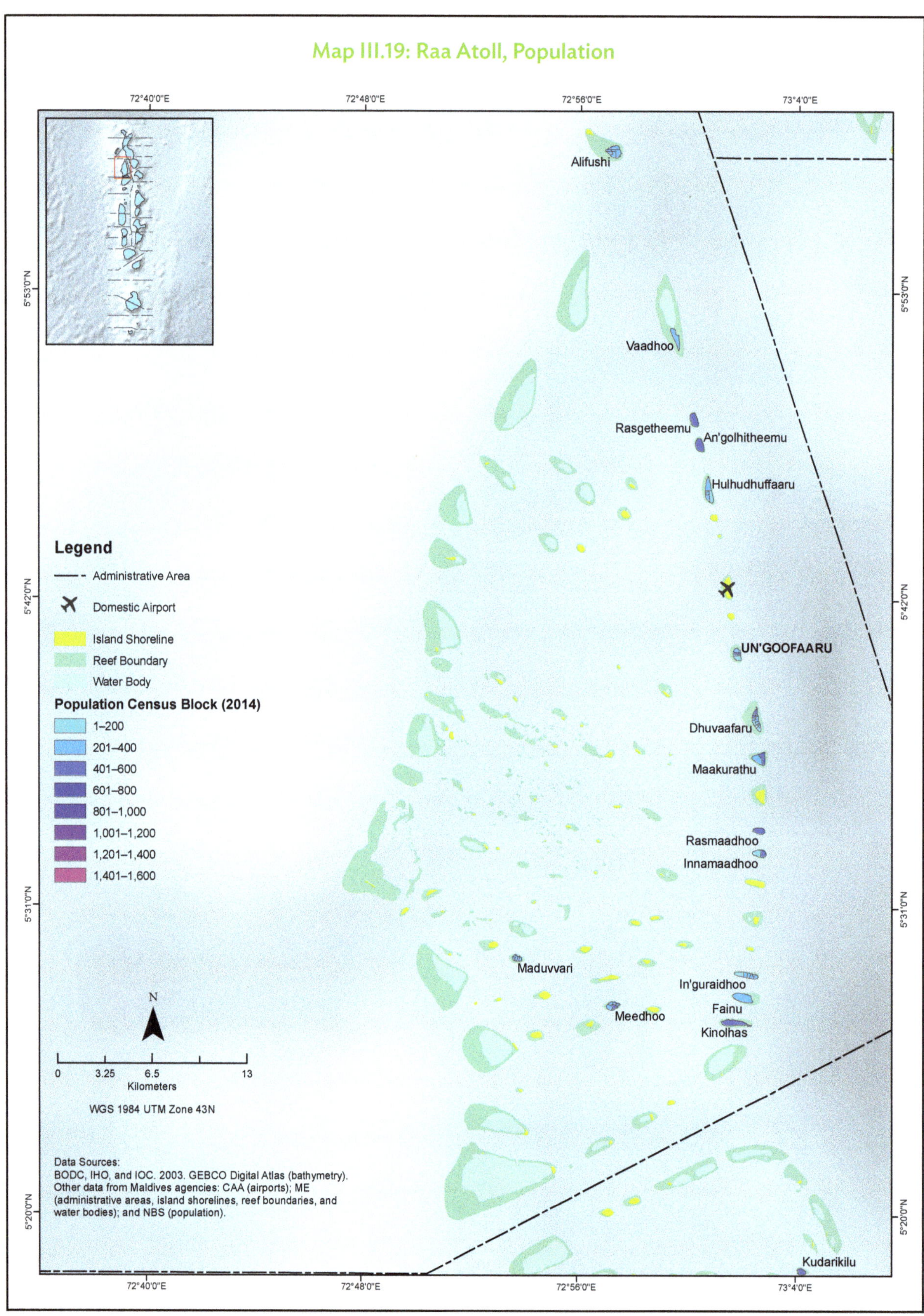
Map III.19: Raa Atoll, Population
72°40'0"E
72°48'0"E
72°56'0"E
73°4'0"E
5°53'0"N
5°42'0"N
5°31'0"N
5°20'0"N
Alifushi
Vaadhoo
Rasgetheemu
An'golhitheemu
Hulhudhuffaaru
UN'GOOFAARU
Dhuvaafaru
Maakurathu
Rasmaadhoo
Innamaadhoo
Maduvvari
In'guraidhoo
Fainu
Meedhoo
Kinolhas
Kudarikilu
Legend
Administrative Area
Domestic Airport
Island Shoreline
Reef Boundary
Water Body
Population Census Block (2014)
1–200
201–400
401–600
601–800
801–1,000
1,001–1,200
1,201–1,400
1,401–1,600
N
0 3.25 6.5 13
Kilometers
WGS 1984 UTM Zone 43N
Data Sources:
BODC, IHO, and IOC. 2003. GEBCO Digital Atlas (bathymetry).
Other data from Maldives agencies: CAA (airports); ME (administrative areas, island shorelines, reef boundaries, and water bodies); and NBS (population).

Map III.20: Shaviyani Atoll, Population

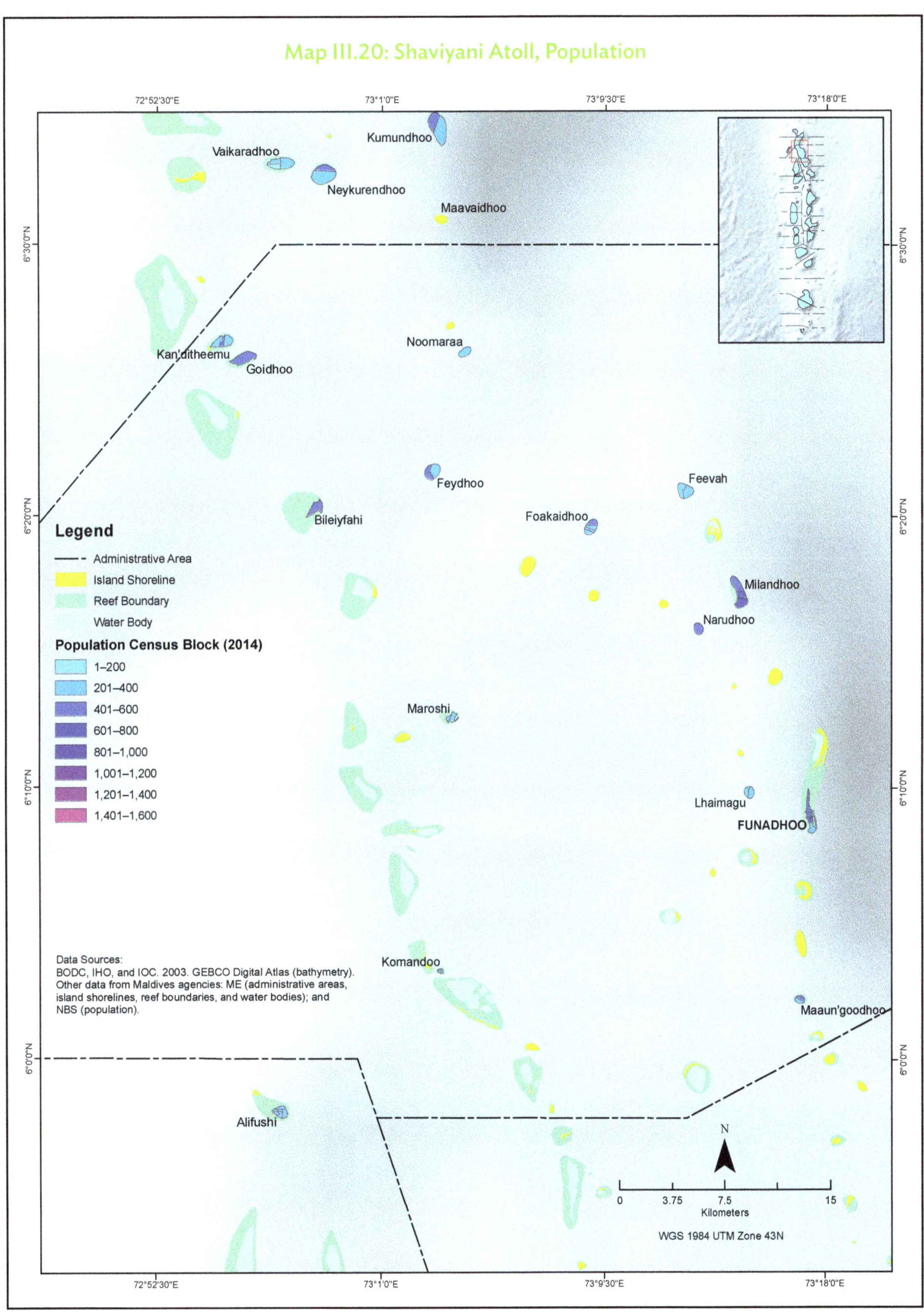

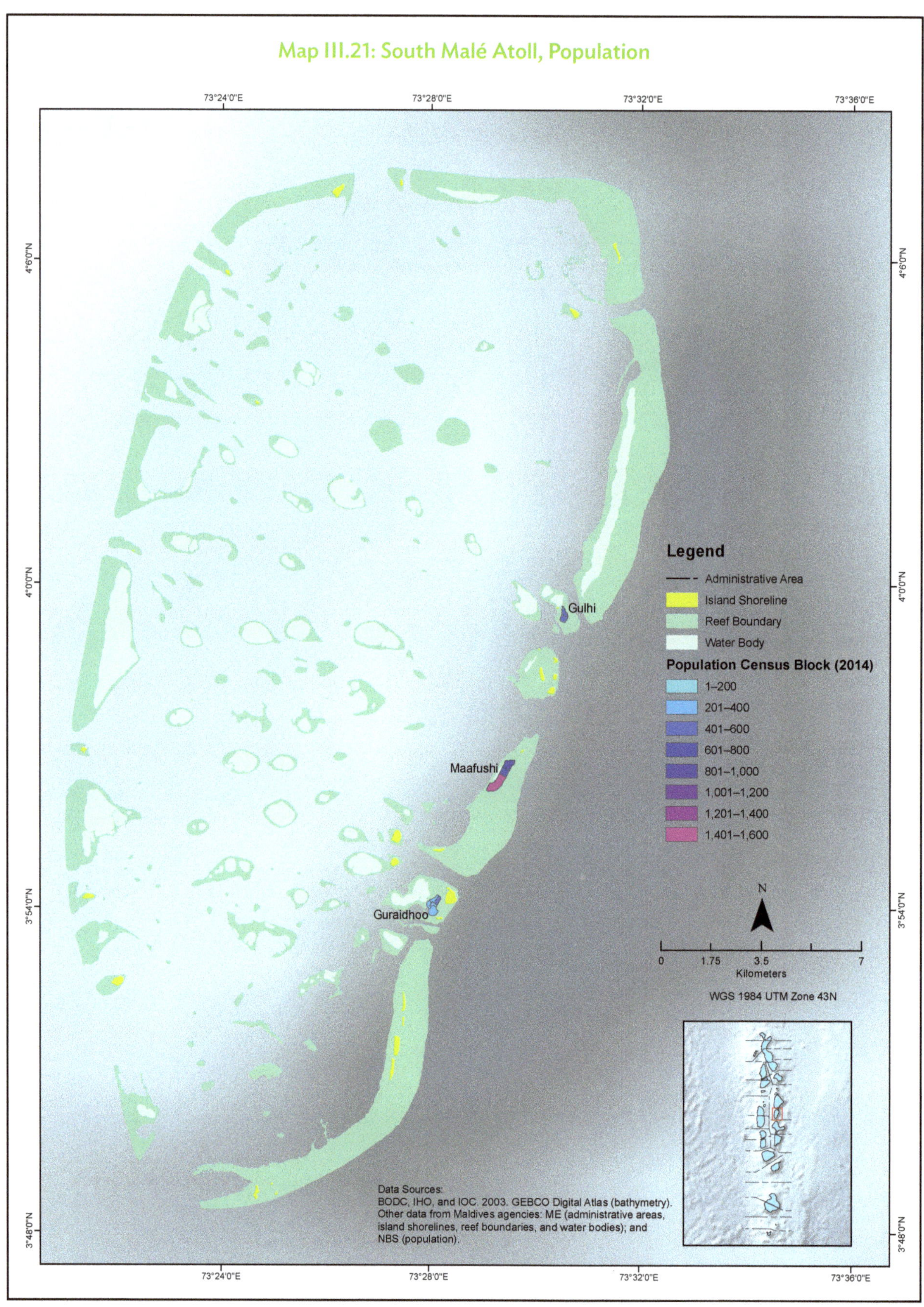
Map III.21: South Malé Atoll, Population
73°24'0"E
73°28'0"E
73°32'0"E
73°36'0"E
4°6'0"N
4°0'0"N
3°54'0"N
3°48'0"N
Gulhi
Maafushi
Guraidhoo
Legend
Administrative Area
Island Shoreline
Reef Boundary
Water Body
Population Census Block (2014)
1–200
201–400
401–600
601–800
801–1,000
1,001–1,200
1,201–1,400
1,401–1,600
N
0 1.75 3.5 7
Kilometers
WGS 1984 UTM Zone 43N
Data Sources:
BODC, IHO, and IOC. 2003. GEBCO Digital Atlas (bathymetry).
Other data from Maldives agencies: ME (administrative areas, island shorelines, reef boundaries, and water bodies); and NBS (population).

Map III.22: Thaa Atoll, Population

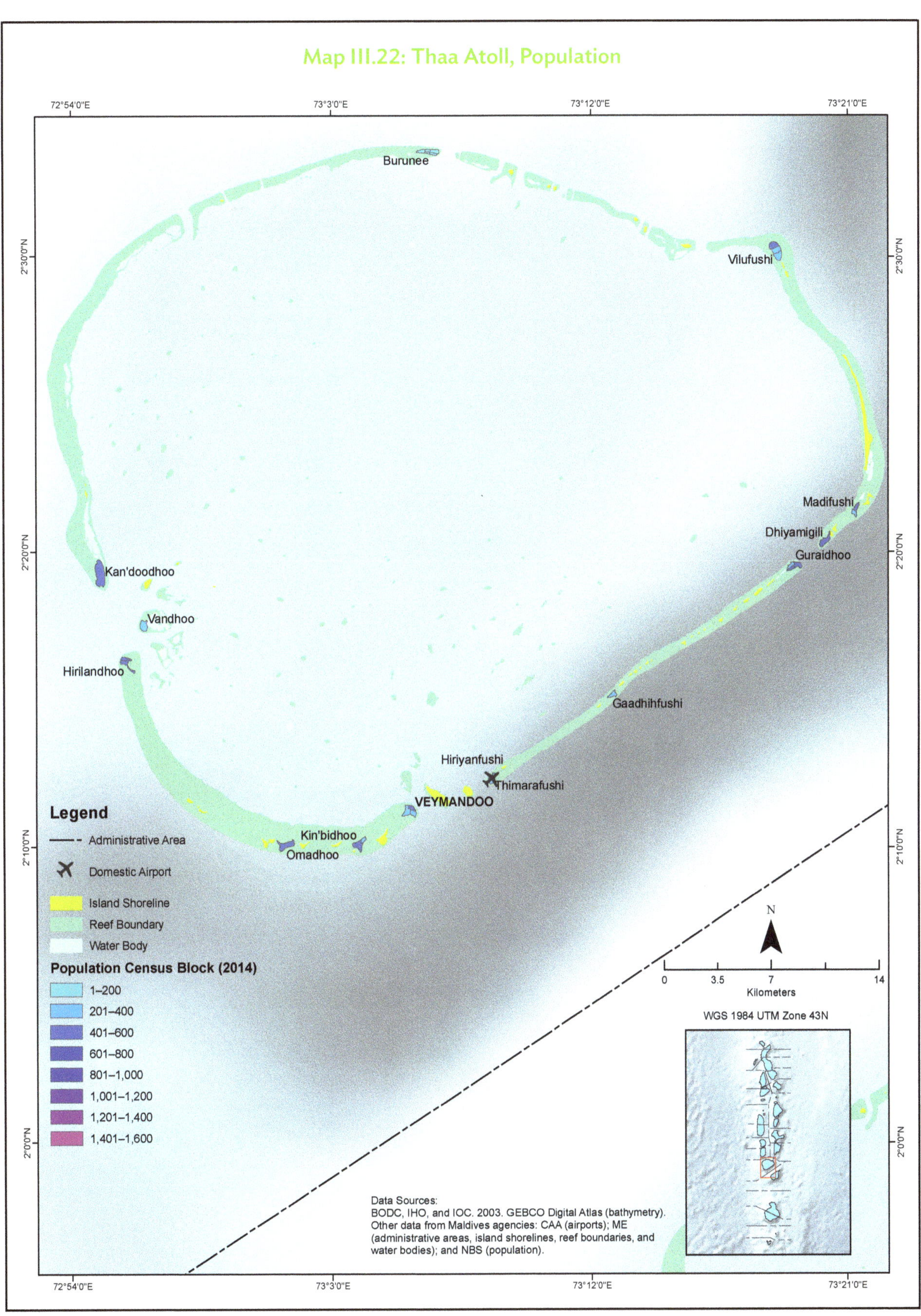

Map III.23: Vaavu Atoll, Population

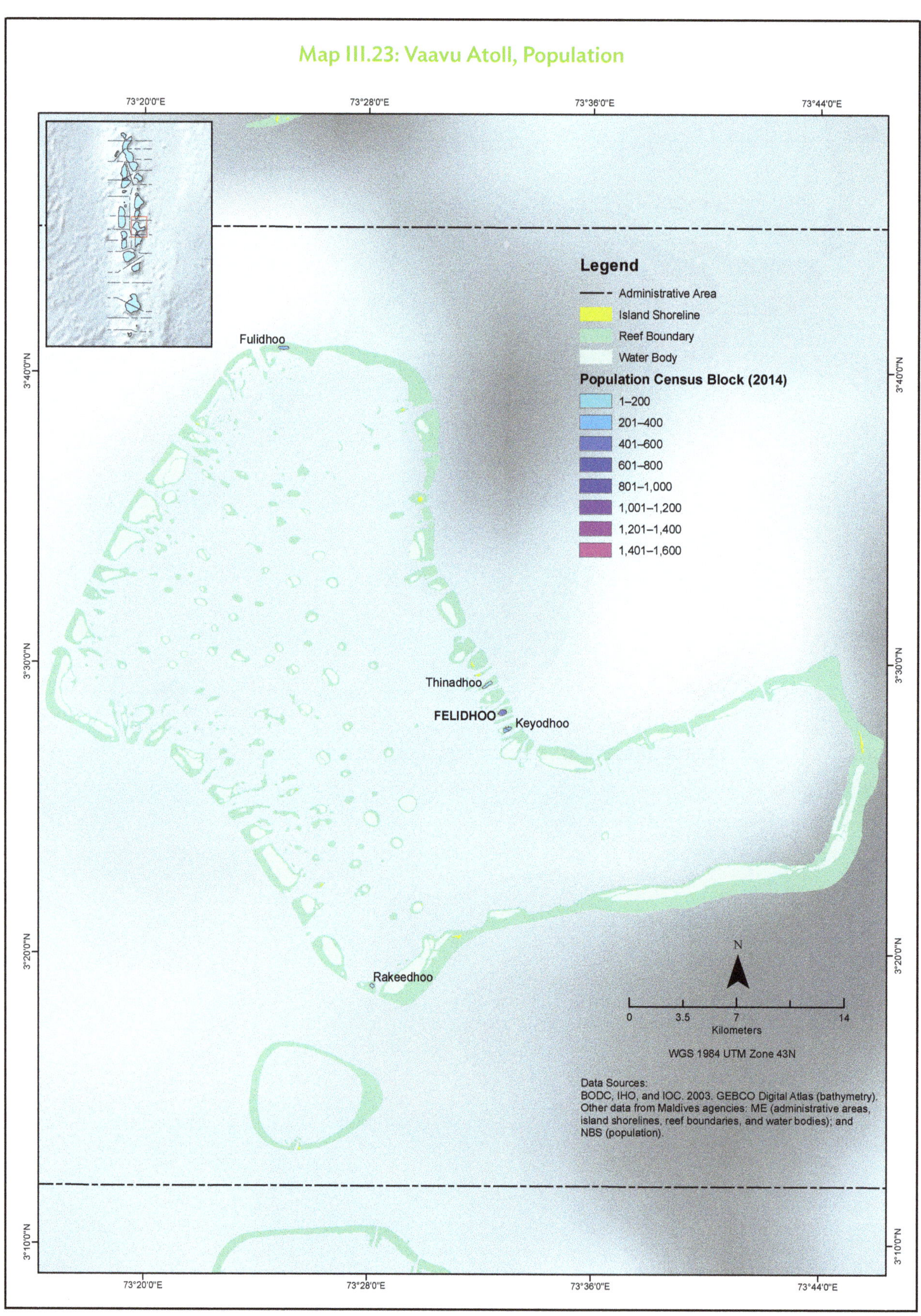

Education

Maldives has a high literacy rate. About 88,000 students are enrolled in the country's 376 government, private, and community schools, most of which are in Kaafu Atoll. Less than 2% of the country's population are illiterate and less than 1% of the youth aged 15–24 cannot read and write (United Nations Educational, Scientific and Cultural Organization's Institute for Statistics 2014).

Schools with the highest student capacities are located in Malé, which is also in Kaafu Atoll and the city with the largest population in Maldives. Most of the schools in Malé are situated in low-lying grounds where they are vulnerable to various hazards related to inundation.

Religion and literacy. All Maldivians practice Islam as a religion as required by law. Almost all people in Maldives are literate (right photo by Mark Fischer, left photo by Adam Jones).

Table III.2: Number of Schools and Student Capacity per Atoll, Maldives

Atoll	Number of Schools	Student Capacity
Alifu Alifu	13	1,681
Alifu Dhaalu	18	2,130
Baa	22	2,614
Dhaalu	11	1,697
Faafu	8	1,421
Gaafu Alifu	14	2,242
Gaafu Dhaalu	17	3,197
Gnaviyani	8	2,465
Haa Alifu	24	3,584
Haa Dhaalu	29	5,625
Kaafu	54	34,440
Laamu	21	3,527
Lhaviyani	7	2,092
Meemu	12	1,217
Noonu	25	2,967
Raa	28	4,704
Seenu	22	5,611
Shaviyani	19	3,680
Thaa	19	2,547
Vaavu	5	347
TOTAL	**376**	**87,788**

Source: Ministry of Education, 2016.

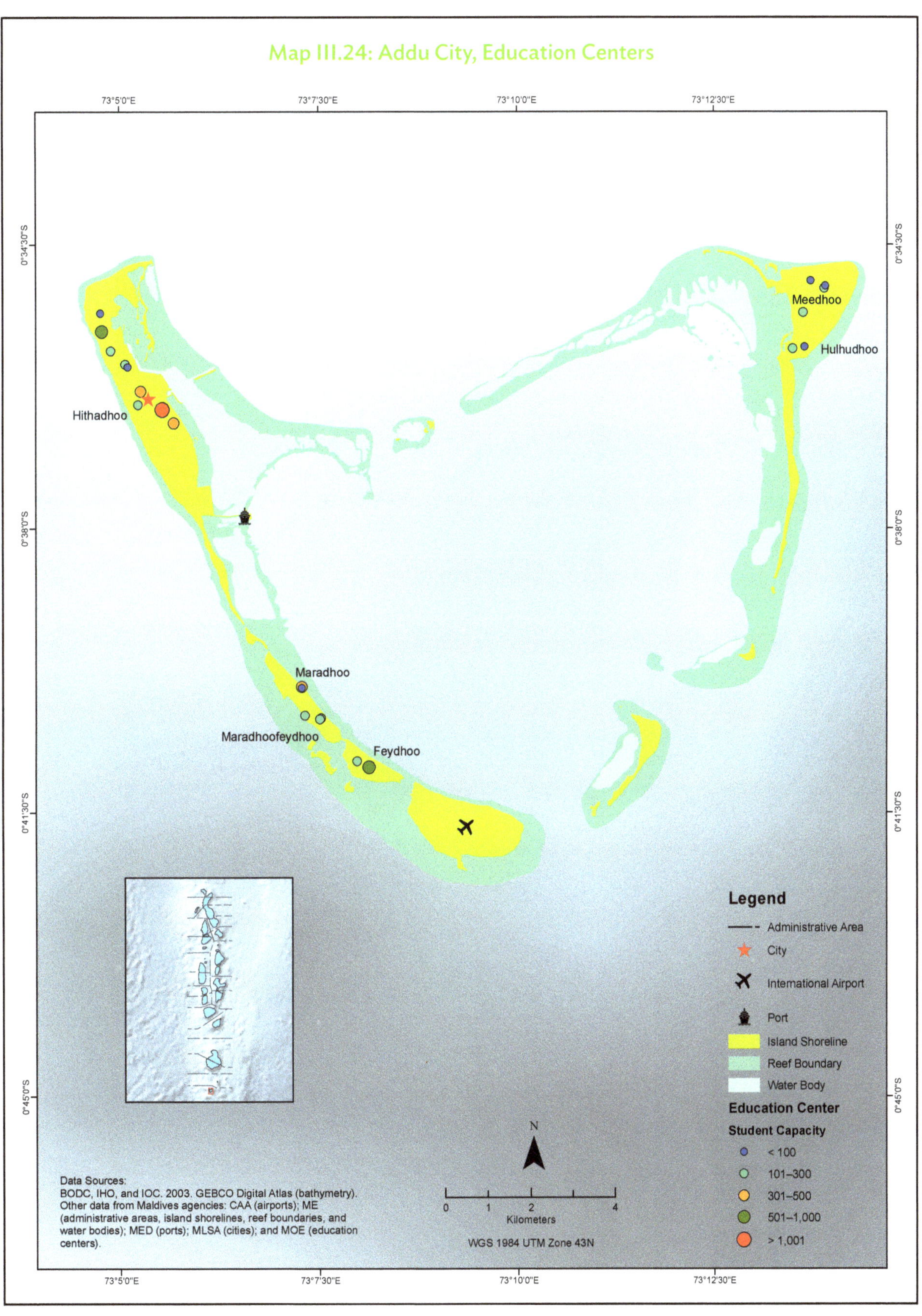
Map III.24: Addu City, Education Centers
73°5'0"E
73°7'30"E
73°10'0"E
73°12'30"E
0°34'30"S
0°38'0"S
0°41'30"S
0°45'0"S
Meedhoo
Hulhudhoo
Hithadhoo
Maradhoo
Maradhoofeydhoo
Feydhoo
Legend
Administrative Area
City
International Airport
Port
Island Shoreline
Reef Boundary
Water Body
Education Center
Student Capacity
< 100
101–300
301–500
501–1,000
> 1,001
N
0 1 2 4
Kilometers
WGS 1984 UTM Zone 43N
Data Sources:
BODC, IHO, and IOC. 2003. GEBCO Digital Atlas (bathymetry).
Other data from Maldives agencies: CAA (airports); ME (administrative areas, island shorelines, reef boundaries, and water bodies); MED (ports); MLSA (cities); and MOE (education centers).

Map III.25: Alifu Alifu Atoll, Education Centers

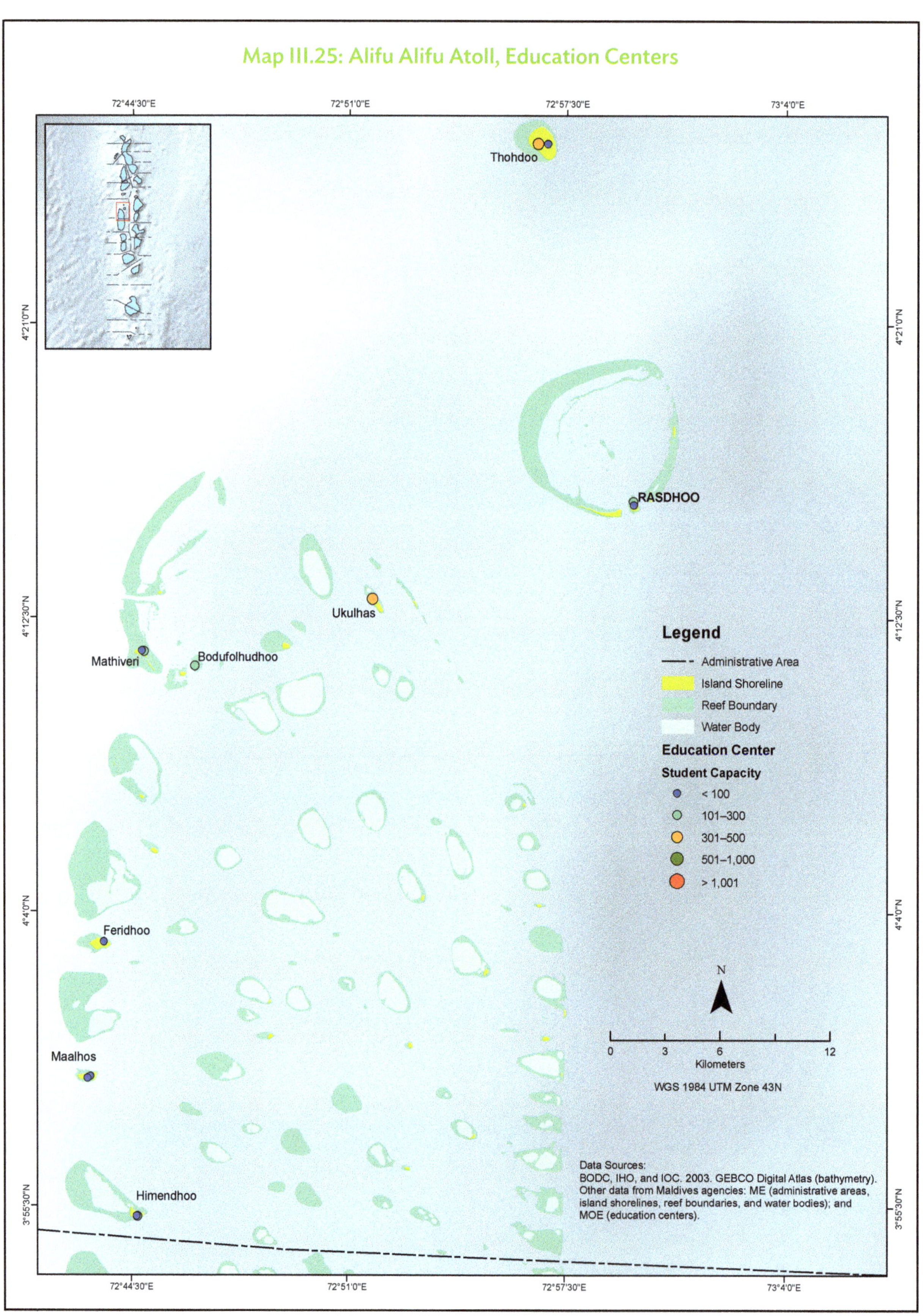

Map III.26: Alifu Dhaalu Atoll, Education Centers

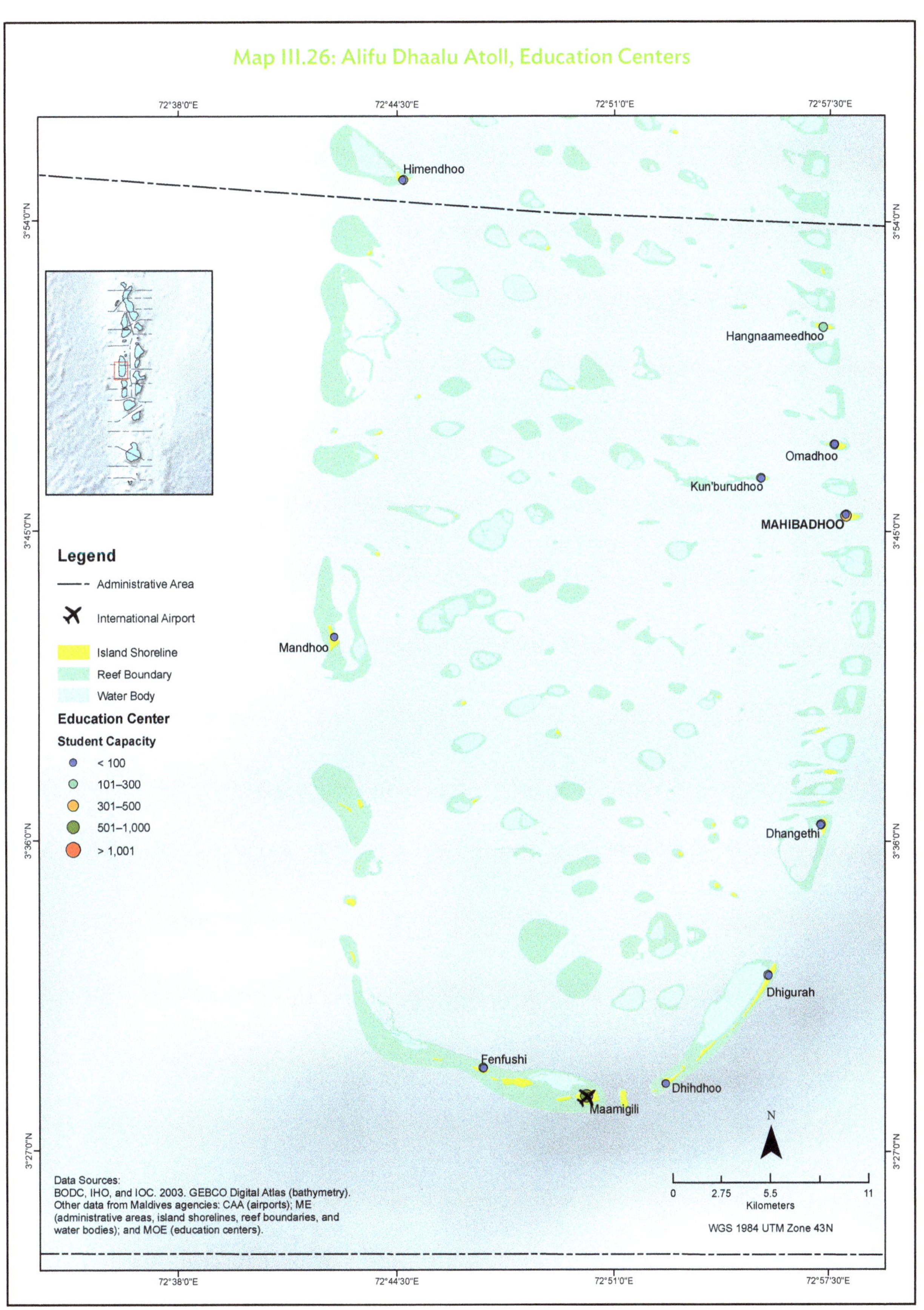

Map III.27: Baa Atoll, Education Centers

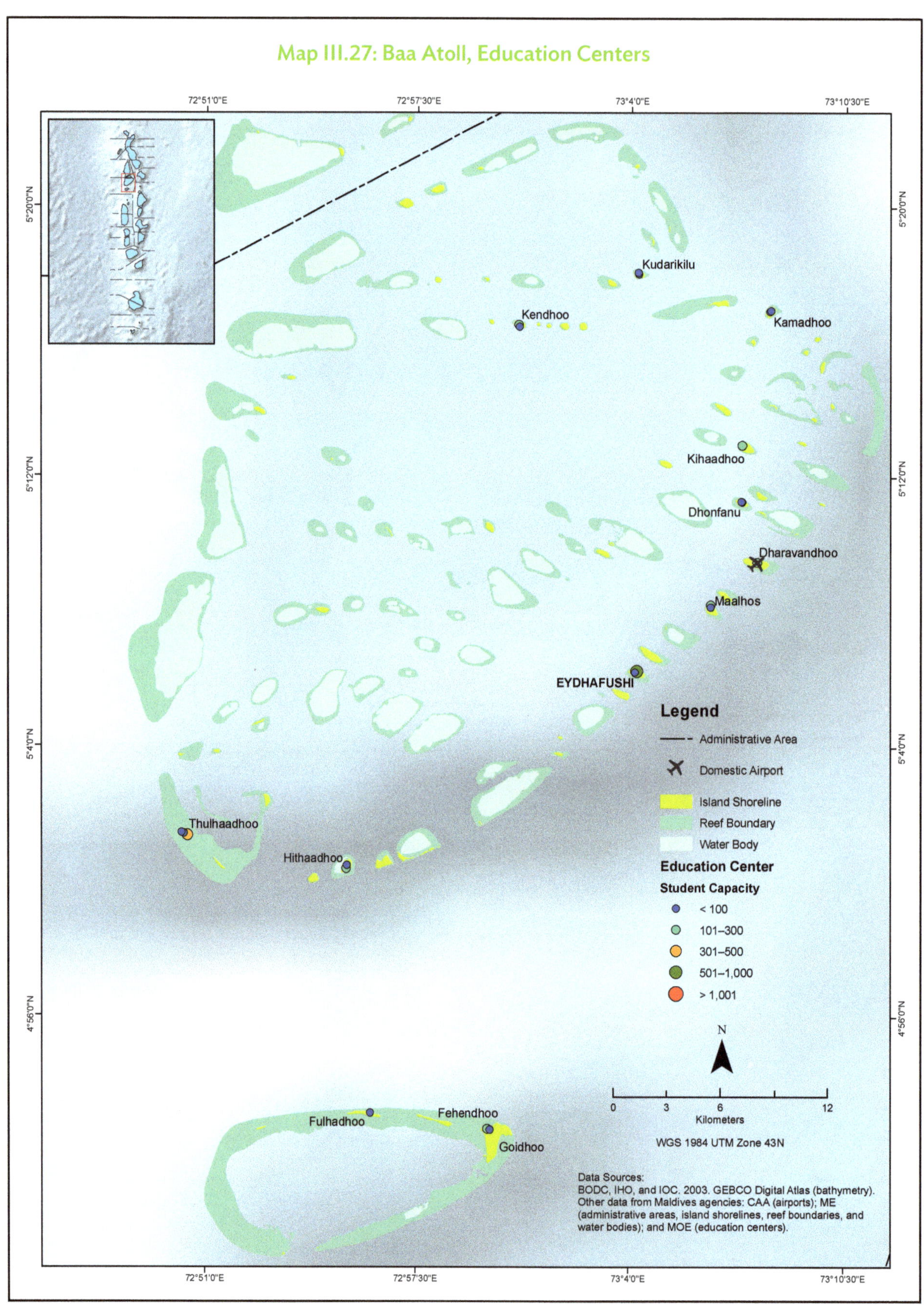

Map III.28: Dhaalu Atoll, Education Centers

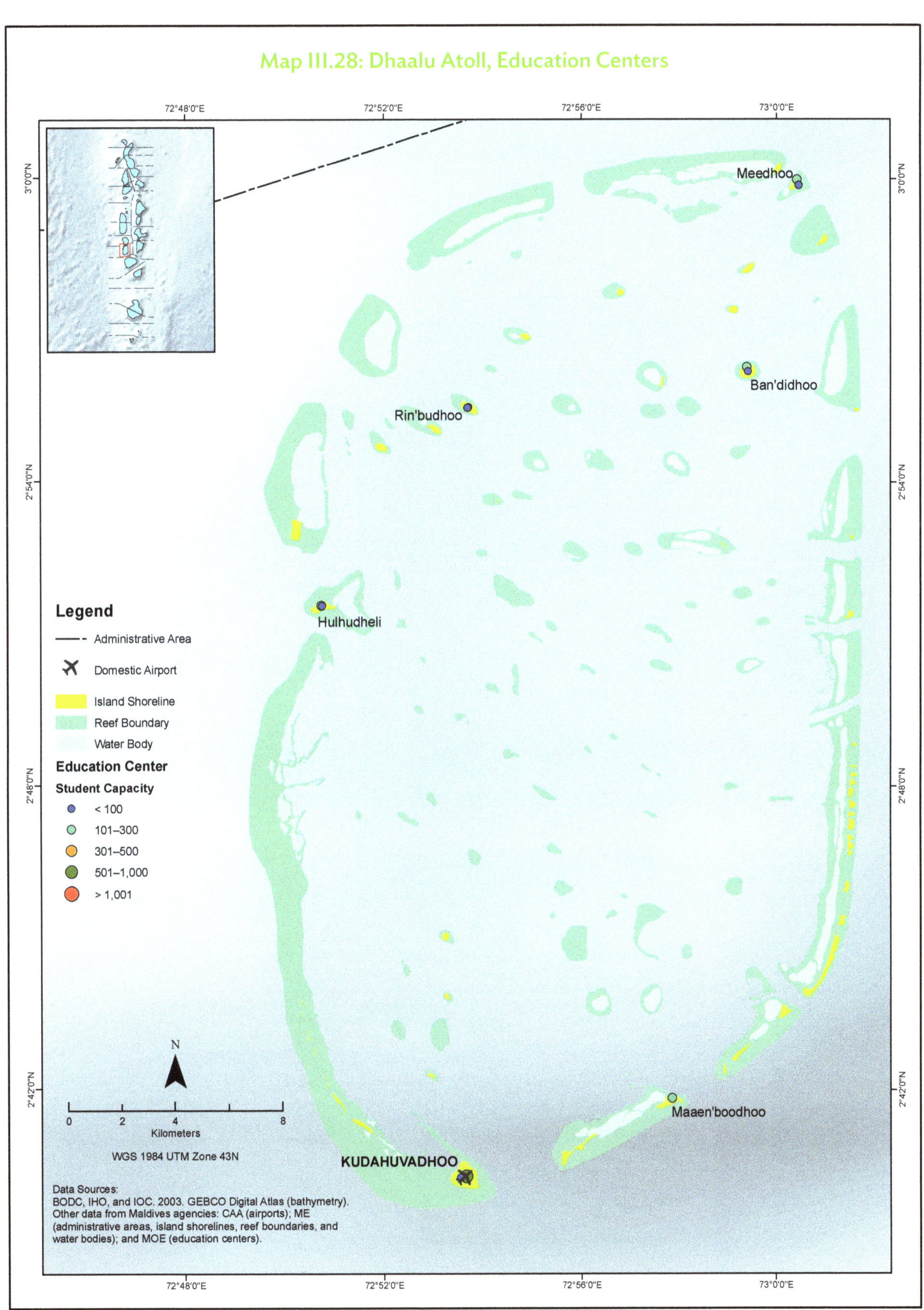

Map III.29: Faafu Atoll, Education Centers

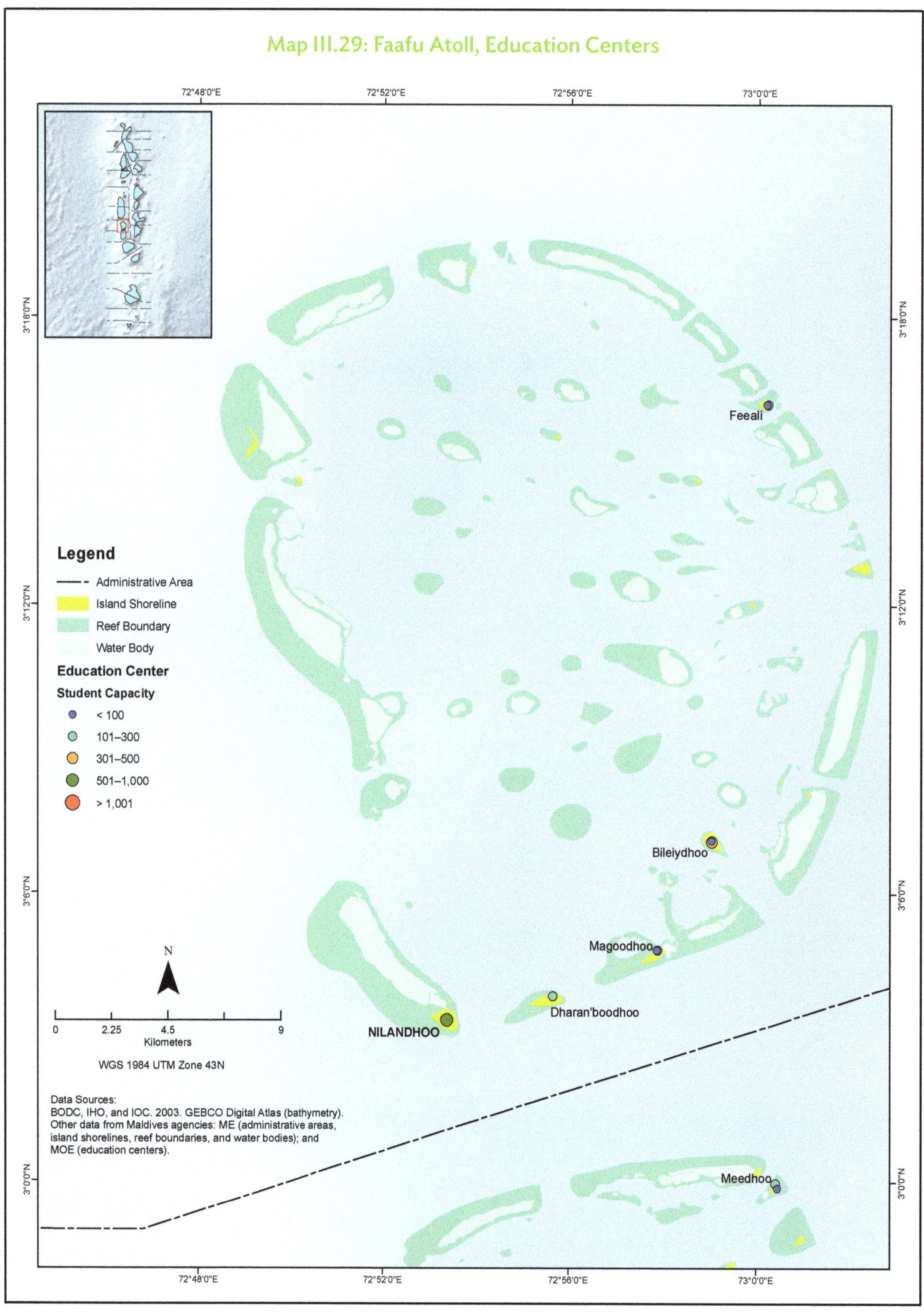

Map III.30: Gaafu Alifu Atoll, Education Centers

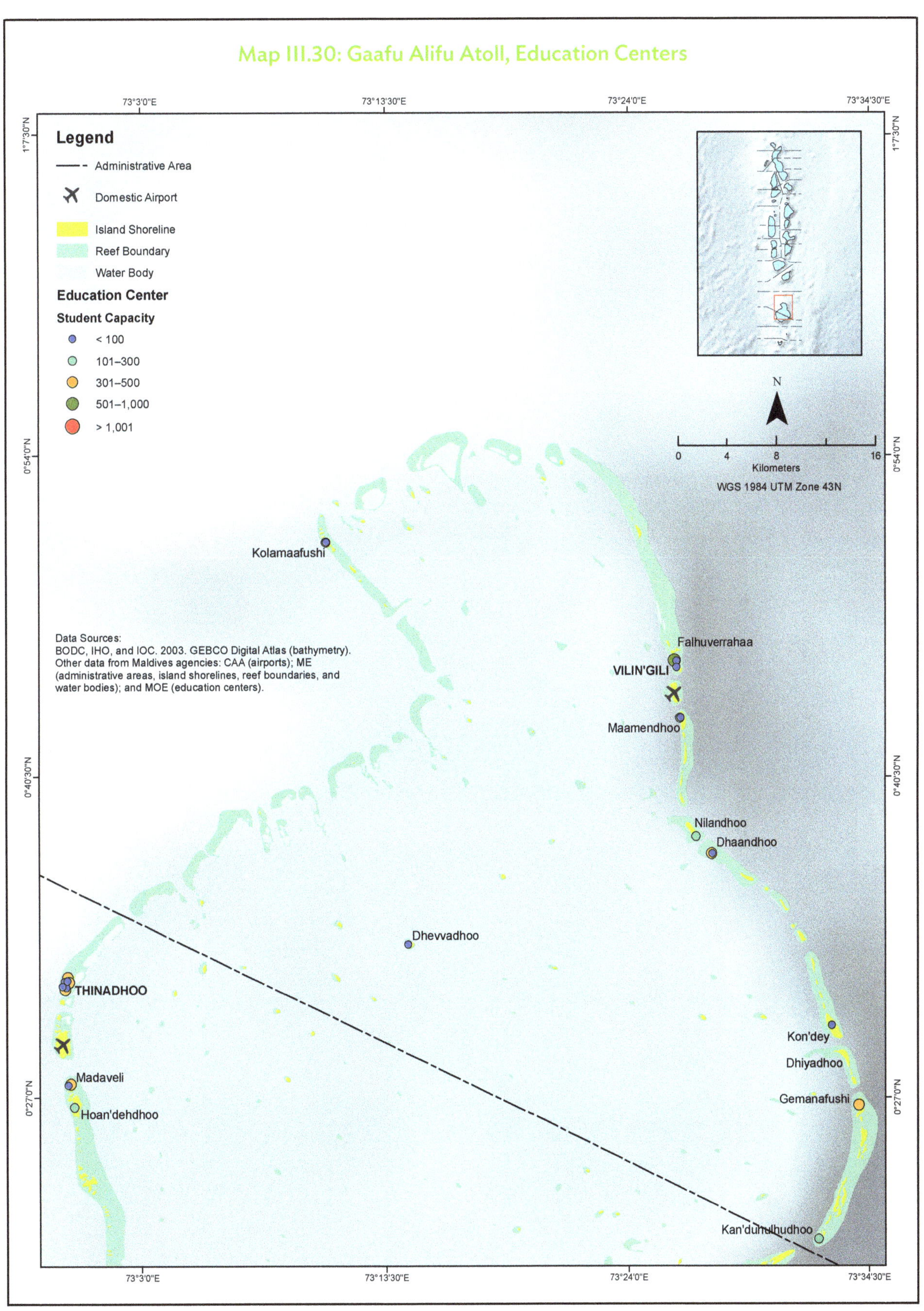

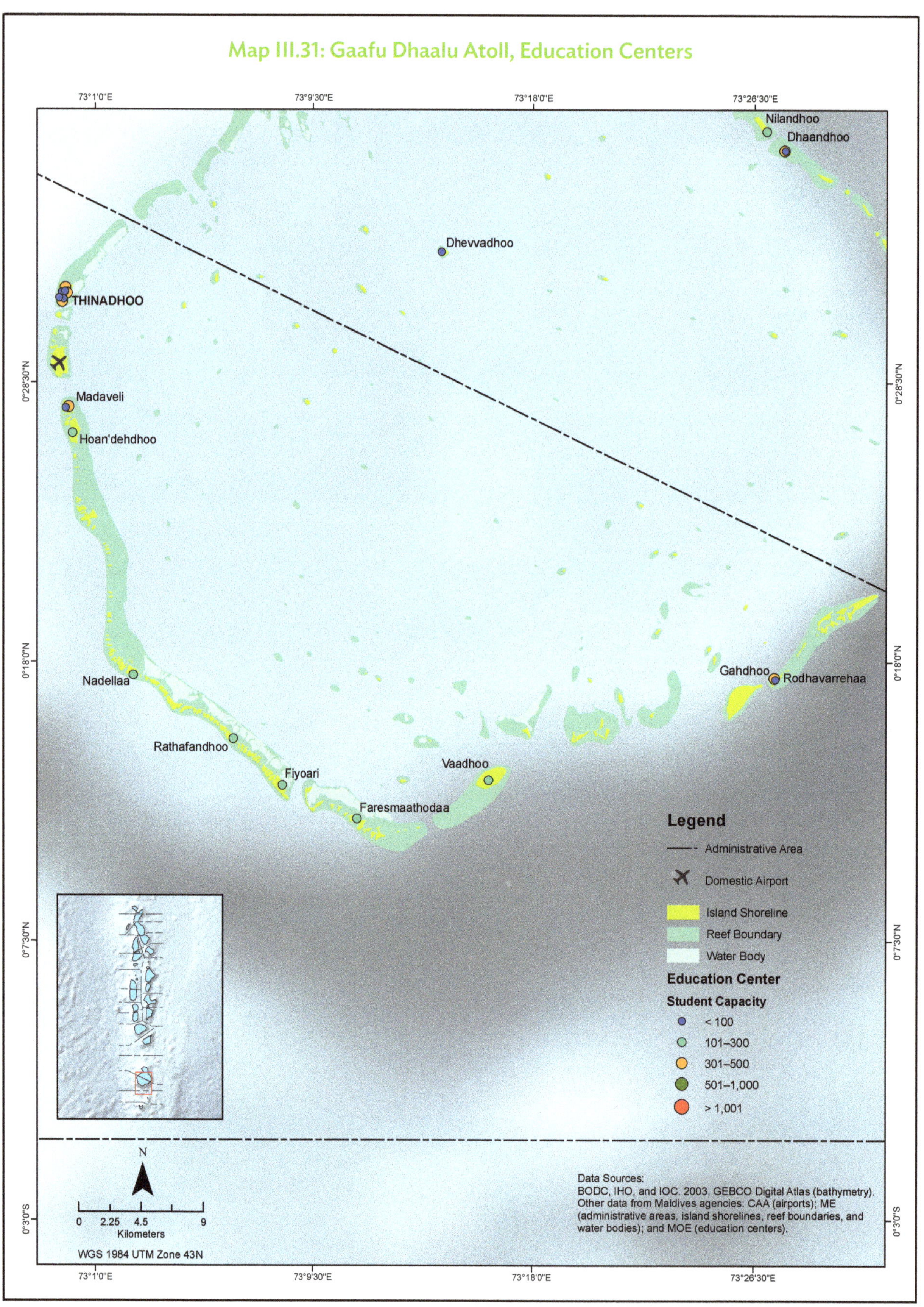
Map III.31: Gaafu Dhaalu Atoll, Education Centers
73°1'0"E
73°9'30"E
73°18'0"E
73°26'30"E
0°28'30"N
0°18'0"N
0°7'30"N
0°3'0"S
Nilandhoo
Dhaandhoo
Dhevvadhoo
THINADHOO
Madaveli
Hoan'dehdhoo
Nadellaa
Rathafandhoo
Fiyoari
Faresmaathodaa
Vaadhoo
Gahdhoo
Rodhavarrehaa
Legend
Administrative Area
Domestic Airport
Island Shoreline
Reef Boundary
Water Body
Education Center
Student Capacity
< 100
101–300
301–500
501–1,000
> 1,001
N
0 2.25 4.5 9
Kilometers
WGS 1984 UTM Zone 43N
Data Sources:
BODC, IHO, and IOC. 2003. GEBCO Digital Atlas (bathymetry).
Other data from Maldives agencies: CAA (airports); ME (administrative areas, island shorelines, reef boundaries, and water bodies); and MOE (education centers).

Map III.32: Gnaviyani Atoll, Education Centers

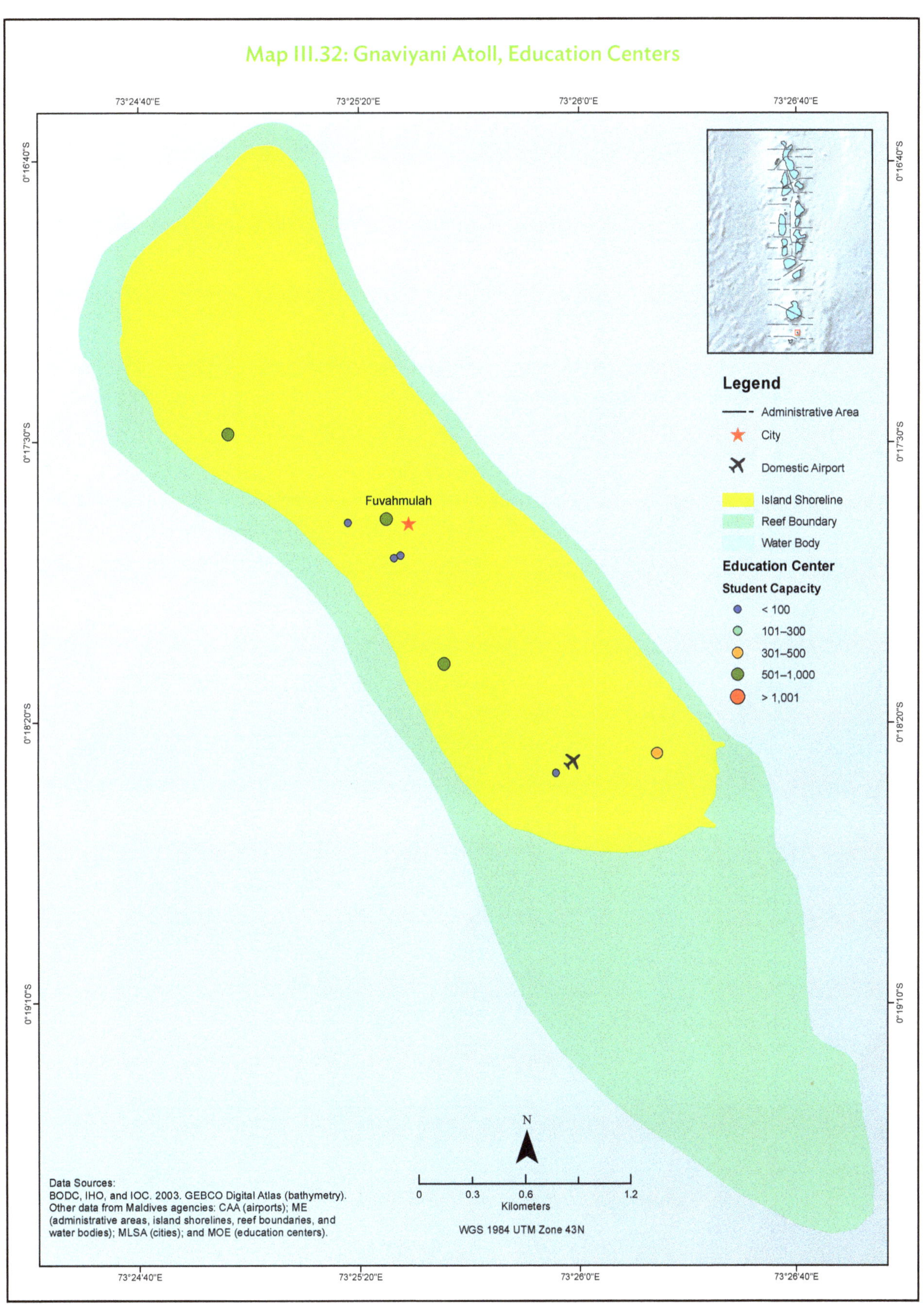

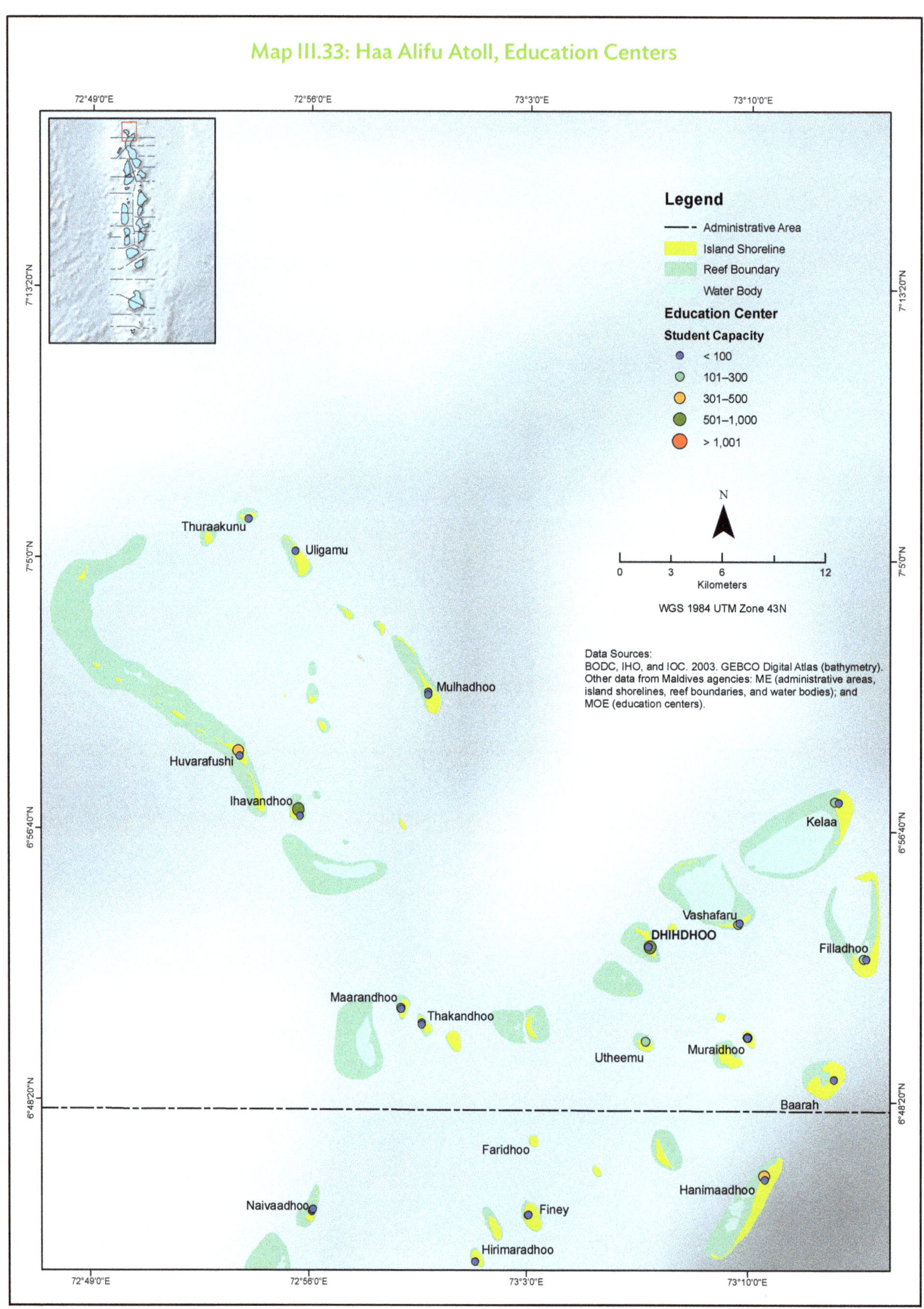
Map III.33: Haa Alifu Atoll, Education Centers
72°49'0"E
72°56'0"E
73°3'0"E
73°10'0"E
7°13'20"N
7°5'0"N
6°56'40"N
6°48'20"N
Legend
Administrative Area
Island Shoreline
Reef Boundary
Water Body
Education Center
Student Capacity
< 100
101–300
301–500
501–1,000
> 1,001
N
0 3 6 12
Kilometers
WGS 1984 UTM Zone 43N
Data Sources:
BODC, IHO, and IOC. 2003. GEBCO Digital Atlas (bathymetry).
Other data from Maldives agencies: ME (administrative areas, island shorelines, reef boundaries, and water bodies); and MOE (education centers).
Thuraakunu
Uligamu
Mulhadhoo
Huvarafushi
Ihavandhoo
Kelaa
Vashafaru
DHIHDHOO
Filladhoo
Maarandhoo
Thakandhoo
Utheemu
Muraidhoo
Baarah
Faridhoo
Hanimaadhoo
Naivaadhoo
Finey
Hirimaradhoo

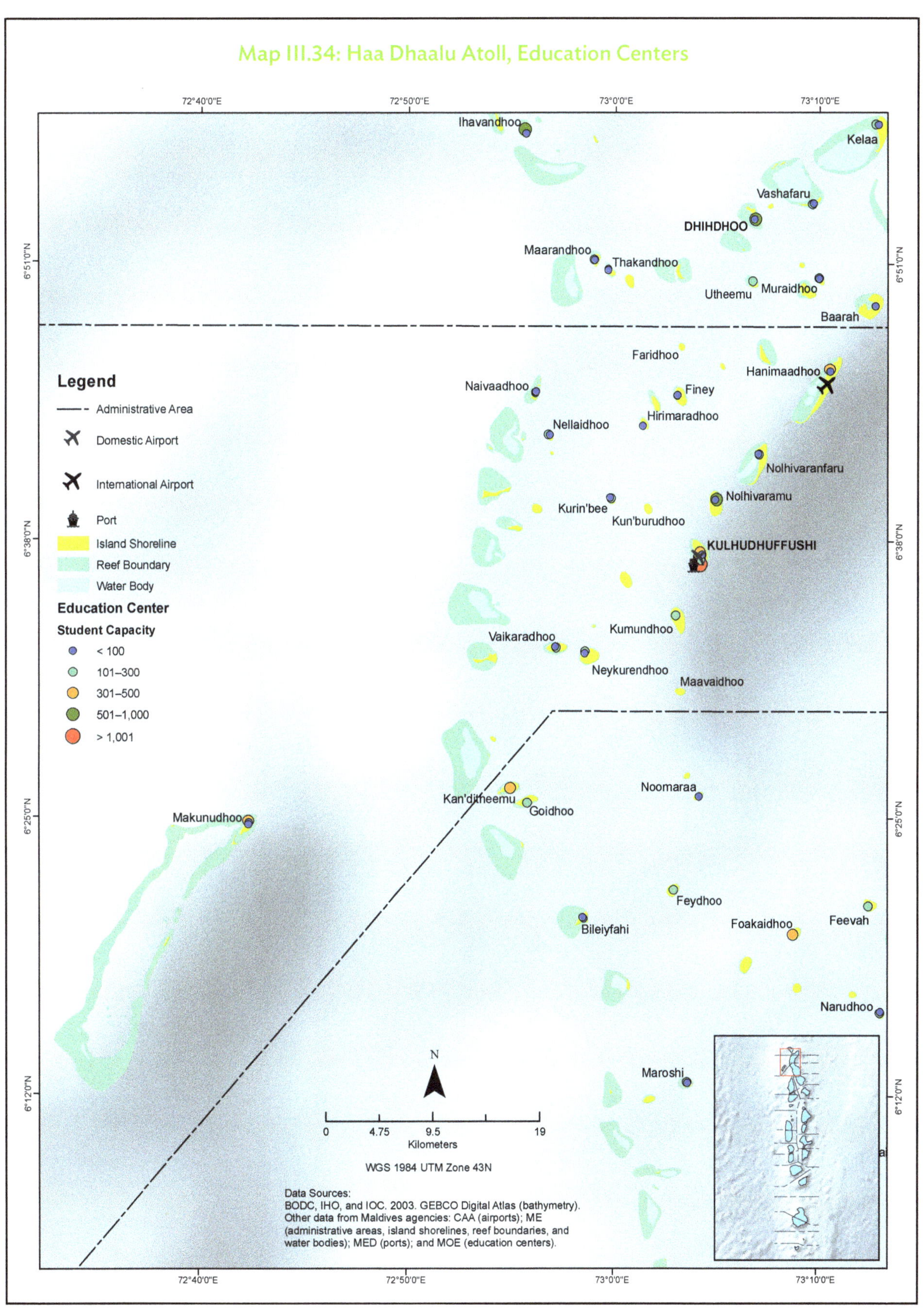
Map III.34: Haa Dhaalu Atoll, Education Centers
72°40'0"E
72°50'0"E
73°0'0"E
73°10'0"E
6°51'0"N
6°38'0"N
6°25'0"N
6°12'0"N
Ihavandhoo
Kelaa
Vashafaru
DHIHDHOO
Maarandhoo
Thakandhoo
Utheemu
Muraidhoo
Baarah
Faridhoo
Hanimaadhoo
Naivaadhoo
Finey
Hirimaradhoo
Nellaidhoo
Nolhivaranfaru
Nolhivaramu
Kurin'bee
Kun'burudhoo
KULHUDHUFFUSHI
Kumundhoo
Vaikaradhoo
Neykurendhoo
Maavaidhoo
Noomaraa
Kan'ditheemu
Goidhoo
Makunudhoo
Feydhoo
Bileiyfahi
Foakaidhoo
Feevah
Narudhoo
Maroshi
Legend
Administrative Area
Domestic Airport
International Airport
Port
Island Shoreline
Reef Boundary
Water Body
Education Center
Student Capacity
< 100
101–300
301–500
501–1,000
> 1,001
N
0
4.75
9.5
19
Kilometers
WGS 1984 UTM Zone 43N
Data Sources:
BODC, IHO, and IOC. 2003. GEBCO Digital Atlas (bathymetry).
Other data from Maldives agencies: CAA (airports); ME (administrative areas, island shorelines, reef boundaries, and water bodies); MED (ports); and MOE (education centers).

Map III.35: Laamu Atoll, Education Centers

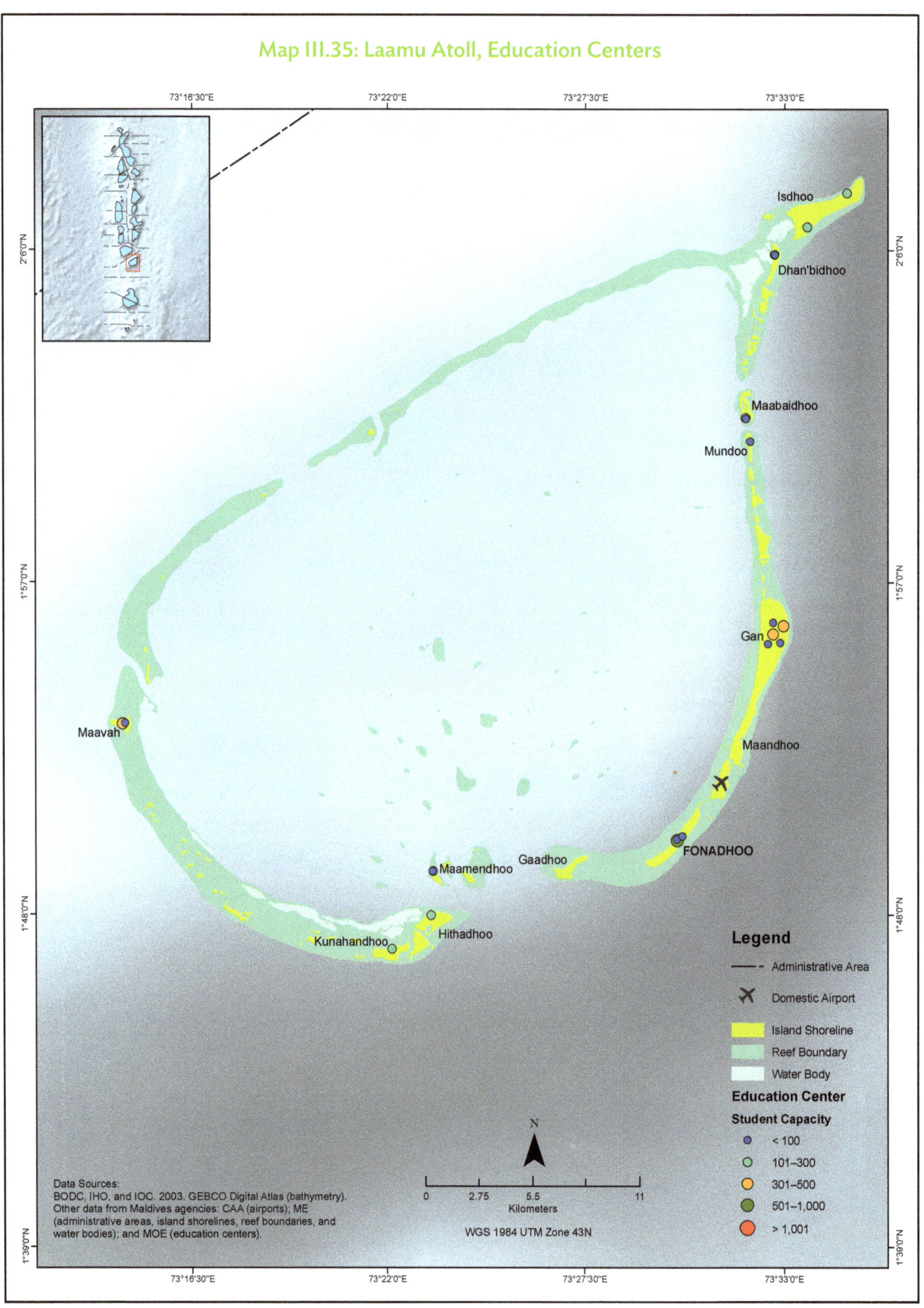

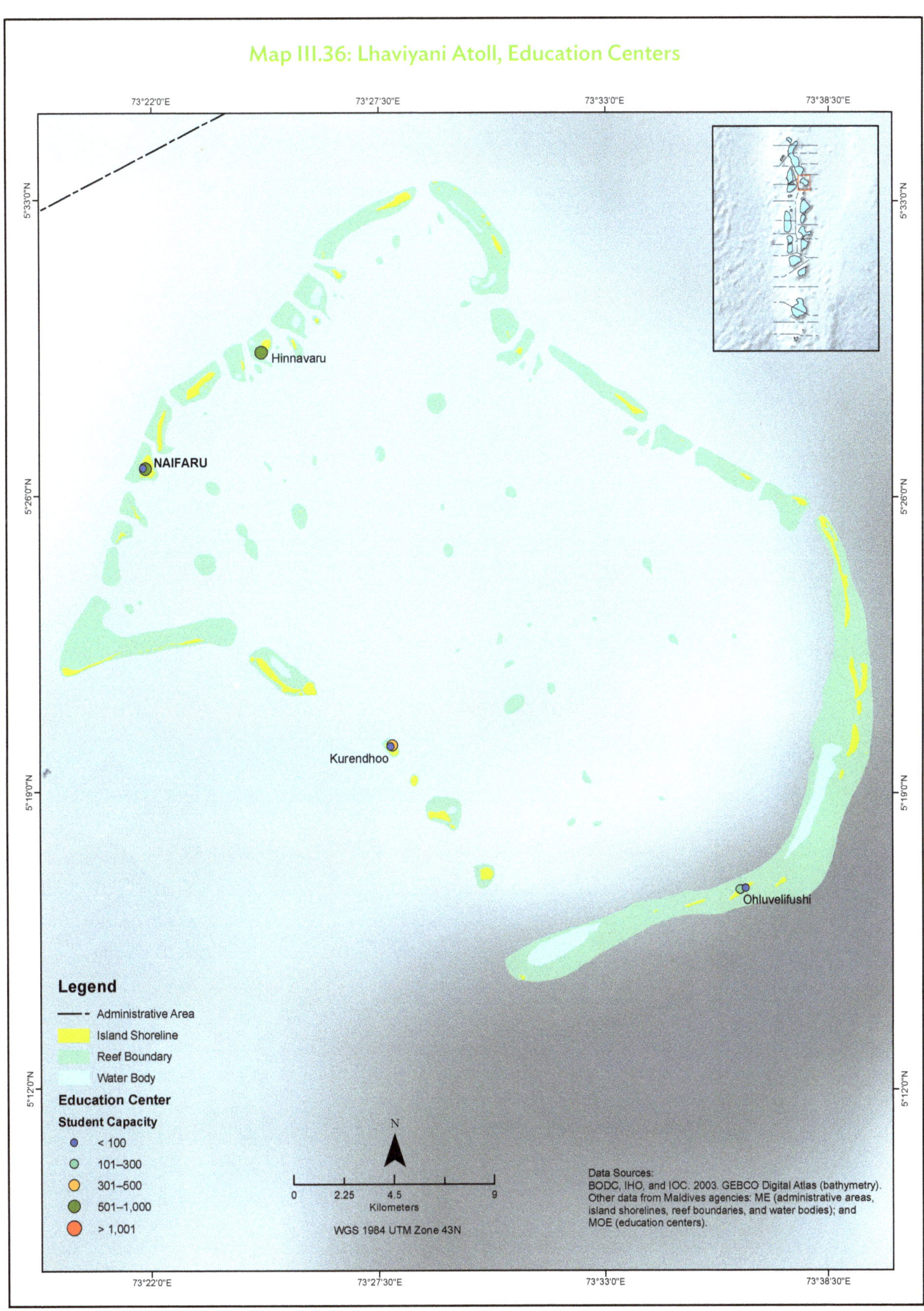
Map III.36: Lhaviyani Atoll, Education Centers
73°22'0"E
73°27'30"E
73°33'0"E
73°38'30"E
5°33'0"N
5°26'0"N
5°19'0"N
5°12'0"N
Hinnavaru
NAIFARU
Kurendhoo
Ohluvelifushi
Legend
Administrative Area
Island Shoreline
Reef Boundary
Water Body
Education Center
Student Capacity
< 100
101–300
301–500
501–1,000
> 1,001
N
0
2.25
4.5
9
Kilometers
WGS 1984 UTM Zone 43N
Data Sources:
BODC, IHO, and IOC. 2003. GEBCO Digital Atlas (bathymetry).
Other data from Maldives agencies: ME (administrative areas, island shorelines, reef boundaries, and water bodies); and MOE (education centers).

Map III.37: Meemu Atoll, Education Centers

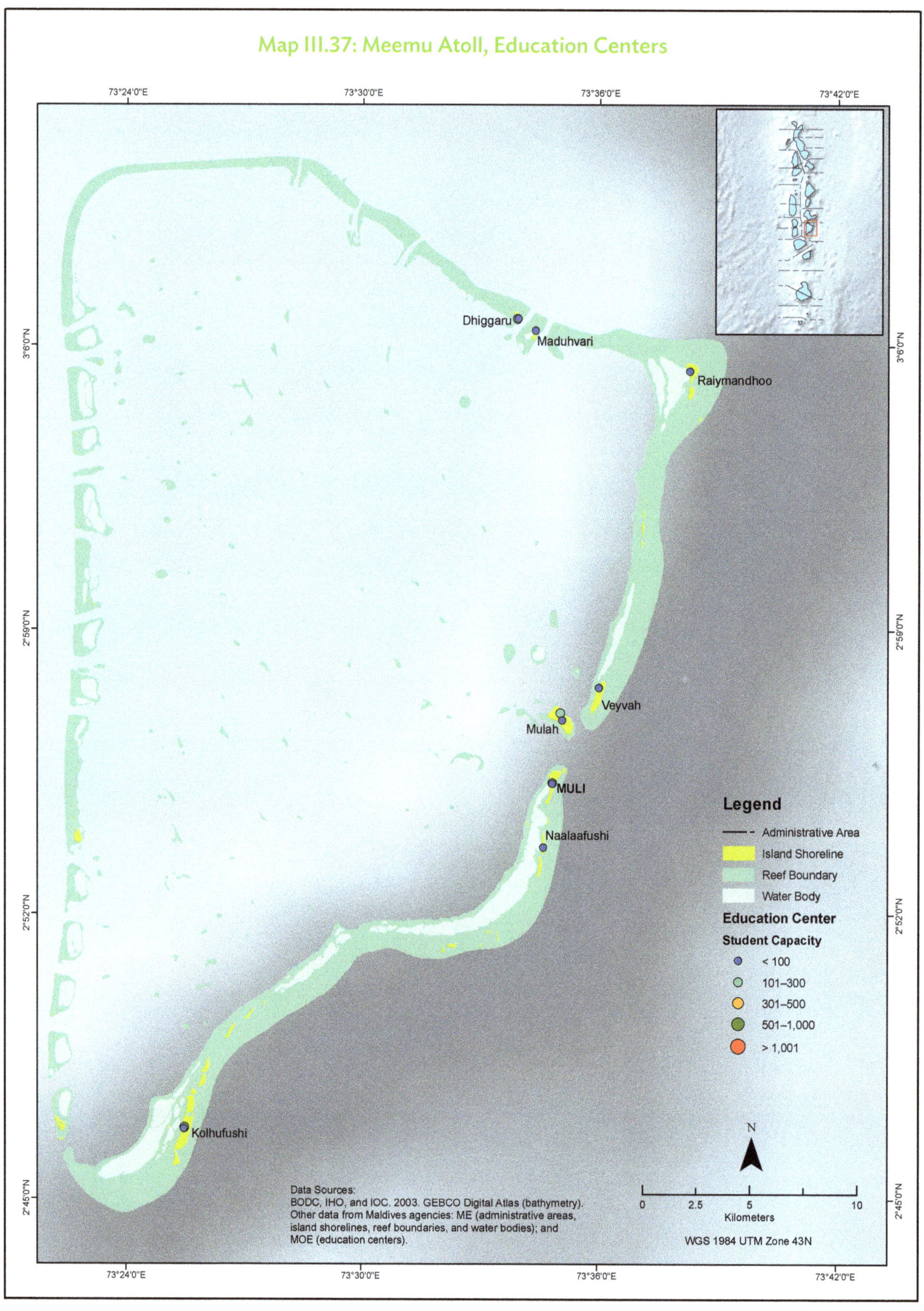

Map III.38: Noonu Atoll, Education Centers

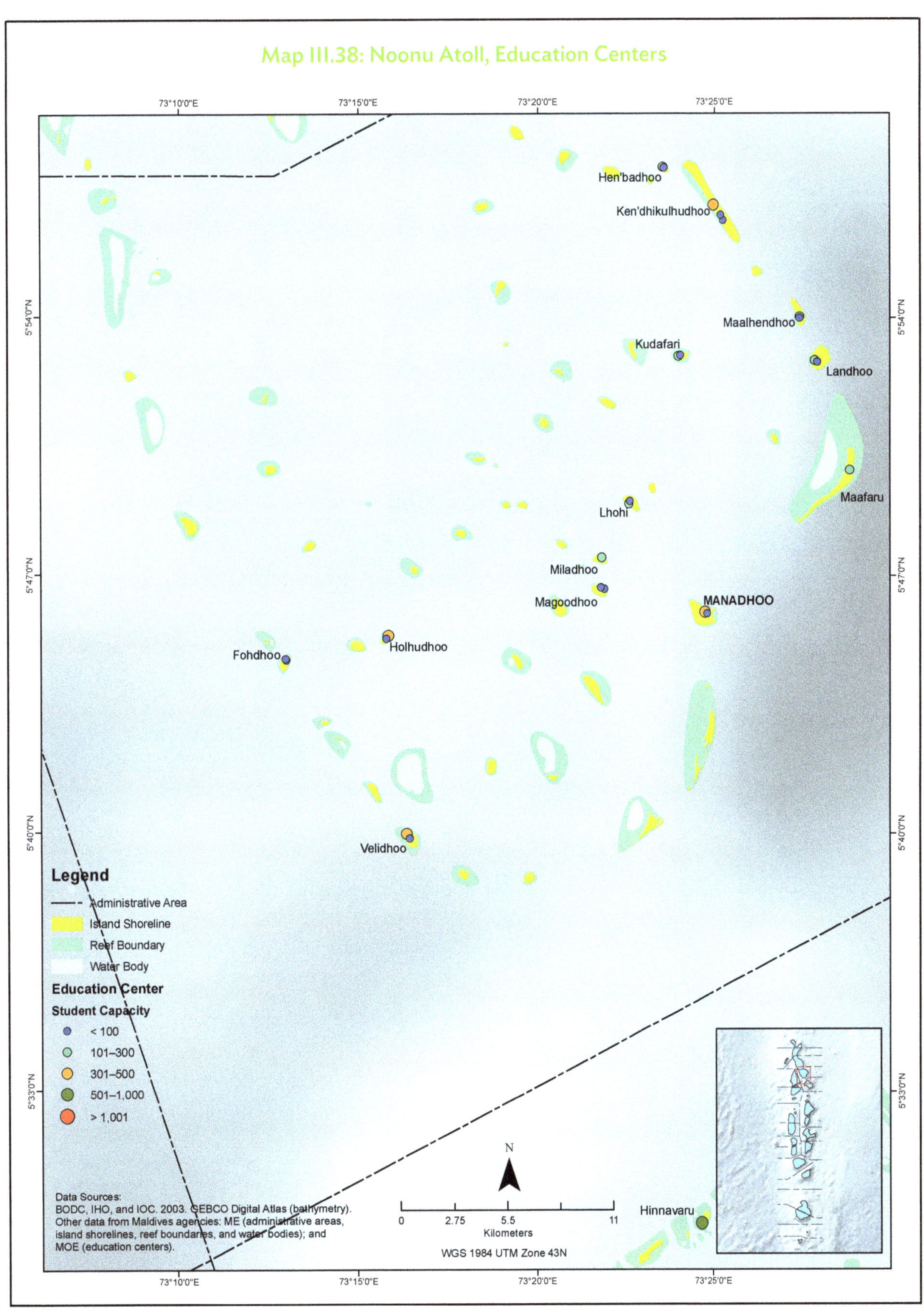

Map III.39: North Malé Atoll, Education Centers

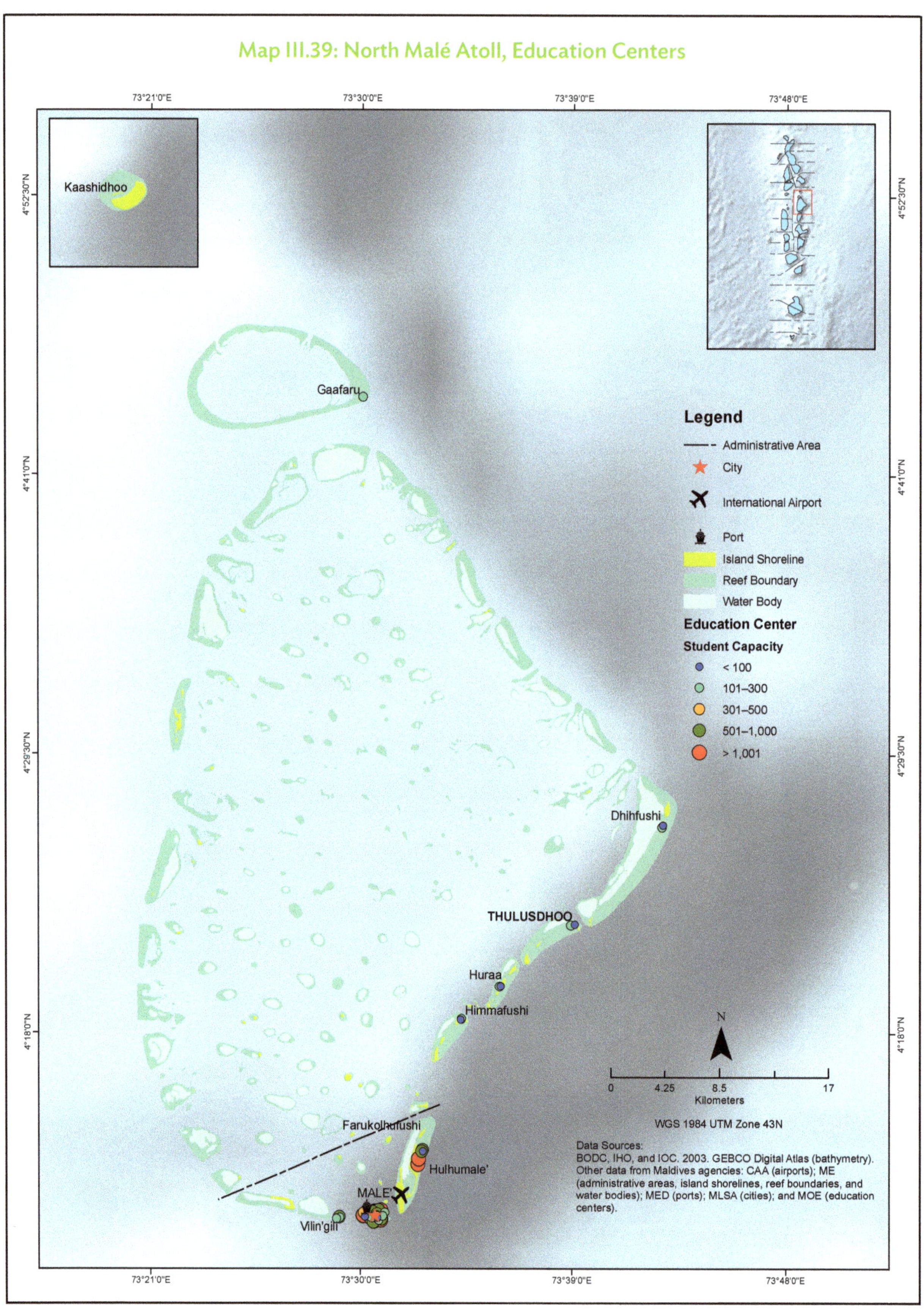

Map III.40: Raa Atoll, Education Centers

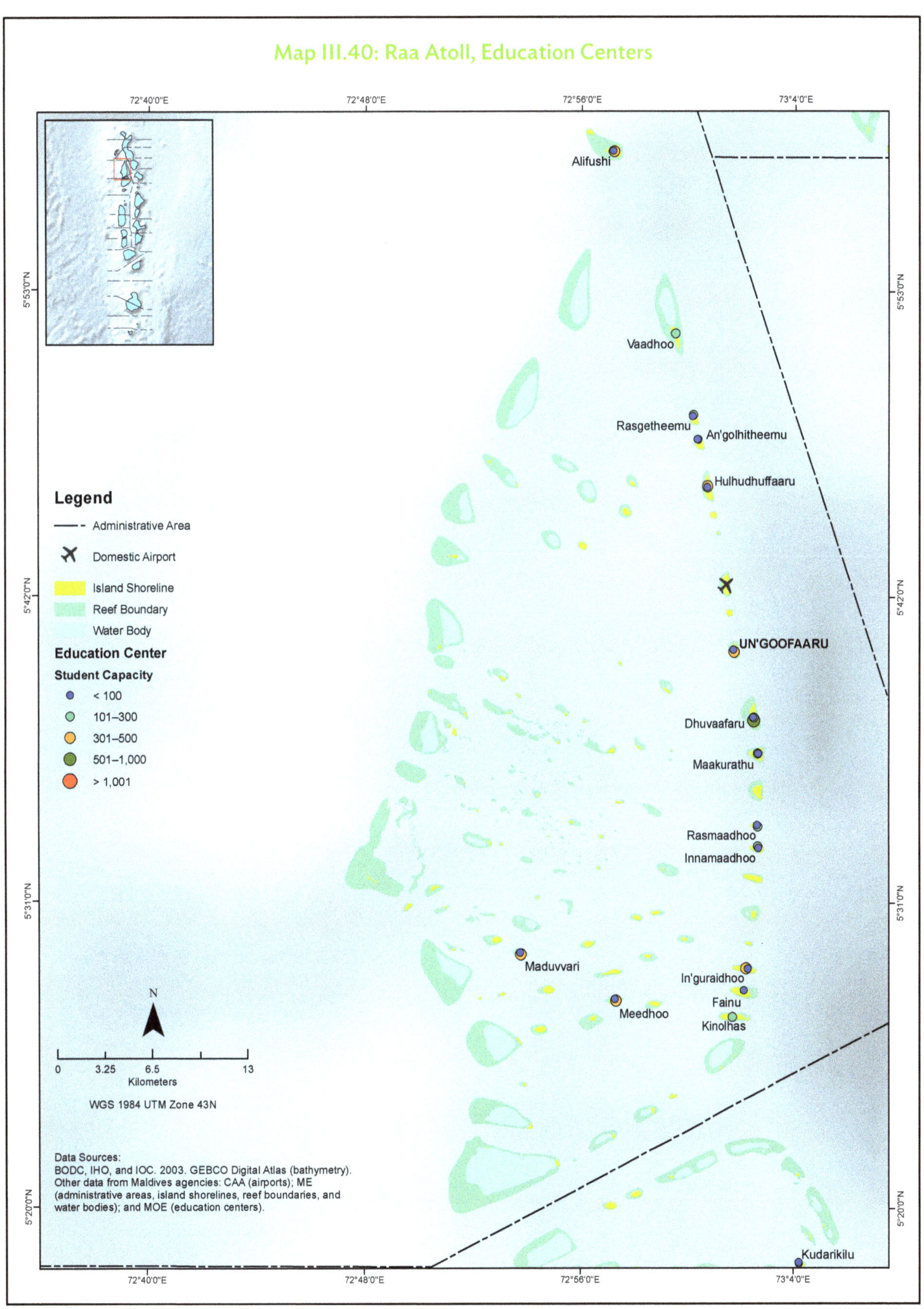

Map III.41: Shaviyani Atoll, Education Centers

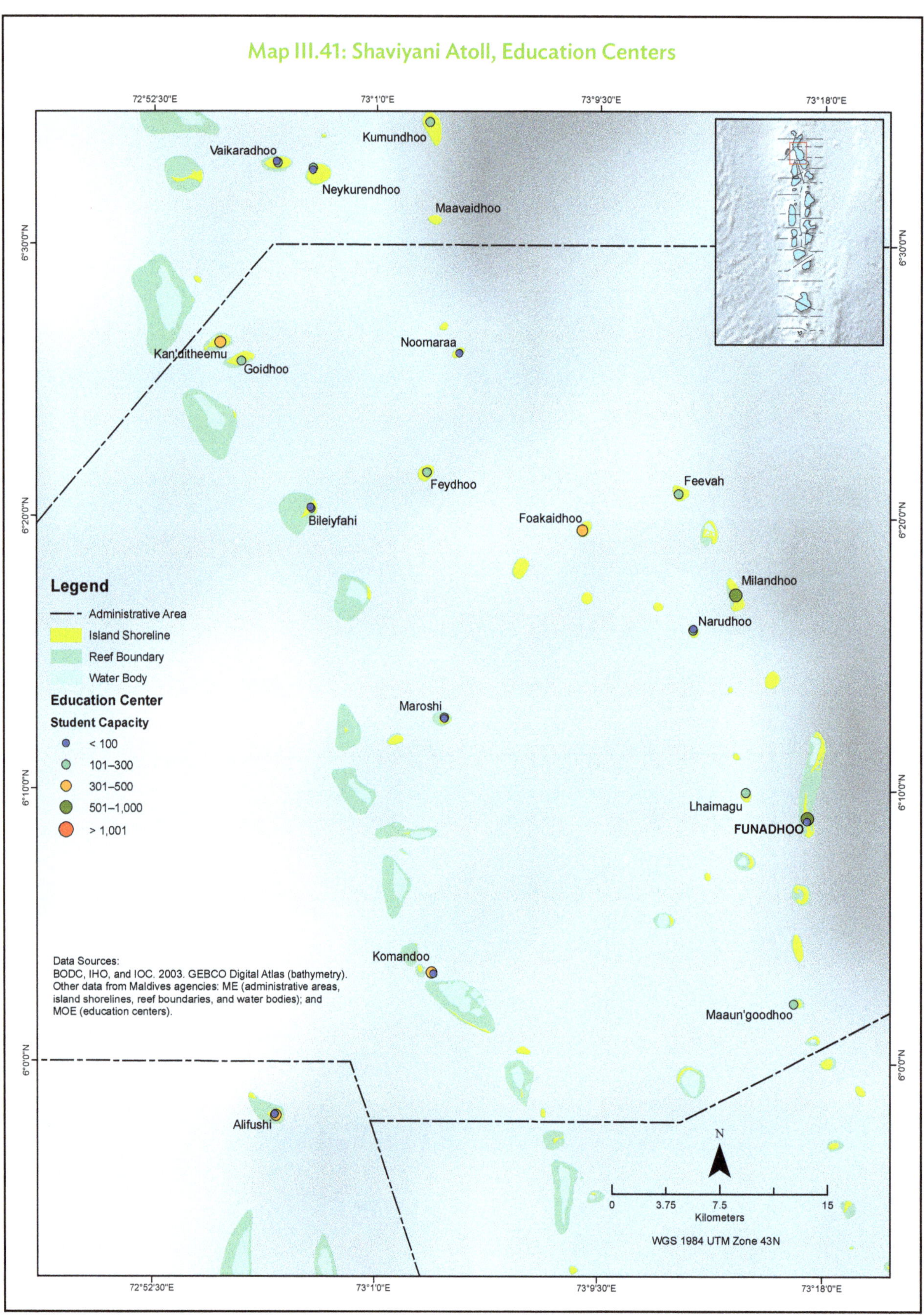

Map III.42: South Malé Atoll, Education Centers

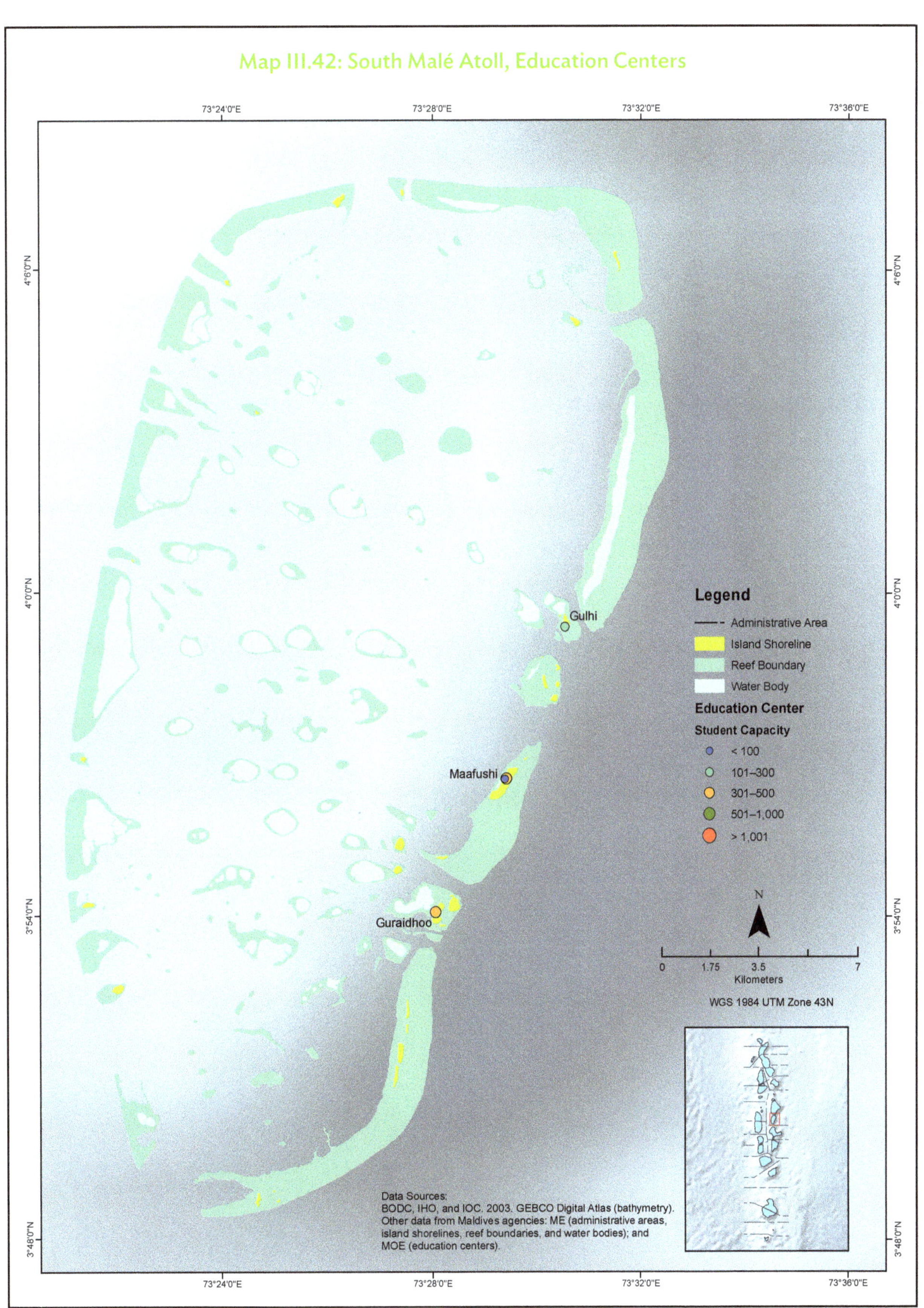

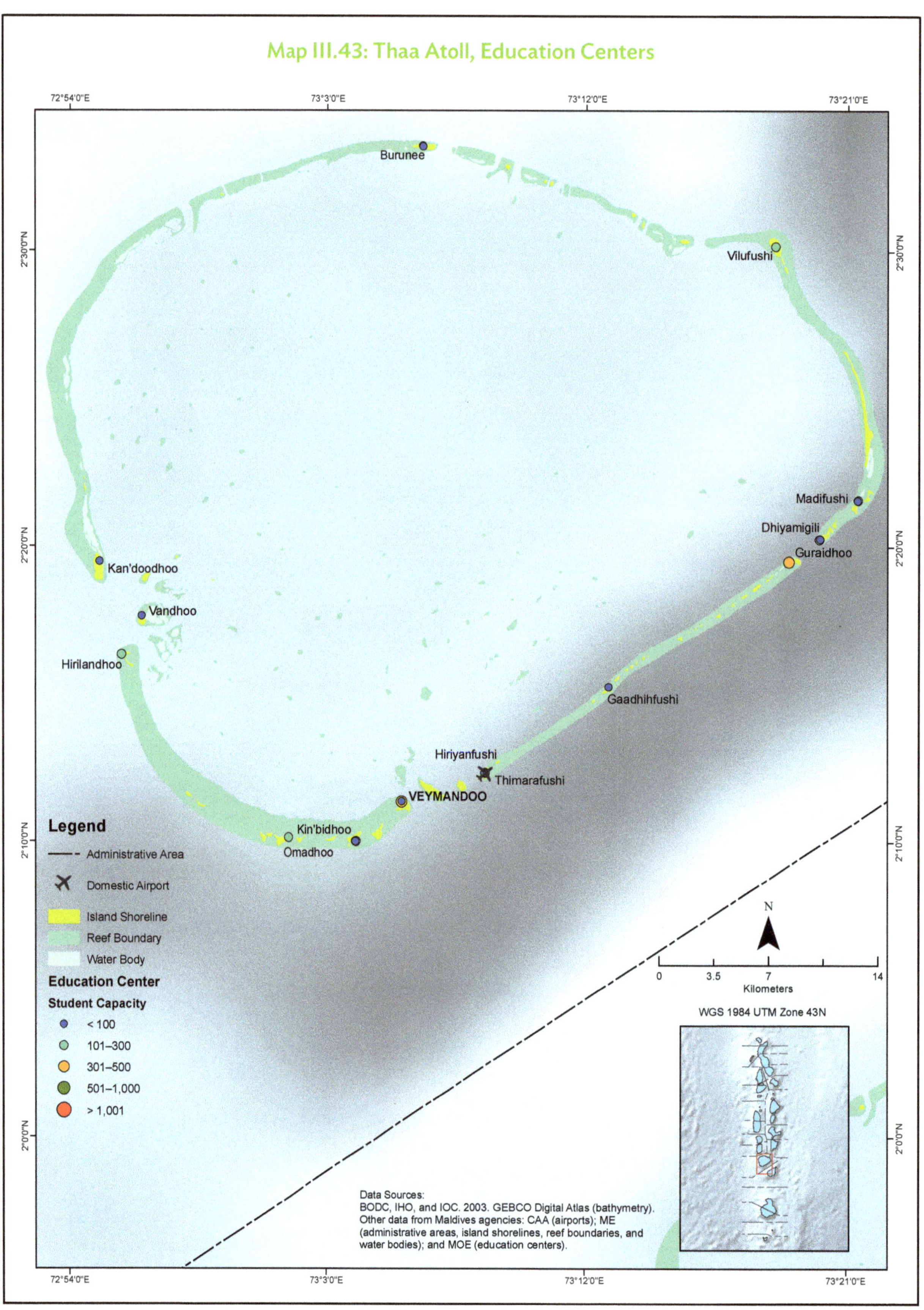
Map III.43: Thaa Atoll, Education Centers
72°54'0"E
73°3'0"E
73°12'0"E
73°21'0"E
2°30'0"N
2°20'0"N
2°10'0"N
2°0'0"N
Burunee
Vilufushi
Madifushi
Dhiyamigili
Guraidhoo
Kan'doodhoo
Vandhoo
Hirilandhoo
Gaadhihfushi
Hiriyanfushi
Thimarafushi
VEYMANDOO
Kin'bidhoo
Omadhoo
Legend
Administrative Area
Domestic Airport
Island Shoreline
Reef Boundary
Water Body
Education Center
Student Capacity
< 100
101–300
301–500
501–1,000
> 1,001
N
0
3.5
7
14
Kilometers
WGS 1984 UTM Zone 43N
Data Sources:
BODC, IHO, and IOC. 2003. GEBCO Digital Atlas (bathymetry).
Other data from Maldives agencies: CAA (airports); ME (administrative areas, island shorelines, reef boundaries, and water bodies); and MOE (education centers).

Map III.44: Vaavu Atoll, Education Centers

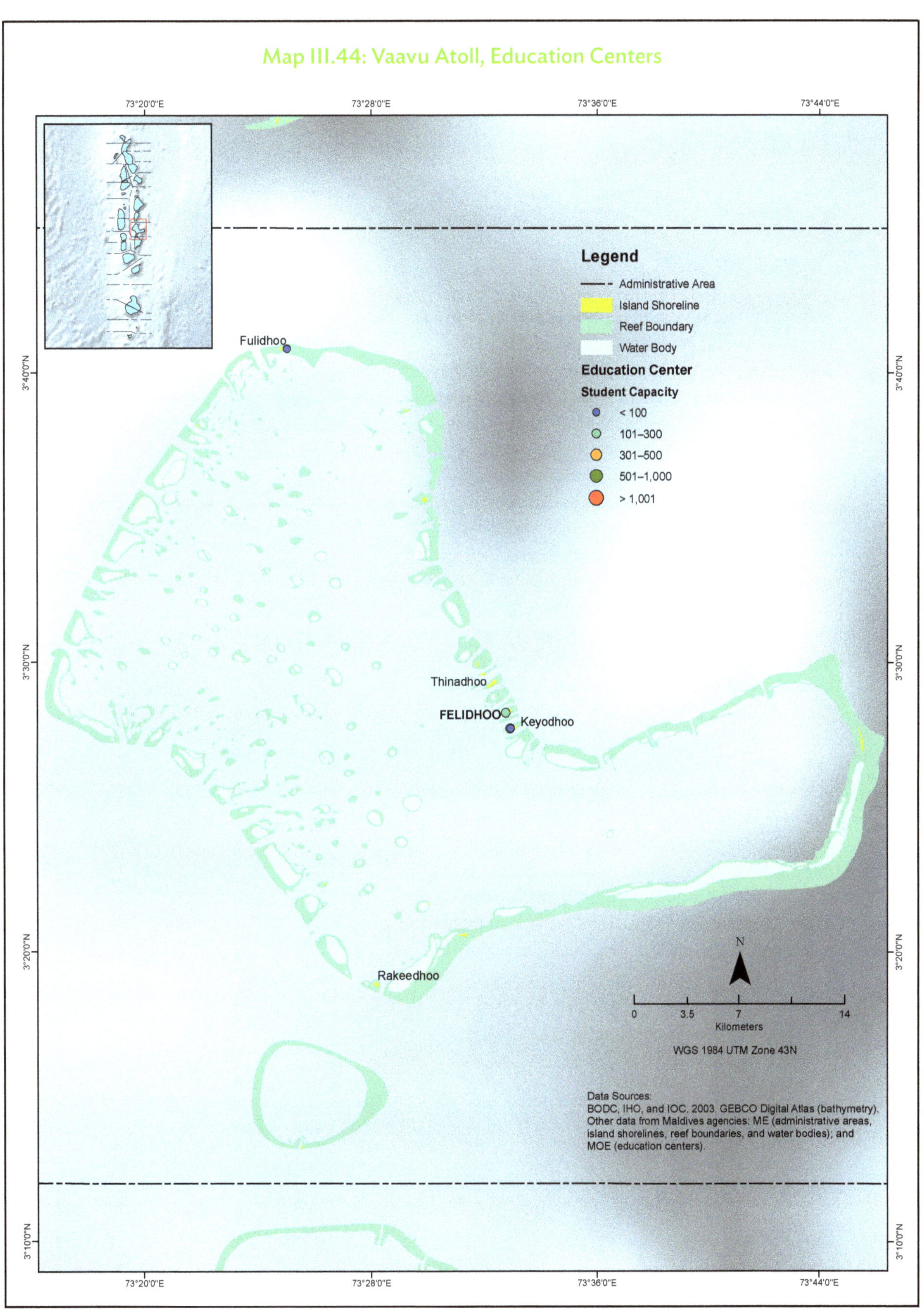

Health

Changing climate—particularly extreme weather conditions—and geologic hazards can affect the health of people, especially persons with disabilities, the elderly, children, and pregnant women. These can cause injuries and even death. Warmer temperature and wetter monsoons, as seen in climate projections for Maldives (*Multihazard Risk Atlas of Maldives: Climate and Geophysical Hazards—Volume II*), could increase the incidences of vector-borne diseases like dengue (Ahmed and Suphachalasai 2014) and water-borne diseases (Intergovernmental Panel on Climate Change 2001). Inundation, as a result of elevated sea levels or rainfall-induced flooding, could generate an environment favorable to breeding of mosquitoes and other disease vectors (Ahmed and Suphachalasai 2014).

This volume maps out health facilities as one indicator of vulnerability to climate and disaster risks. Based on 2016 data from the Ministry of Health, Maldives has a total of 184 hospitals classified into health centers, private hospitals, government hospitals, atoll hospitals, and regional hospitals. These hospitals are distributed mostly among the highly populated atolls of Raa, Kaafu, and Haa Alifu.

Table III.3: Types of Hospitals in Maldives

Hospital Type	Number
Health center	161
Atoll hospital	13
Regional hospital	6
Government hospital	2
Private hospital	2

Source: Ministry of Health, 2016.

Table III.4: Number of Hospitals per Atoll, Maldives

ATOLL	Number of Hospitals
Alifu Dhaalu Atoll	1
Alifu Alifu Atoll	8
Alifu Dhaalu Atoll	9
Baa Atoll	13
Dhaalu Atoll	7
Faafu Atoll	5
Gaafu Alifu Atoll	8
Gaafu Dhaalu Atoll	9
Gnaviyani Atoll	1
Haa Alifu Atoll	14
Haa Dhaalu Atoll	13
Kaafu Atoll	14
Laamu Atoll	11
Lhaviyani Atoll	4
Meemu Atoll	8
Noonu Atoll	12
Raa Atoll	15
Seenu Atoll	4
Shaviyani Atoll	14
Thaa Atoll	13
Vaavu Atoll	1
TOTAL	**184**

Source: Ministry of Health, 2016.

Map III.45: Maldives, Healthcare Facilities

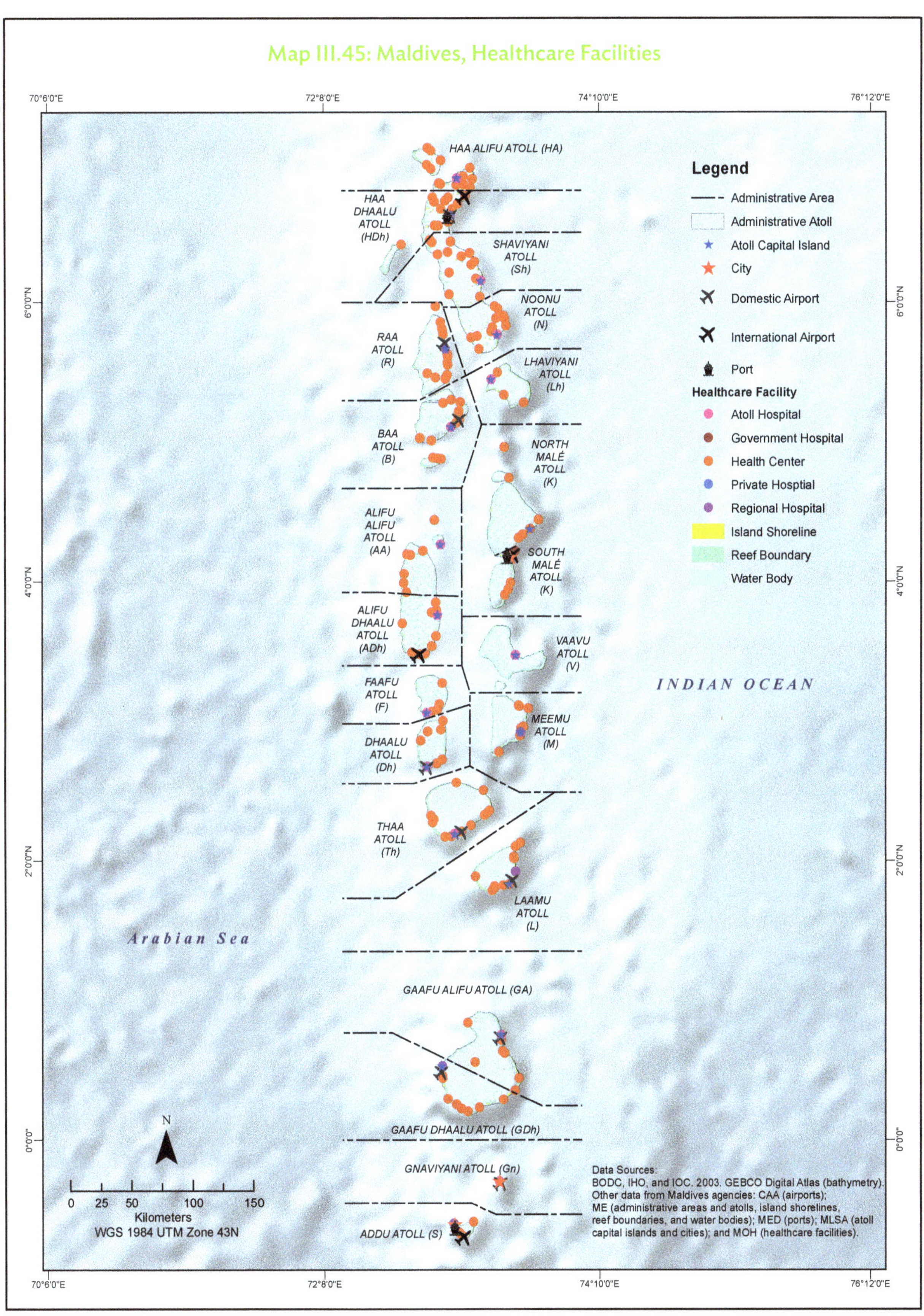

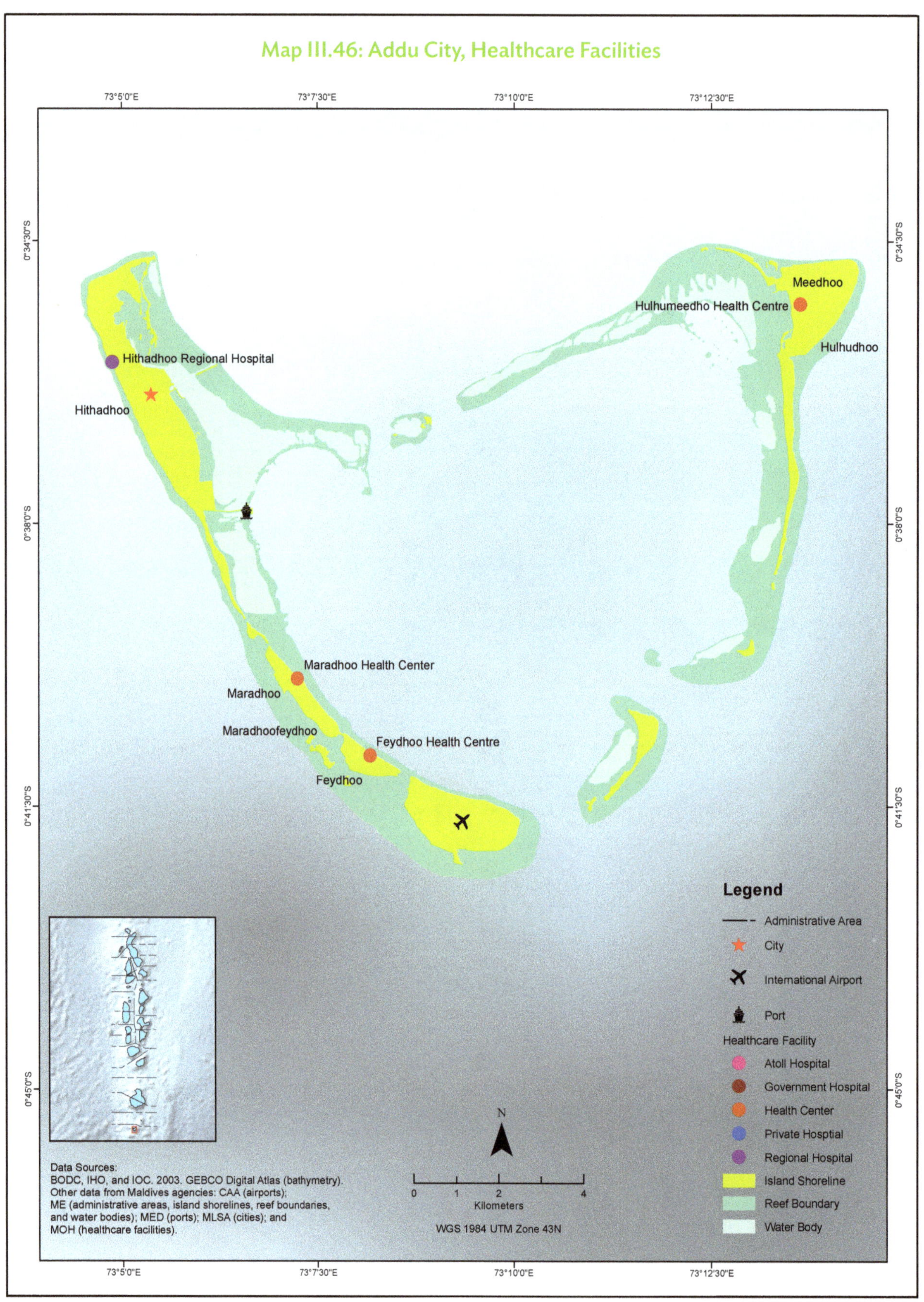
Map III.46: Addu City, Healthcare Facilities
73°5'0"E
73°7'30"E
73°10'0"E
73°12'30"E
0°34'30"S
0°38'0"S
0°41'30"S
0°45'0"S
Meedhoo
Hulhumeedho Health Centre
Hulhudhoo
Hithadhoo Regional Hospital
Hithadhoo
Maradhoo Health Center
Maradhoo
Maradhoofeydhoo
Feydhoo Health Centre
Feydhoo
Legend
Administrative Area
City
International Airport
Port
Healthcare Facility
Atoll Hospital
Government Hospital
Health Center
Private Hosptial
Regional Hospital
Island Shoreline
Reef Boundary
Water Body
N
0 1 2 4
Kilometers
WGS 1984 UTM Zone 43N
Data Sources:
BODC, IHO, and IOC. 2003. GEBCO Digital Atlas (bathymetry).
Other data from Maldives agencies: CAA (airports);
ME (administrative areas, island shorelines, reef boundaries, and water bodies); MED (ports); MLSA (cities); and
MOH (healthcare facilities).

Map III.47: Alifu Alifu Atoll, Healthcare Facilities

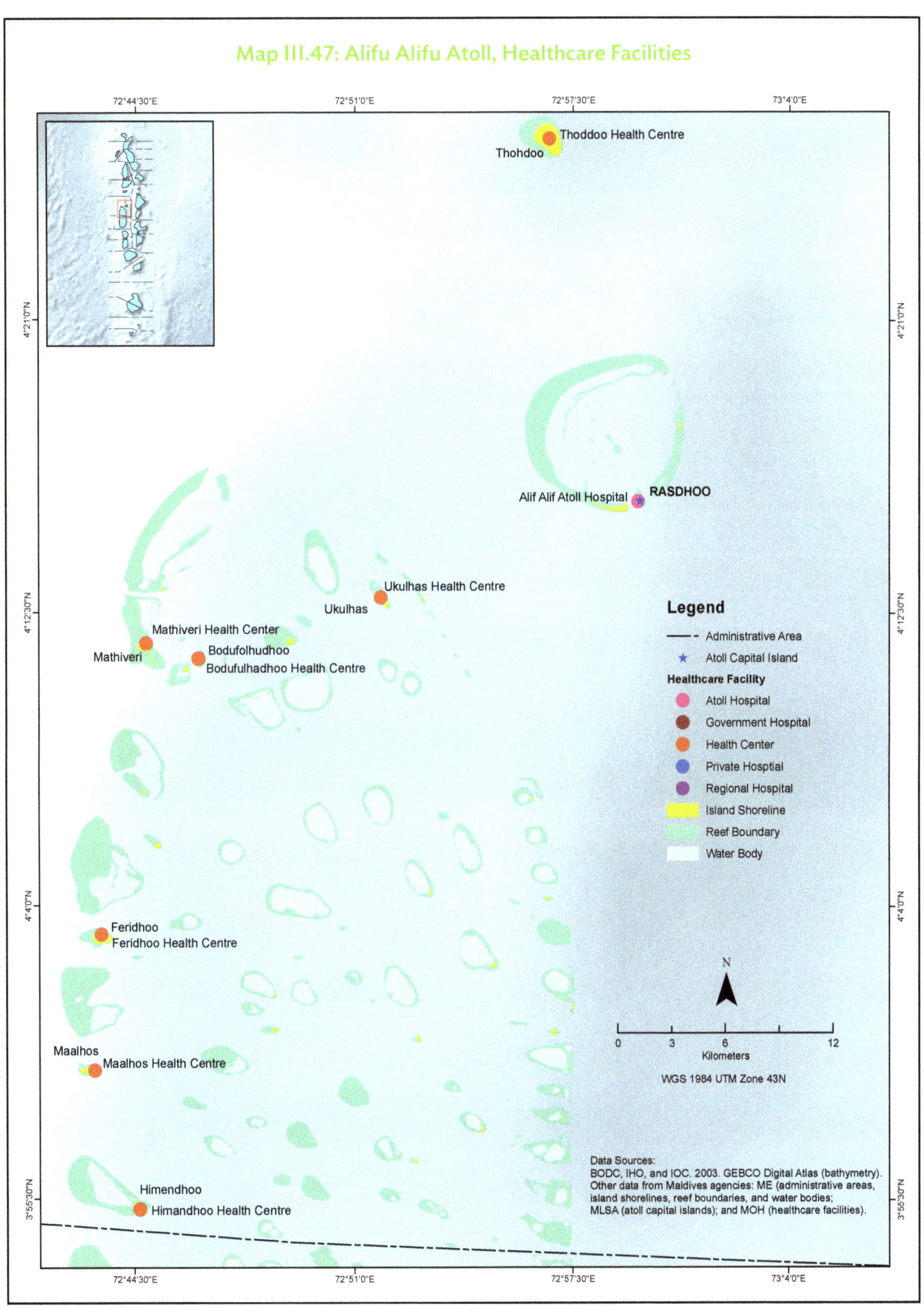

Map III.48: Alifu Dhaalu Atoll, Healthcare Facilities

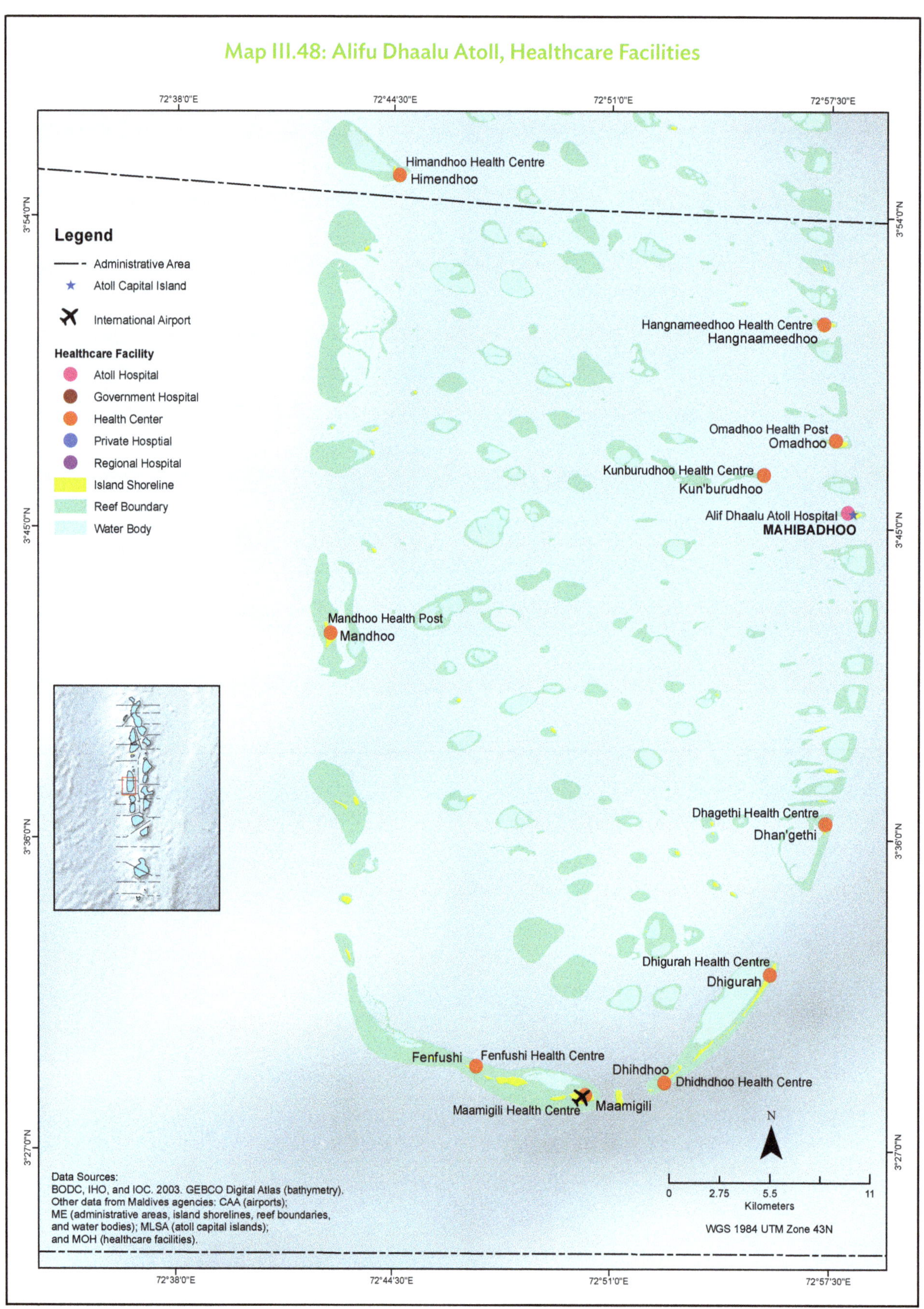

Map III.49: Baa Atoll, Healthcare Facilities

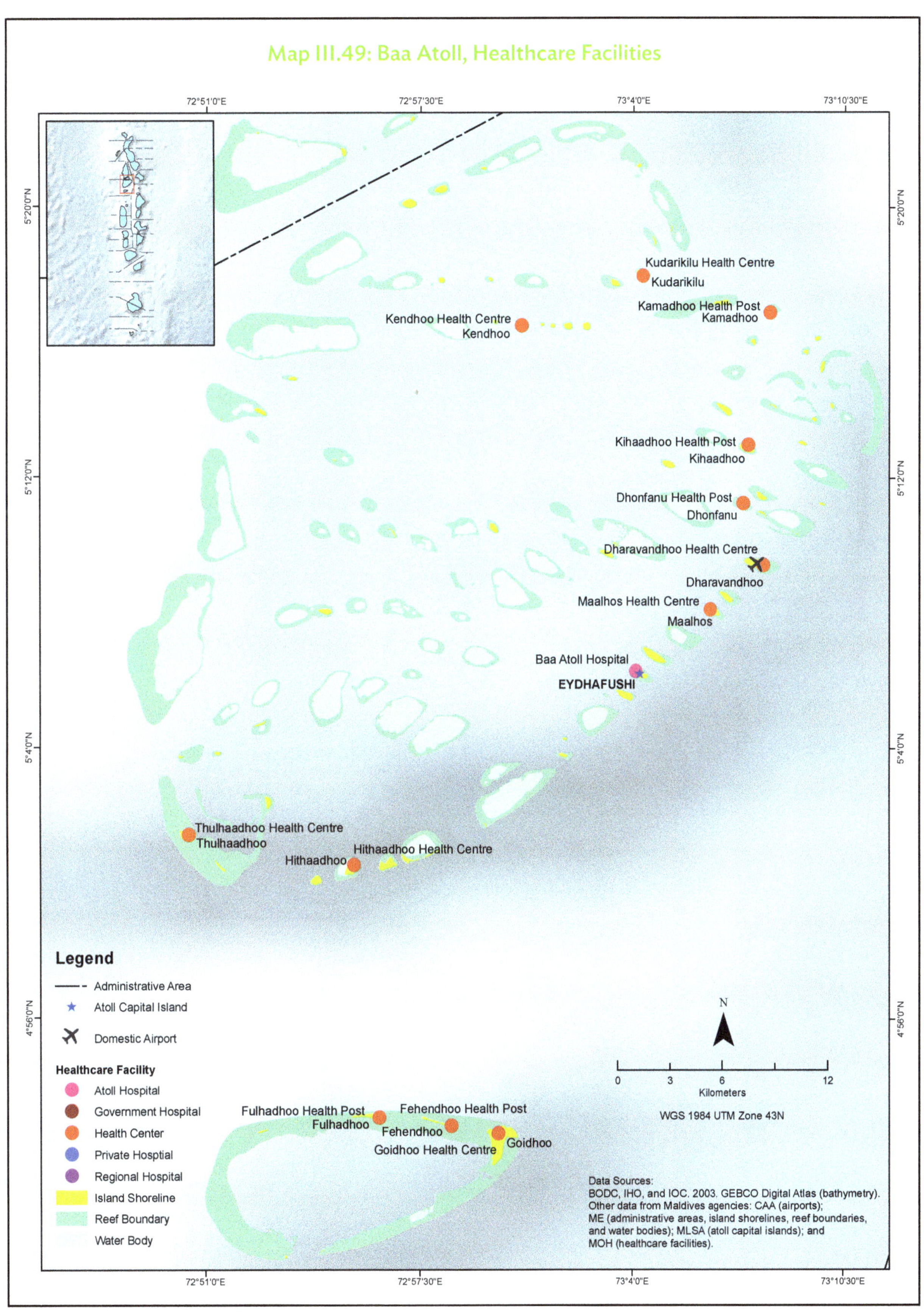

Map III.50: Dhaalu Atoll, Healthcare Facilities

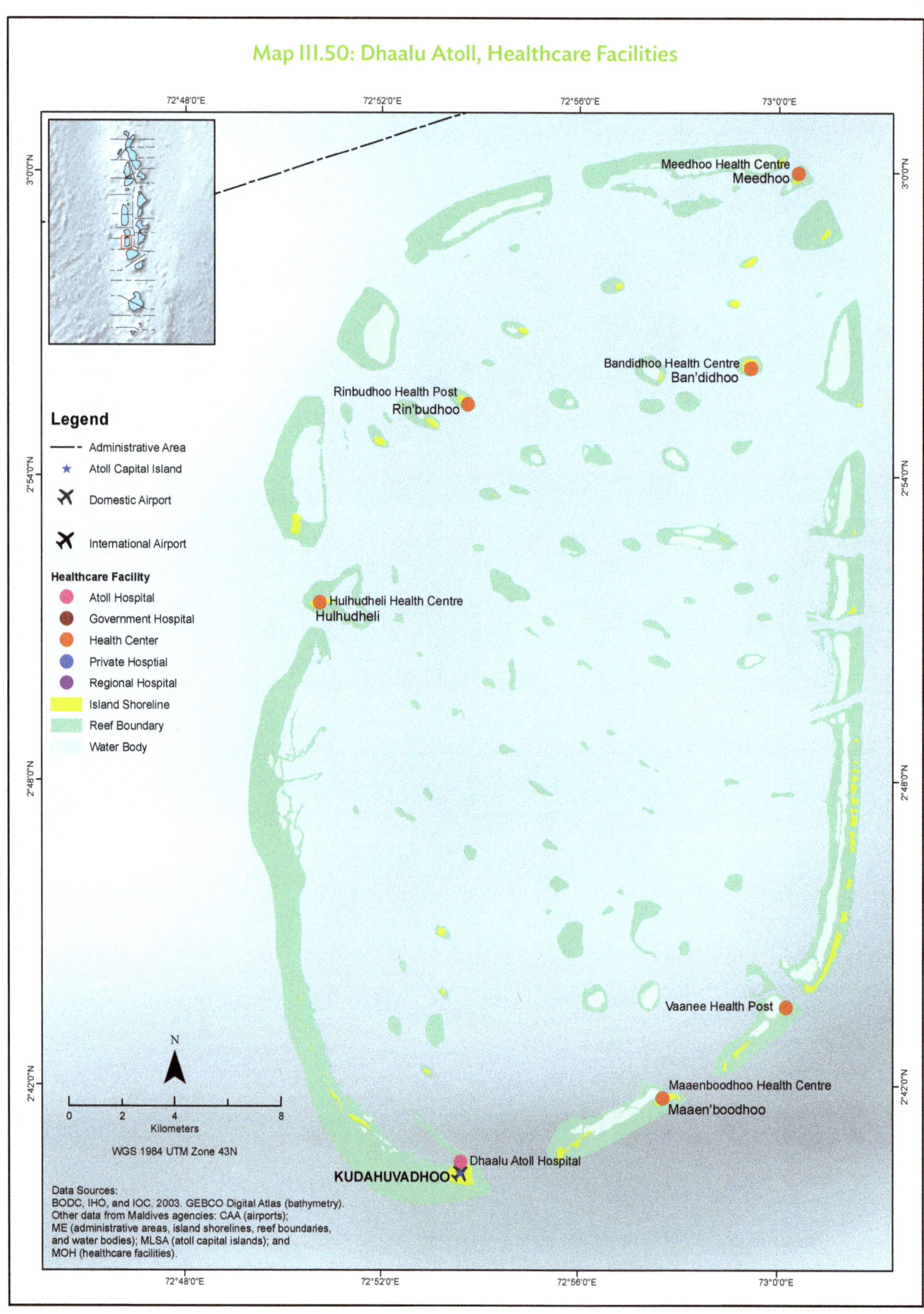

Map III.51: Faafu Atoll, Healthcare Facilities

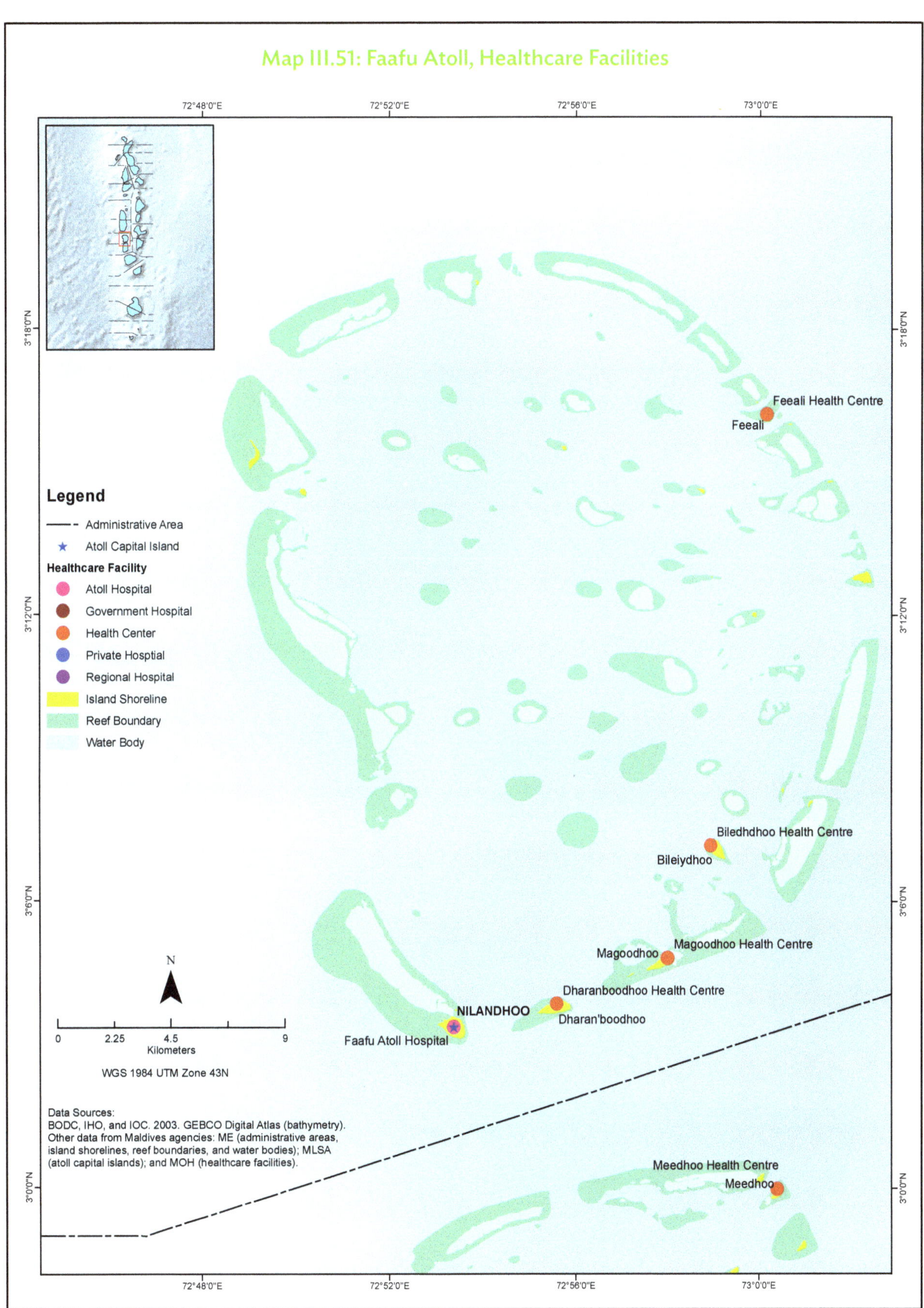

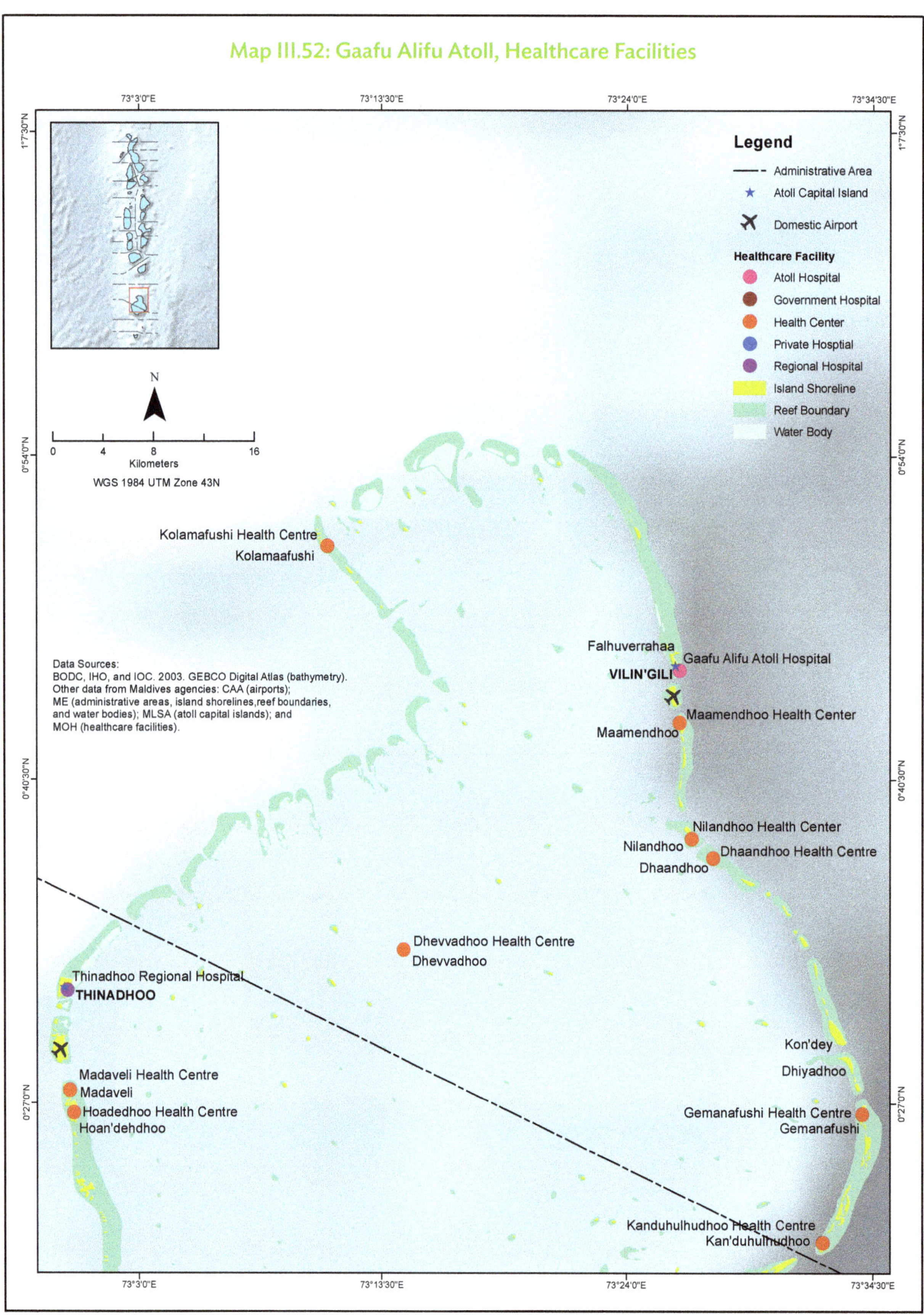
Map III.52: Gaafu Alifu Atoll, Healthcare Facilities
73°3'0"E
73°13'30"E
73°24'0"E
73°34'30"E
1°7'30"N
0°54'0"N
0°40'30"N
0°27'0"N
Legend
Administrative Area
Atoll Capital Island
Domestic Airport
Healthcare Facility
Atoll Hospital
Government Hospital
Health Center
Private Hosptial
Regional Hospital
Island Shoreline
Reef Boundary
Water Body
N
0 4 8 16
Kilometers
WGS 1984 UTM Zone 43N
Data Sources:
BODC, IHO, and IOC. 2003. GEBCO Digital Atlas (bathymetry).
Other data from Maldives agencies: CAA (airports);
ME (administrative areas, island shorelines,reef boundaries,
and water bodies); MLSA (atoll capital islands); and
MOH (healthcare facilities).
Kolamafushi Health Centre
Kolamaafushi
Falhuverrahaa
Gaafu Alifu Atoll Hospital
VILIN'GILI
Maamendhoo Health Center
Maamendhoo
Nilandhoo Health Center
Nilandhoo
Dhaandhoo Health Centre
Dhaandhoo
Dhevvadhoo Health Centre
Dhevvadhoo
Thinadhoo Regional Hospital
THINADHOO
Madaveli Health Centre
Madaveli
Hoadedhoo Health Centre
Hoan'dehdhoo
Kon'dey
Dhiyadhoo
Gemanafushi Health Centre
Gemanafushi
Kanduhulhudhoo Health Centre
Kan'duhulhudhoo

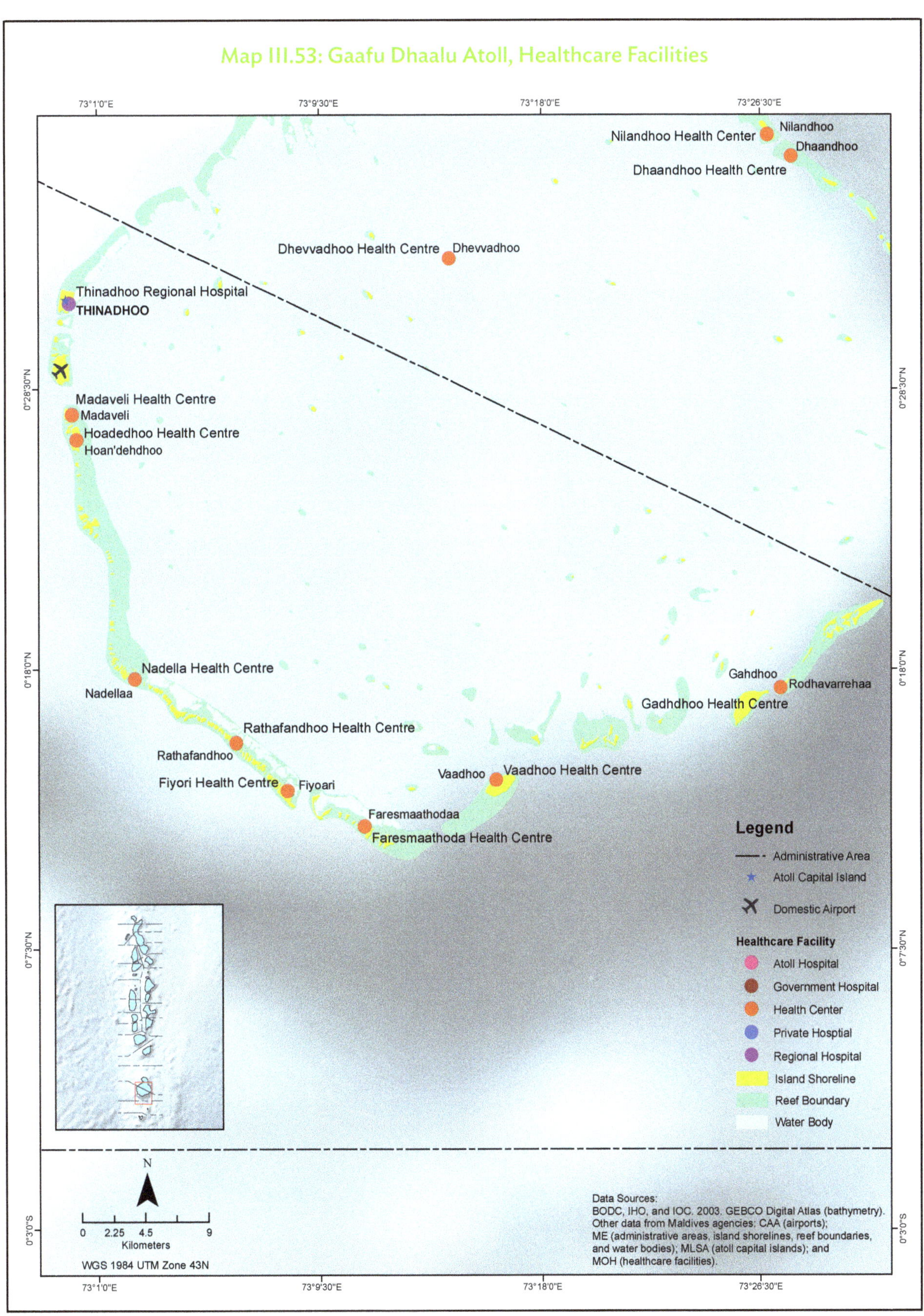
Map III.53: Gaafu Dhaalu Atoll, Healthcare Facilities
73°1'0"E
73°9'30"E
73°18'0"E
73°26'30"E
0°28'30"N
0°18'0"N
0°7'30"N
0°3'0"S
Nilandhoo
Nilandhoo Health Center
Dhaandhoo
Dhaandhoo Health Centre
Dhevvadhoo Health Centre
Dhevvadhoo
Thinadhoo Regional Hospital
THINADHOO
Madaveli Health Centre
Madaveli
Hoadedhoo Health Centre
Hoan'dehdhoo
Nadella Health Centre
Nadellaa
Rathafandhoo Health Centre
Rathafandhoo
Fiyori Health Centre
Fiyoari
Faresmaathodaa
Faresmaathoda Health Centre
Vaadhoo
Vaadhoo Health Centre
Gahdhoo
Rodhavarrehaa
Gadhdhoo Health Centre
Legend
Administrative Area
Atoll Capital Island
Domestic Airport
Healthcare Facility
Atoll Hospital
Government Hospital
Health Center
Private Hosptial
Regional Hospital
Island Shoreline
Reef Boundary
Water Body
N
0 2.25 4.5 9
Kilometers
WGS 1984 UTM Zone 43N
Data Sources:
BODC, IHO, and IOC. 2003. GEBCO Digital Atlas (bathymetry).
Other data from Maldives agencies: CAA (airports);
ME (administrative areas, island shorelines, reef boundaries, and water bodies); MLSA (atoll capital islands); and
MOH (healthcare facilities).

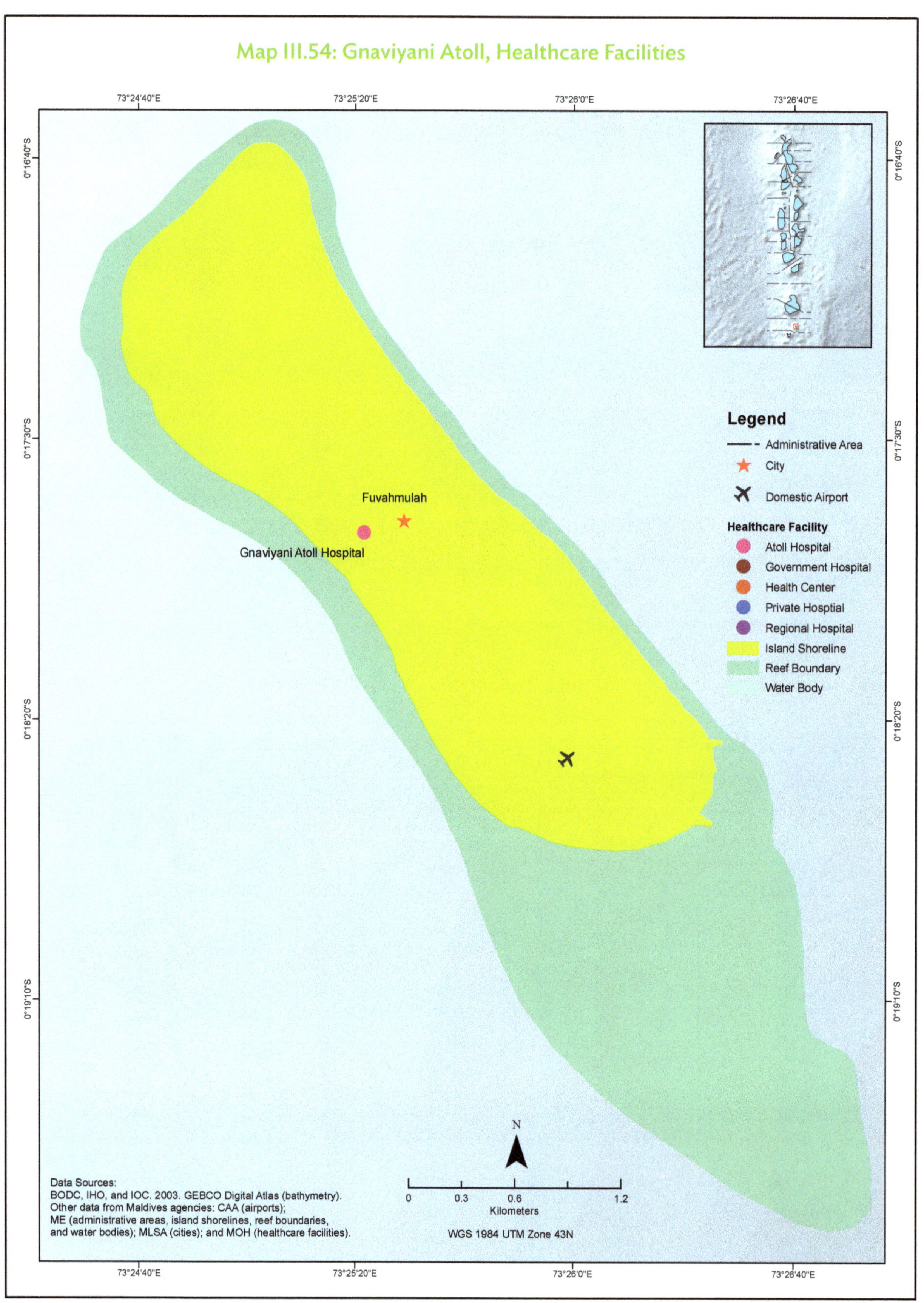
Map III.54: Gnaviyani Atoll, Healthcare Facilities
73°24'40"E
73°25'20"E
73°26'0"E
73°26'40"E
0°16'40"S
0°17'30"S
0°18'20"S
0°19'10"S
Fuvahmulah
Gnaviyani Atoll Hospital
Legend
Administrative Area
City
Domestic Airport
Healthcare Facility
Atoll Hospital
Government Hospital
Health Center
Private Hosptial
Regional Hospital
Island Shoreline
Reef Boundary
Water Body
N
0 0.3 0.6 1.2
Kilometers
WGS 1984 UTM Zone 43N
Data Sources:
BODC, IHO, and IOC. 2003. GEBCO Digital Atlas (bathymetry).
Other data from Maldives agencies: CAA (airports);
ME (administrative areas, island shorelines, reef boundaries,
and water bodies); MLSA (cities); and MOH (healthcare facilities).

Map III.55: Haa Alifu Atoll, Healthcare Facilities

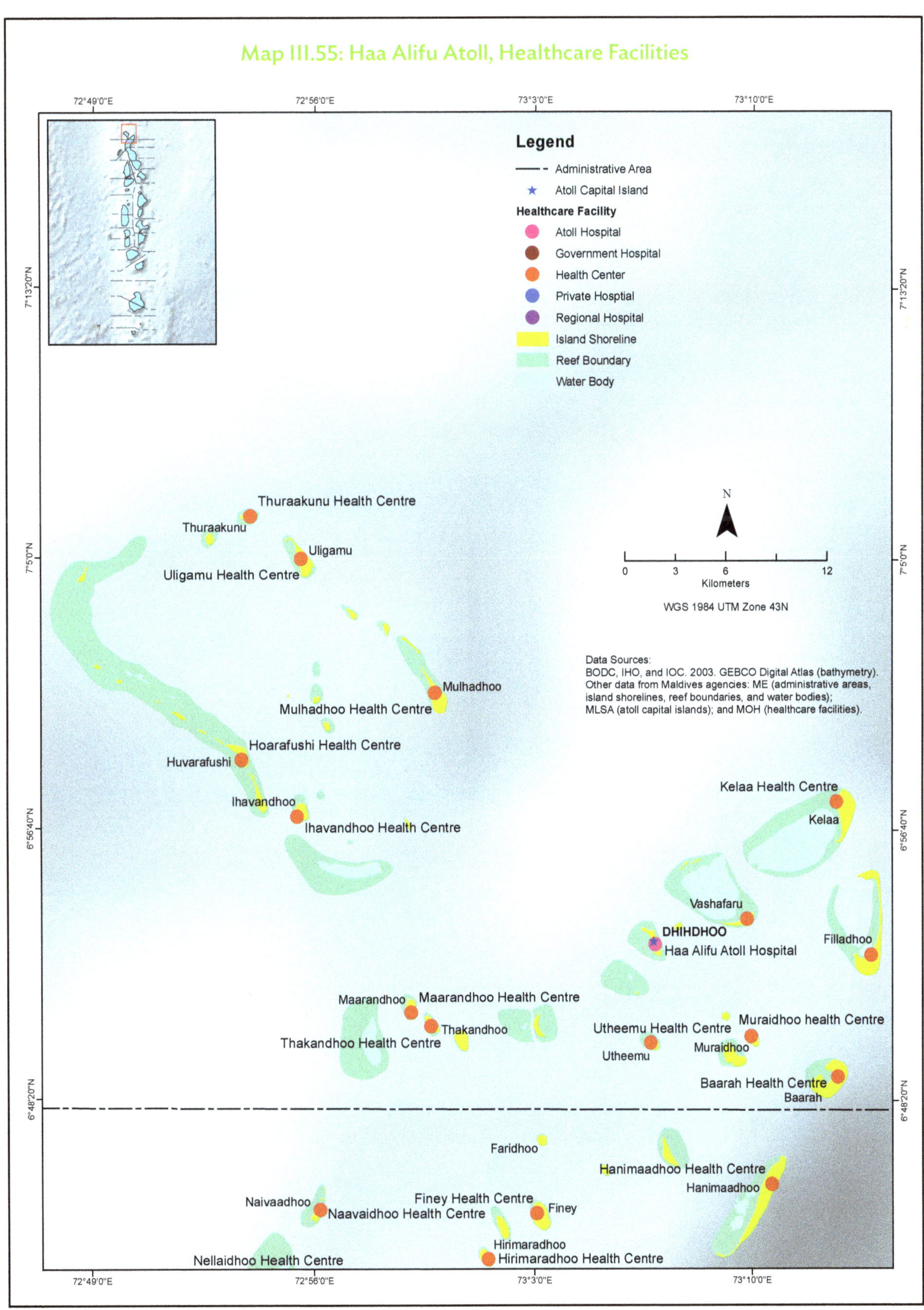

Map III.56: Haa Dhaalu Atoll, Healthcare Facilities

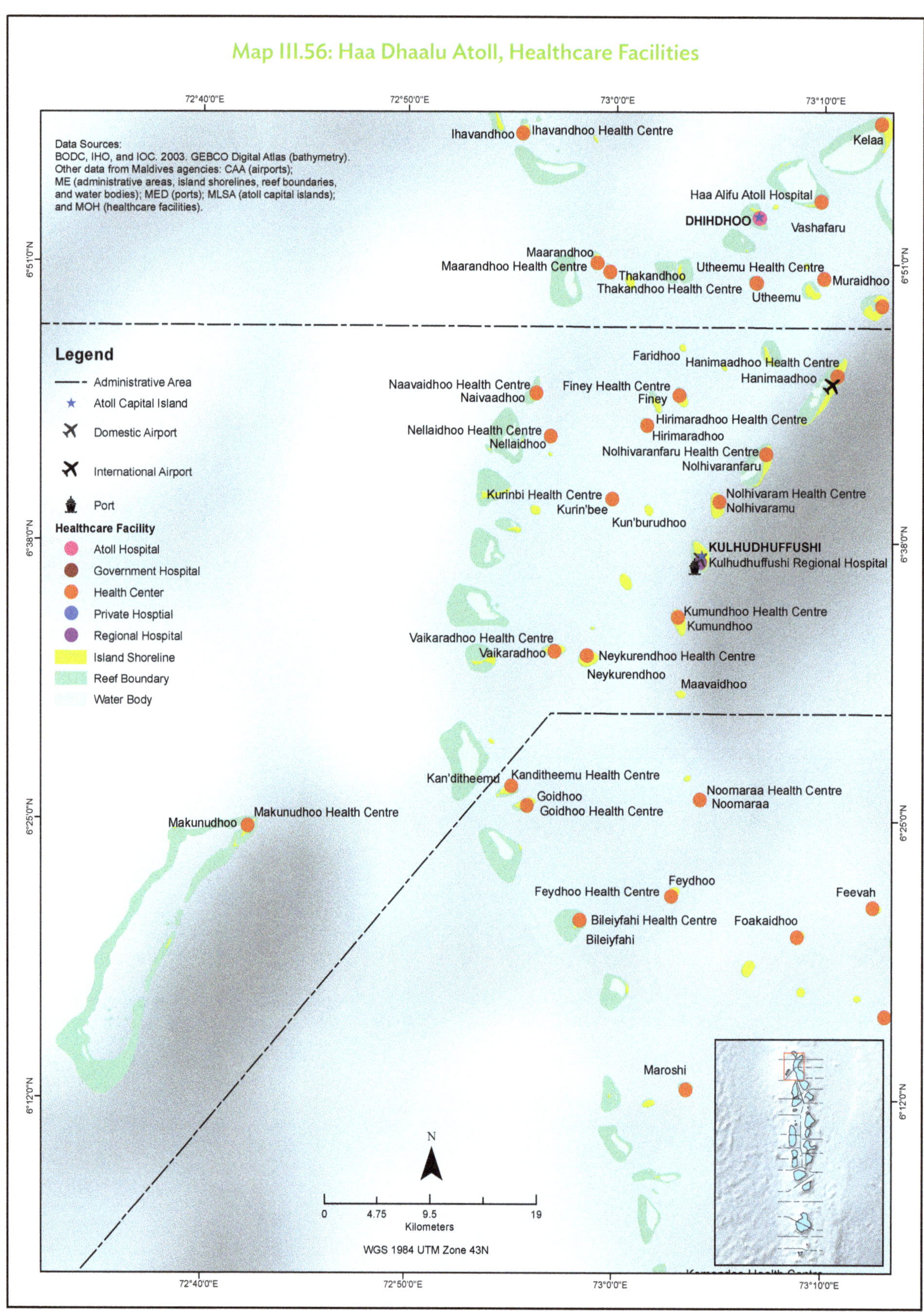

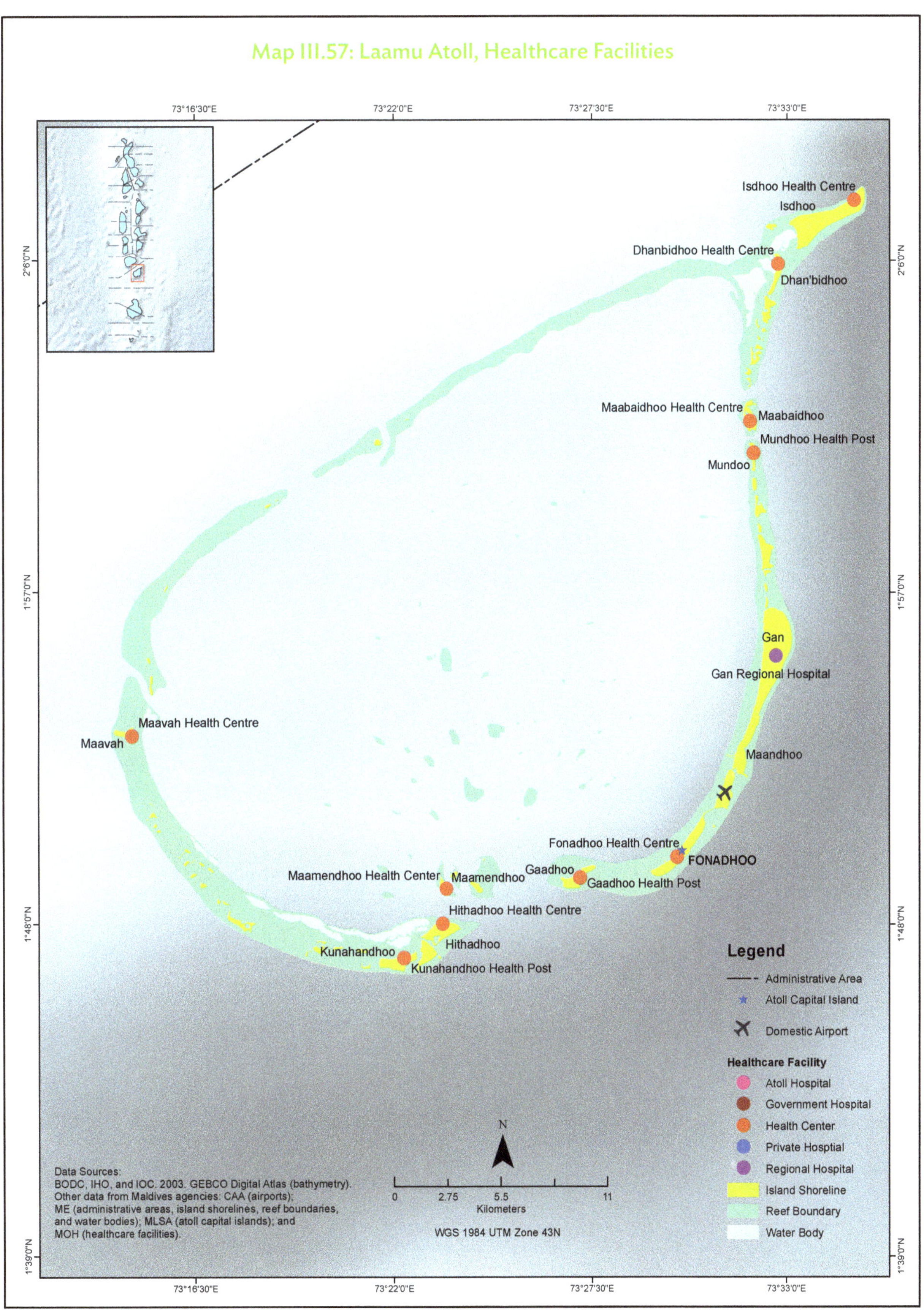
Map III.57: Laamu Atoll, Healthcare Facilities
73°16'30"E
73°22'0"E
73°27'30"E
73°33'0"E
2°6'0"N
1°57'0"N
1°48'0"N
1°39'0"N
Isdhoo Health Centre
Isdhoo
Dhanbidhoo Health Centre
Dhan'bidhoo
Maabaidhoo Health Centre
Maabaidhoo
Mundhoo Health Post
Mundoo
Gan
Gan Regional Hospital
Maavah Health Centre
Maavah
Maandhoo
Fonadhoo Health Centre
FONADHOO
Gaadhoo
Gaadhoo Health Post
Maamendhoo Health Center
Maamendhoo
Hithadhoo Health Centre
Hithadhoo
Kunahandhoo
Kunahandhoo Health Post
Legend
Administrative Area
Atoll Capital Island
Domestic Airport
Healthcare Facility
Atoll Hospital
Government Hospital
Health Center
Private Hosptial
Regional Hospital
Island Shoreline
Reef Boundary
Water Body
N
0 2.75 5.5 11
Kilometers
WGS 1984 UTM Zone 43N
Data Sources:
BODC, IHO, and IOC. 2003. GEBCO Digital Atlas (bathymetry).
Other data from Maldives agencies: CAA (airports);
ME (administrative areas, island shorelines, reef boundaries, and water bodies); MLSA (atoll capital islands); and
MOH (healthcare facilities).

Map III.58: Lhaviyani Atoll, Healthcare Facilities

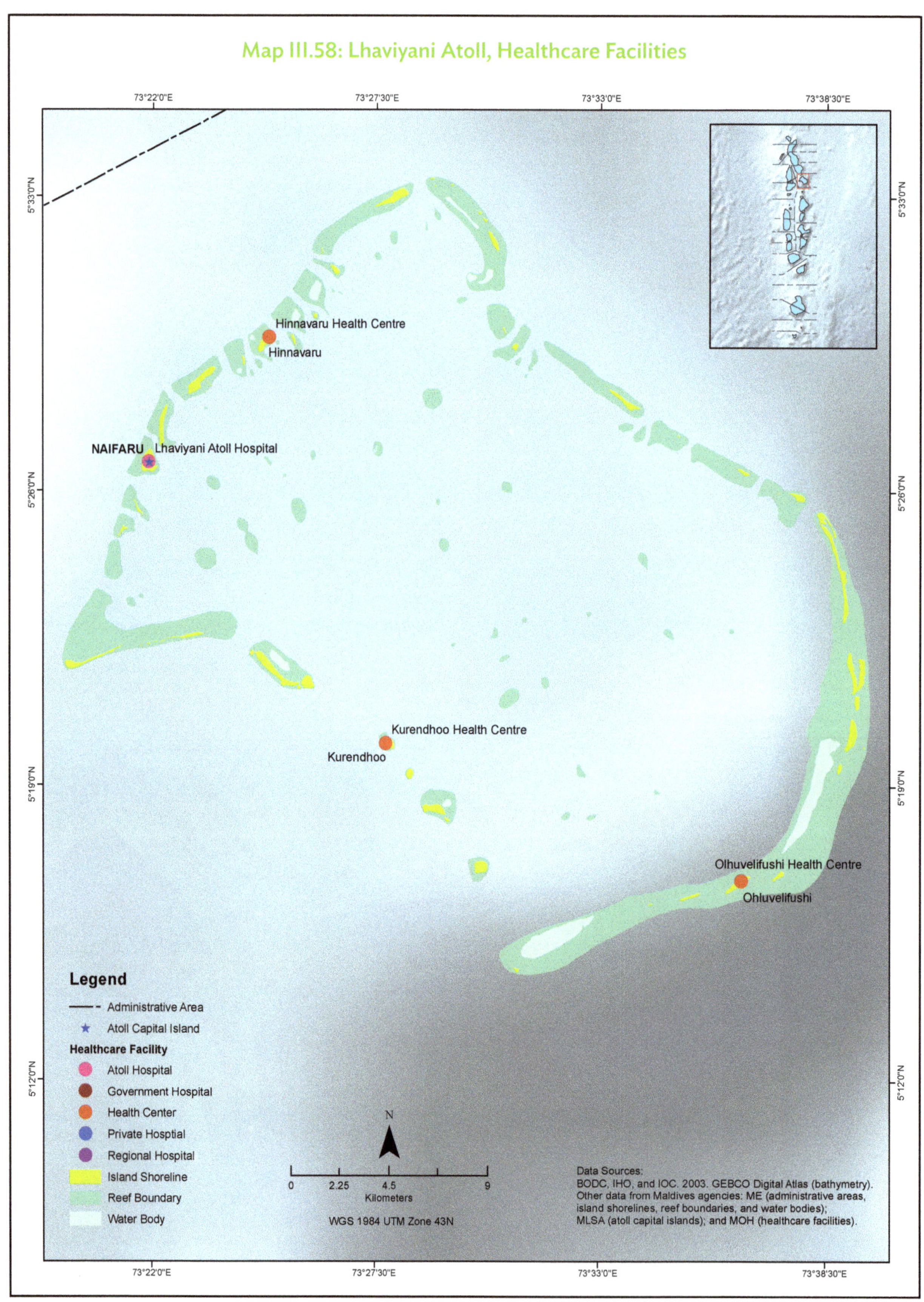

Map III.59: Meemu Atoll, Healthcare Facilities

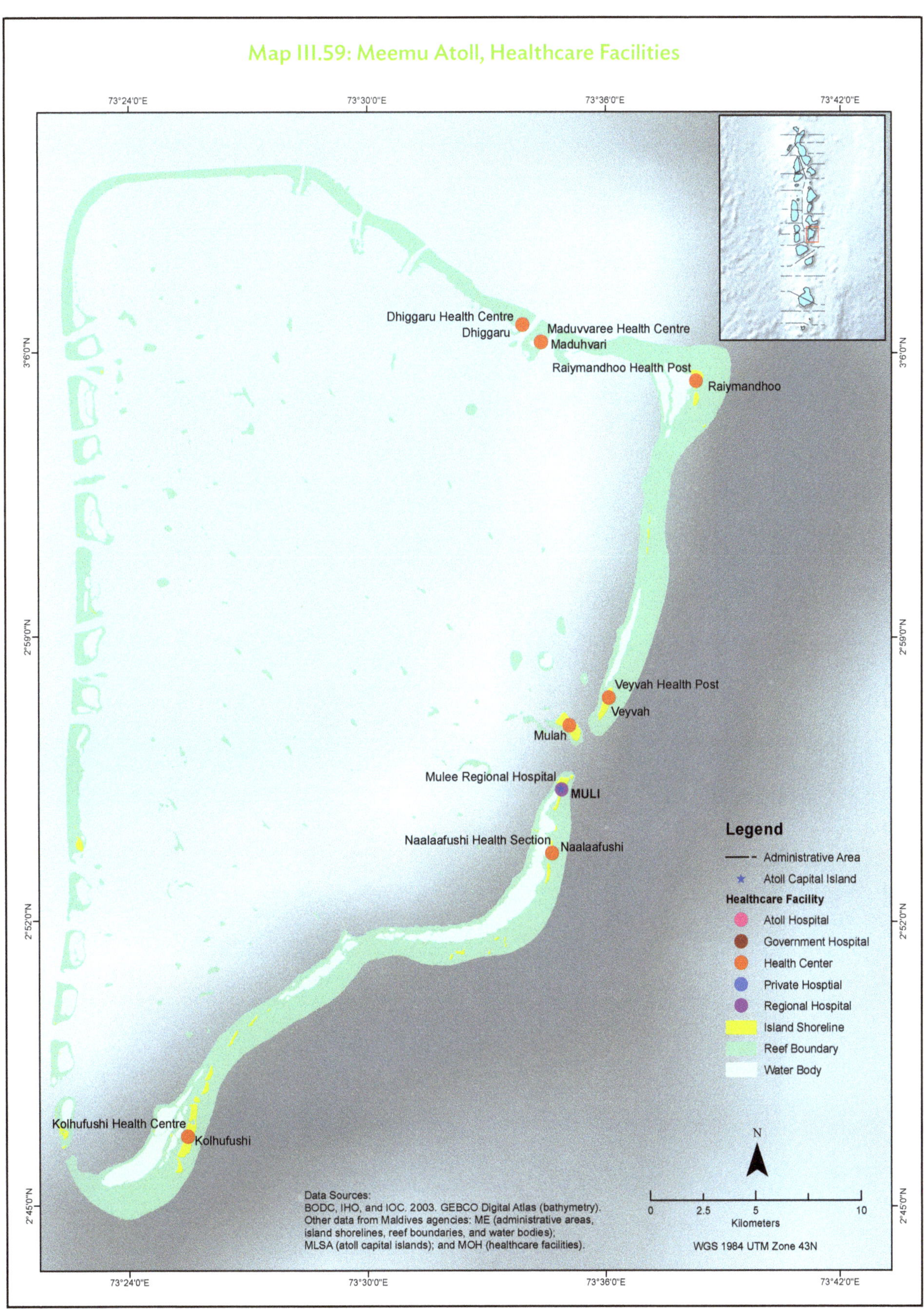

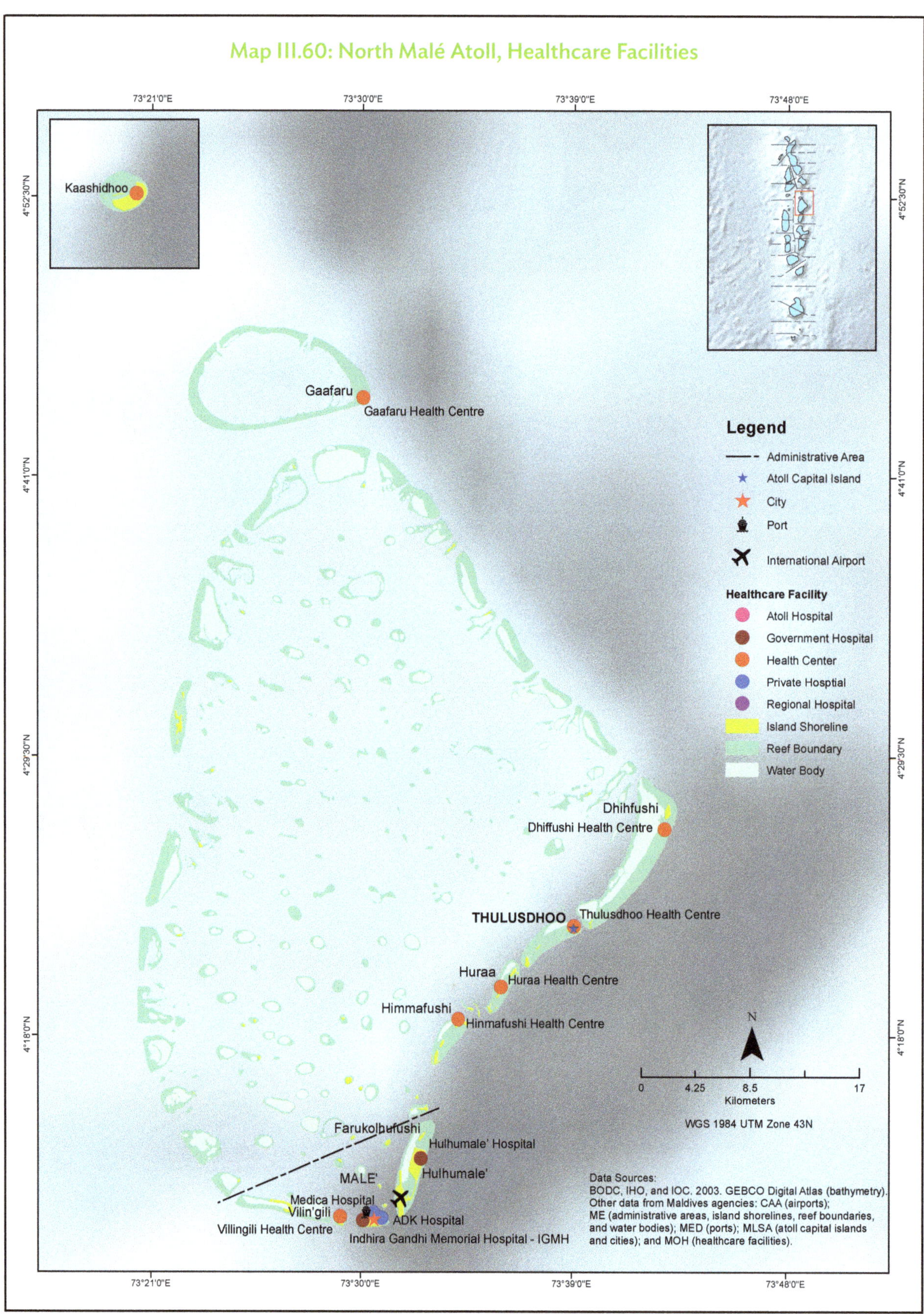
Map III.60: North Malé Atoll, Healthcare Facilities
73°21'0"E
73°30'0"E
73°39'0"E
73°48'0"E
4°52'30"N
4°41'0"N
4°29'30"N
4°18'0"N
Kaashidhoo
Gaafaru
Gaafaru Health Centre
Dhihfushi
Dhiffushi Health Centre
THULUSDHOO
Thulusdhoo Health Centre
Huraa
Huraa Health Centre
Himmafushi
Hinmafushi Health Centre
Farukolhufushi
Hulhumale' Hospital
Hulhumale'
MALE'
Medica Hospital
Vilin'gili
Villingili Health Centre
ADK Hospital
Indhira Gandhi Memorial Hospital - IGMH
Legend
Administrative Area
Atoll Capital Island
City
Port
International Airport
Healthcare Facility
Atoll Hospital
Government Hospital
Health Center
Private Hosptial
Regional Hospital
Island Shoreline
Reef Boundary
Water Body
N
0
4.25
8.5
17
Kilometers
WGS 1984 UTM Zone 43N
Data Sources:
BODC, IHO, and IOC. 2003. GEBCO Digital Atlas (bathymetry).
Other data from Maldives agencies: CAA (airports);
ME (administrative areas, island shorelines, reef boundaries, and water bodies); MED (ports); MLSA (atoll capital islands and cities); and MOH (healthcare facilities).

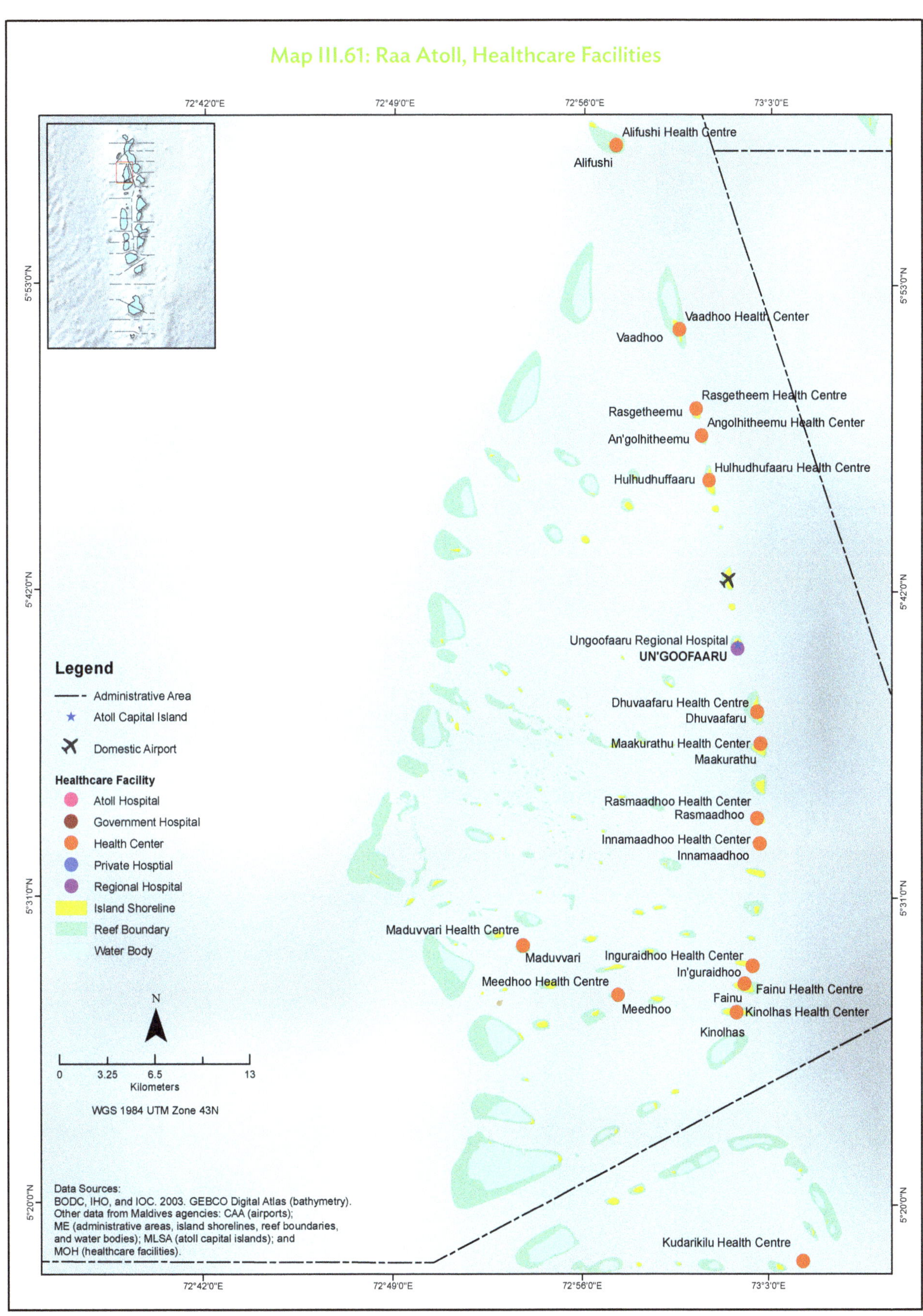
Map III.61: Raa Atoll, Healthcare Facilities
72°42'0"E
72°49'0"E
72°56'0"E
73°3'0"E
5°53'0"N
5°42'0"N
5°31'0"N
5°20'0"N
Alifushi Health Centre
Alifushi
Vaadhoo Health Center
Vaadhoo
Rasgetheem Health Centre
Rasgetheemu
Angolhitheemu Health Center
An'golhitheemu
Hulhudhufaaru Health Centre
Hulhudhuffaaru
Ungoofaaru Regional Hospital
UN'GOOFAARU
Dhuvaafaru Health Centre
Dhuvaafaru
Maakurathu Health Center
Maakurathu
Rasmaadhoo Health Center
Rasmaadhoo
Innamaadhoo Health Center
Innamaadhoo
Maduvvari Health Centre
Maduvvari
Inguraidhoo Health Center
In'guraidhoo
Meedhoo Health Centre
Meedhoo
Fainu Health Centre
Fainu
Kinolhas Health Center
Kinolhas
Kudarikilu Health Centre
Legend
Administrative Area
Atoll Capital Island
Domestic Airport
Healthcare Facility
Atoll Hospital
Government Hospital
Health Center
Private Hosptial
Regional Hospital
Island Shoreline
Reef Boundary
Water Body
N
0
3.25
6.5
13
Kilometers
WGS 1984 UTM Zone 43N
Data Sources:
BODC, IHO, and IOC. 2003. GEBCO Digital Atlas (bathymetry).
Other data from Maldives agencies: CAA (airports);
ME (administrative areas, island shorelines, reef boundaries,
and water bodies); MLSA (atoll capital islands); and
MOH (healthcare facilities).

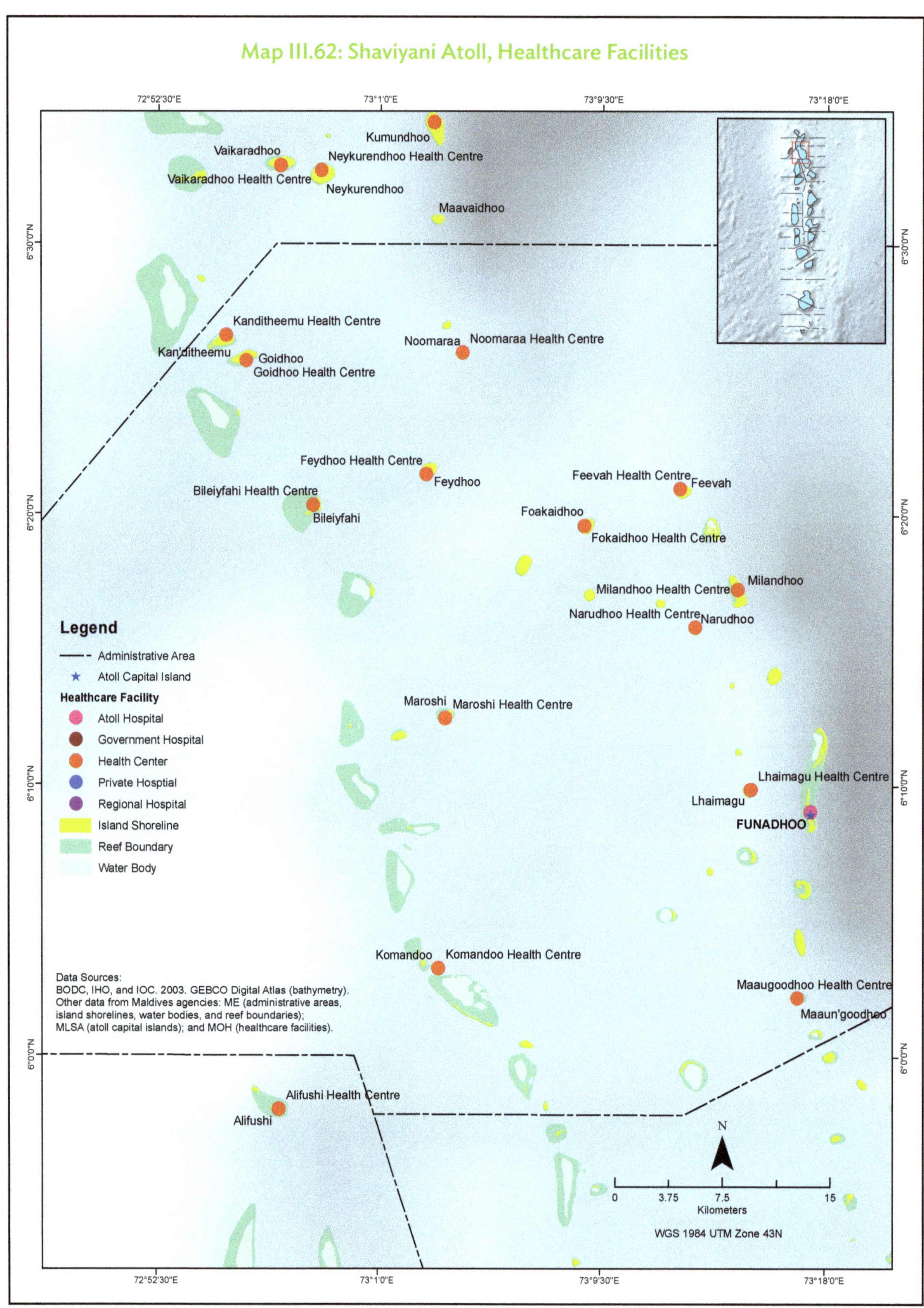

Map III.62: Shaviyani Atoll, Healthcare Facilities
72°52'30"E
73°1'0"E
73°9'30"E
73°18'0"E
6°30'0"N
6°20'0"N
6°10'0"N
6°0'0"N
Kumundhoo
Vaikaradhoo
Neykurendhoo Health Centre
Vaikaradhoo Health Centre
Neykurendhoo
Maavaidhoo
Kanditheemu Health Centre
Kan'ditheemu
Goidhoo
Goidhoo Health Centre
Noomaraa
Noomaraa Health Centre
Feydhoo Health Centre
Feydhoo
Bileiyfahi Health Centre
Bileiyfahi
Feevah Health Centre
Feevah
Foakaidhoo
Fokaidhoo Health Centre
Milandhoo
Milandhoo Health Centre
Narudhoo Health Centre
Narudhoo
Maroshi
Maroshi Health Centre
Lhaimagu Health Centre
Lhaimagu
FUNADHOO
Komandoo
Komandoo Health Centre
Maaugoodhoo Health Centre
Maaun'goodhoo
Alifushi Health Centre
Alifushi
Legend
Administrative Area
Atoll Capital Island
Healthcare Facility
Atoll Hospital
Government Hospital
Health Center
Private Hosptial
Regional Hospital
Island Shoreline
Reef Boundary
Water Body
Data Sources:
BODC, IHO, and IOC. 2003. GEBCO Digital Atlas (bathymetry).
Other data from Maldives agencies: ME (administrative areas, island shorelines, water bodies, and reef boundaries);
MLSA (atoll capital islands); and MOH (healthcare facilities).
N
0
3.75
7.5
15
Kilometers
WGS 1984 UTM Zone 43N

Map III.63: South Malé Atoll, Healthcare Facilities

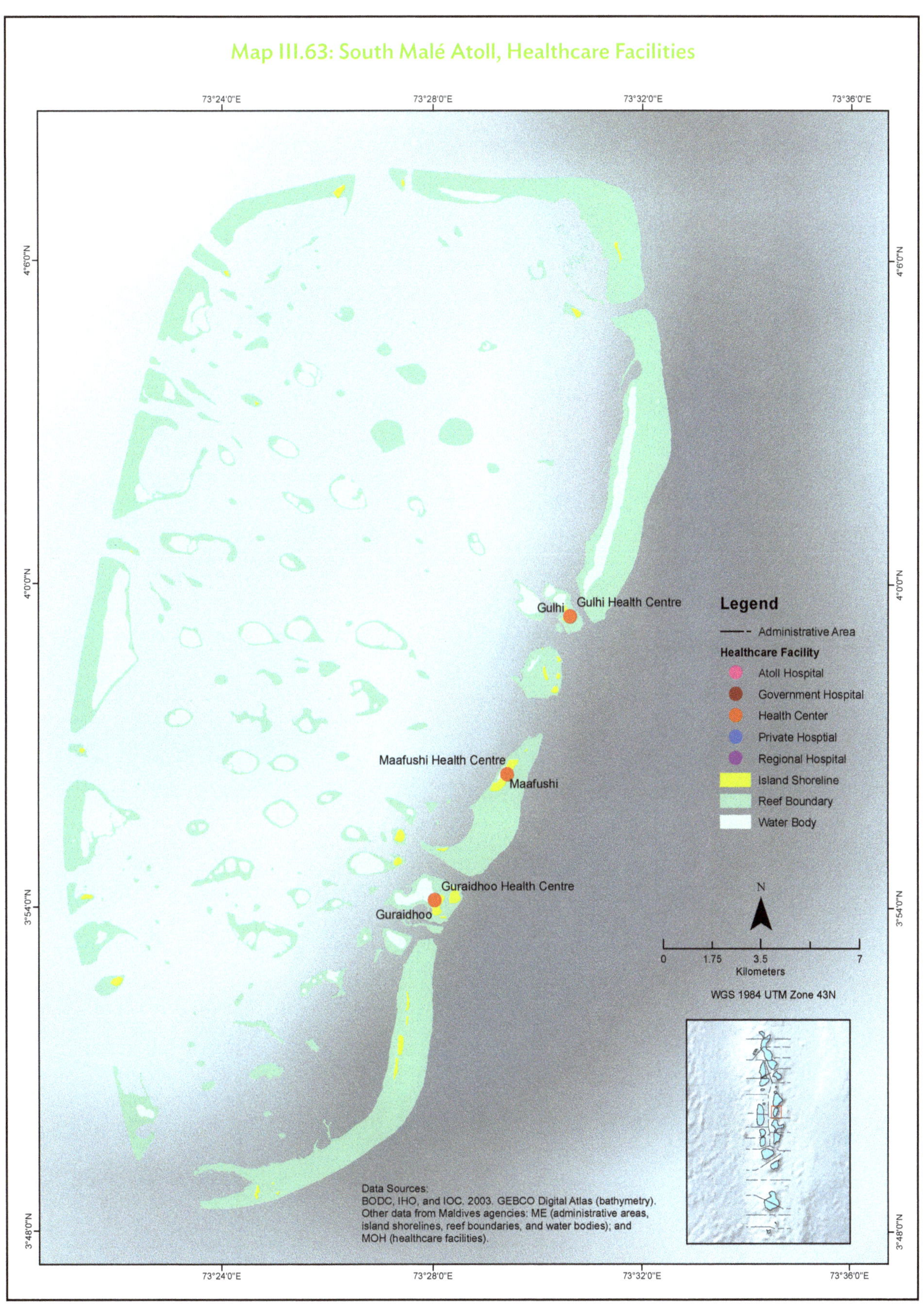

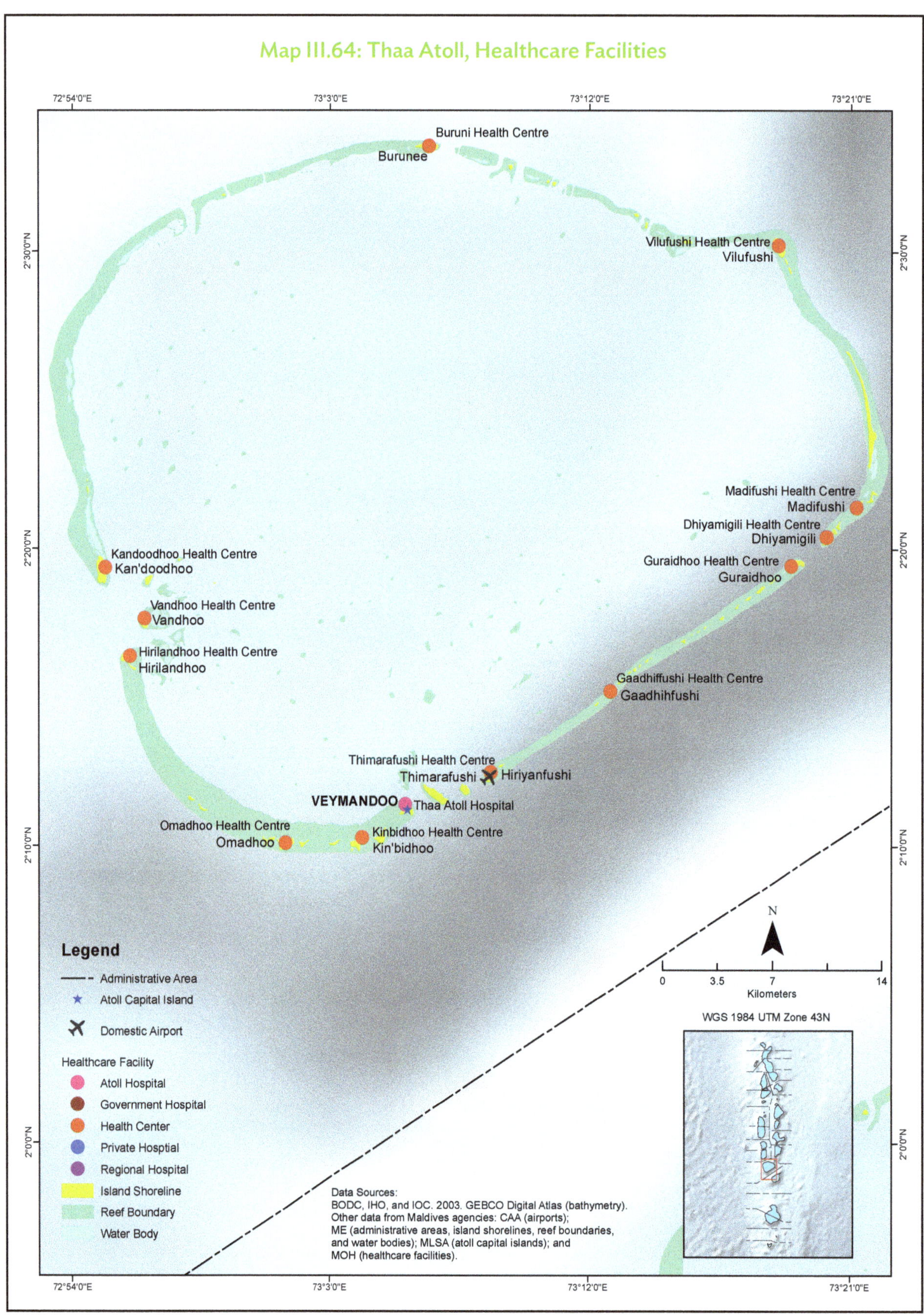
Map III.64: Thaa Atoll, Healthcare Facilities
72°54'0"E
73°3'0"E
73°12'0"E
73°21'0"E
2°30'0"N
2°20'0"N
2°10'0"N
2°0'0"N
Buruni Health Centre
Burunee
Vilufushi Health Centre
Vilufushi
Madifushi Health Centre
Madifushi
Dhiyamigili Health Centre
Dhiyamigili
Guraidhoo Health Centre
Guraidhoo
Kandoodhoo Health Centre
Kan'doodhoo
Vandhoo Health Centre
Vandhoo
Hirilandhoo Health Centre
Hirilandhoo
Gaadhiffushi Health Centre
Gaadhihfushi
Thimarafushi Health Centre
Thimarafushi
Hiriyanfushi
VEYMANDOO
Thaa Atoll Hospital
Omadhoo Health Centre
Omadhoo
Kinbidhoo Health Centre
Kin'bidhoo
Legend
Administrative Area
Atoll Capital Island
Domestic Airport
Healthcare Facility
Atoll Hospital
Government Hospital
Health Center
Private Hosptial
Regional Hospital
Island Shoreline
Reef Boundary
Water Body
N
0
3.5
7
14
Kilometers
WGS 1984 UTM Zone 43N
Data Sources:
BODC, IHO, and IOC. 2003. GEBCO Digital Atlas (bathymetry).
Other data from Maldives agencies: CAA (airports);
ME (administrative areas, island shorelines, reef boundaries,
and water bodies); MLSA (atoll capital islands); and
MOH (healthcare facilities).

Map III.65: Vaavu Atoll, Healthcare Facilities

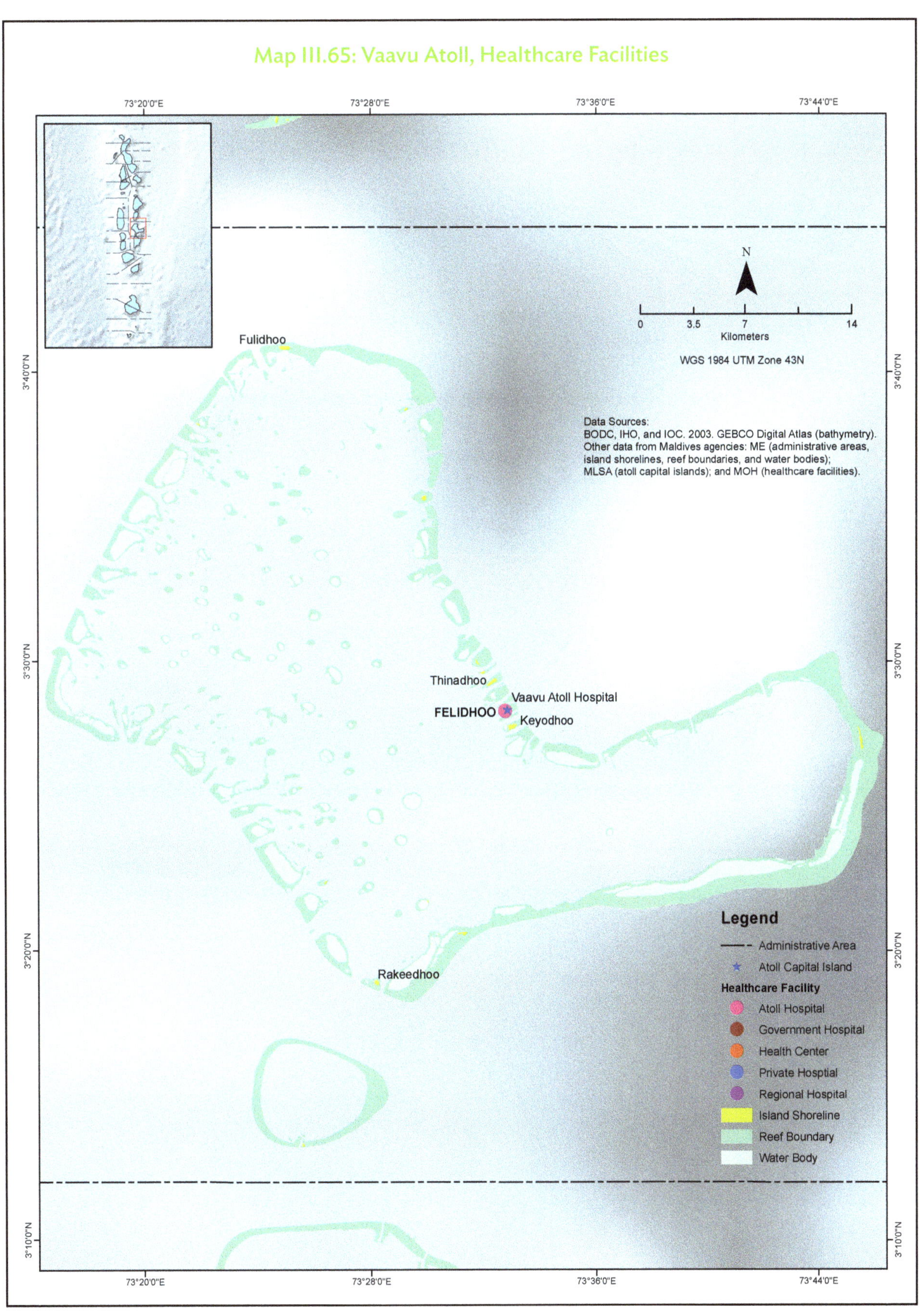

Lives Centered at Sea

Life in Maldives revolves around the sea, which provides Maldivians with plenty of resources for livelihood including fisheries and tourism.

Rich aquatic resources made fisheries the main source of livelihood in Maldives, until tourism took over. Up to now, fishing contributes to national productivity with tuna, grouper, and snapper among the popular species of fish caught and sold in the market as food.

The natural beauty of its beaches has enticed tourists to visit Maldives for over 4 decades. Although the sea and tropical climate provide for the success of the tourism sector, they also pose concerns for changes in the country's sea level and future climate.

Fisheries and tourism are susceptible to the changing climate. Future warmer temperatures can affect fish production as higher sea surface temperature leads to coral bleaching, and tourists wanting to see the rich aquatic life in Maldives may no longer be as satisfied (Hosterman and Smith 2014). A wetter climate can also affect the influx of tourists.

In the future, a wetter climate could be less favorable for tourism activities (*Multihazard Risk Atlas of Maldives: Climate and Geophysical Hazards—Volume II*, p. 32). This may be observed especially during the southwest monsoon in June, July, and August in northern Maldives (*Multihazard Risk Atlas of Maldives: Climate and Geophysical Hazards—Volume II*, p. 47). The beaches are also threatened by coastal erosion (*Multihazard Risk Atlas of Maldives: Biodiversity—Volume IV*, p. 69). Additionally, low-lying islands are prone to inundation.

An atoll as seen from the sky. A portion of a Maldivian island is captured on an aerial photograph (photo by Sue Todd).

Tourism

Naturally beautiful beaches and the presence of island resorts make tourism the main economic sector in Maldives. Tourists can choose to visit one of the 120 islands with operating resorts, ranging from small resorts with only 14 beds to larger ones with more than 900 beds. Most resorts are located in and around Kaafu Atoll (43 resorts), Alifu Dhaalu Atoll (17 resorts), and Baa Atoll (12 resorts) due to the proximity to Malé City and international airports.

Baa, Lhaviyani, Haa Alifu, and Shaviyani atolls will experience an increase in average rainfall rate, based on climate projection results (*Multihazard Risk Atlas of Maldives: Climate and Geophysical Hazards—Volume II*, p. 32). The resorts in these atolls may experience higher rainfall rate in the future.

Table III.5: Number of Resorts per Atoll, Maldives

Atoll	Number of Resorts
Kaafu	43
Alifu Dhaalu	17
Baa	12
Alifu Alifu	11
Lhaviyani	7
Dhaalu	5
Noonu	5
Gaafu Alifu	4
Gaafu Dhaalu	3
Raa	3
Addu City	2
Meemu	2
Vaavu	2
Faafu	1
Haa Alifu	1
Laamu	1
Thaa	1

Source: Ministry of Tourism, 2017.

Table III.6: Resorts in Maldives

Atoll and Island	Resort	Area (ha)	Number of Beds
Addu City			
Herethera	Canareef Resort Maldives	81.40	542
Vilin'gili	Shangri La's Vilingili Resort & Spa, Maldives	54.44	284
Alifu Alifu Atoll			
Bathalaa	Bathala Island Resort	3.11	90
Ellaidhoo	Ellaidhoo Maldives by Cinnamon	5.90	224
Fesdhoo	W Retreat and Spa-Maldives	4.83	164
Gangehi	Gangehi Island Resort	2.15	72
Halaveli	Constance Halaveli Resort	4.75	172
Kan'dholhudhoo	Kandolhu Island Maldives	4.85	60
Kudafolhudhoo	Nika Island Resort and Spa	5.06	104
Kuramathi	Kuramathi Maldives	30.97	790
Maayafushi	Maayafushi Tourist Resort	3.16	150
Velidhoo	Velidhoo Island Resort	9.22	200
Veligan'du	Veligandu Island Resort	5.29	182
Alifu Dhaalu Atoll			
Angaagau	Angaga Island Resort and Spa	4.39	180
Athurugau	Diamonds Athuruga Beach & Water Villas	3.12	146
Dhihdhoofinolhu	Lux South Ari Atoll, Maldives	15.99	394
Dhihfushi	Holiday Island Resort and Spa	13.14	284
Huvahendhoo	Lily Beach Resort	6.62	250
Kudarah	Kudarah Island Resort	3.60	102
Maafushivaru	Twin Island Resort	3.01	98
Mahchafushi	Centara Grand Island Resort & Spa Maldives	4.24	224
Mirihi	Mirihi Island Resort	3.47	76
Moofushi	Constance Moofushi Resort	3.72	220
Nalaguraidhoo	Sun Island Resort and Spa	54.12	924
Rangaleefinolhu	Conrad Maldives Rangali Island	6.49	304
Theluveligaa	Drift Thelu Veliga Retreat	1.11	60
Thun'dufushi	Diamonds Thudufushi Beach & Water Villas	3.96	144
Vakarufalhi	Vakarufalhi Island Resort	5.23	150
Vilamendhoo	Vilamendhoo Island Resort	12.83	368
Vilin'gilivaru	Ranveli Island Resort	2.98	112
Baa Atoll			
Dhigufaruvinagan'du	Dhigufaru Island Resort	3.14	80
Dhunikolhu	Coco Palm Dhunikolhu	14.73	196

continued on next page

Table III.6 *continued*

Atoll and Island	Resort	Area (ha)	Number of Beds
Fonimagoodhoo	Reethi Beach Resort	10.38	248
Horubadhoo	Royal Island Resort and Spa	16.82	304
Kanufushi	Finolhu Baa Atoll Maldives	11.74	272
Kihaadhuffaru	Kihaad Resort	13.13	236
Kihavahhuravalhi	Anantara Kihavah Villas	11.81	172
Kunfunadhoo	Soneva Fushi Resort	43.45	228
Lan'daagiraavaru	Four Seasons Resort Maldives at Landaa Giraavaru	20.03	232
Milaidhoo	Milaidhoo Island Maldives	5.27	88
Muhdhoo	Dusit Thani Maldives	19.03	208
Voavah	Four Seasons Private Island Maldives at Voavah	2.87	26
Dhaalu Atoll			
Dhoores	aaaVeee Nature's Paradise	8.90	66
En'boodhoofushi	Niyama Maldives	12.94	274
Meedhuhfushi	Sun Aqua Vilu Reef Maldives	6.61	200
Velavaru	Angsana Resort & Spa Maldives - Velavaru	7.24	238
Vommuli	The St. Regis Vommuli Resort, Maldives	9.66	184
Faafu Atoll			
Filitheyo	Filitheyo Island Resort	21.35	250
Gaafu Alifu Atoll			
Falhumaafushi	The Residence Maldives	7.21	108
Funamauddaa	Robinson Club Maldives	10.92	242
Han'dahaa	Park Hyatt Maldives, Hadahaa	9.63	100
Meradhoo	Jumeirah Dhevanafushi	6.05	74
Gaafu Dhaalu Atoll			
Havoddaa	Amari Havodda Maldives	10.90	240
Konottaa	Outrigger Konotta Maldives Resort	8.77	110
Maguhdhuvaa	Ayada Maldives	14.56	200
Haa Alifu Atoll			
Alidhoo	Alidhoo Island Resort	17.19	200
Manafaru	JA Manafaru	14.95	174
Kaafu Atoll			
Asdhoo	Asdu Sun Island	3.58	60
Baros	Baros Maldives	4.46	150
Biyaadhoo	Biyaadhoo Island Resort	10.15	192
Boduban'dos	Bandos Maldives	19.64	430
Bodufinolhu	Fun Island Resort & Spa	10.79	150
Boduhithi	Coco Bodu Hithi	6.47	200
Bolifushi	Jumeirah Vittaveli	1.79	178

continued on next page

Table III.6 *continued*

Atoll and Island	Resort	Area (ha)	Number of Beds
Dhigufinolhu	Anantara Resort and Spa Maldives	4.80	220
En'boodhoo	Embudhu Village	5.32	236
En'boodhoofinolhu	Taj Exotica Resort and Spa Maldives	4.79	134
Eriadhoo	Eriyadhoo Island Resort	3.13	152
Fihaalhohi	Fihaalhohi Island Resort	8.90	300
Furanafushi	Sheraton Maldives Full Moon Resort & Spa	11.81	352
Gasfinolhu	Club Med Finolhu Villas	3.41	100
Giraavaru	Centara Ras Fushi Resort & Spa	2.96	220
Helen'geli	Oblu By Atmosphere at Helengeli	7.66	236
Hen'badhoo	Vivanta by Taj - Coral Reef, Maldives	3.53	128
Ihuru	Angsana Resort & Spa Maldives Ihuru	3.44	90
Kan'doomaafushi	Holiday Inn Resort Kandooma Maldives	14.75	322
Kanifinolhu	Club Med Kanifinolhu	14.49	492
Kanuoiyhuraa	Cinnamon Dhonveli Maldives	7.61	296
Kuda huraa	Four Seasons Resort Maldives at Kuda Huraa	7.34	212
Kudaban'dos	Malahini Kuda Bandos	4.65	164
Kudahithi	Coco Privé Kuda Hithi Island	1.58	14
Lankanfinolhu	Paradise Island Resort and Spa	19.82	568
Lankanfuhsi	Gili Lankanfuhsi	8.61	94
Lhohifushi	Adaaran Select Hudhuranfushi	19.25	354
Maadhoo	Ozen By Atmosphere At Maadhoo	4.17	198
Mahaanaelhihuraa	Rihiveli	4.20	100
Makunudhoo	Makunudhoo Island	3.18	72
Makunufushi	Cocoa Island	2.53	70
Medhufinolhu	One&Only Reethi Rah, Maldives	42.69	268
Meerufenfushi	Meeru Island Resort	33.52	572
Nakaiychaafushi	Huvafenfushi Resort and Spa	4.30	98
Olhuhali	Olhuveli Beach and Spa Resort	2.39	332
Rannaalhi	Adaaran Club Rannalhi	4.91	260
Thulhaagiri	Thulhagiri Island Resort & Spa	3.27	168
Vaadhoo	Adaaran Prestige Vadoo	1.58	100
Vahbinfaru	Banyan Tree Maldives Vabbinfaru	3.92	96
Velassaru	Velassaru Maldives	8.34	258
Veligan'duhuraa	Naladhu	2.26	138
Vihamanaafushi	Kurumba Maldives	13.71	362
Ziyaaraiyfushi	Summer Island Maldives	5.04	314
Laamu Atoll			
Olhuveli	Six Senses Laamu	18.10	194

continued on next page

Table III.6 *continued*

Atoll and Island	Resort	Area (ha)	Number of Beds
Lhaviyani Atoll			
Huravalhi	Hurawalhi Island Resort	8.13	180
Kanifushi	Atmosphere Kanifushi Maldives	30.03	300
Kanuhuraa	Kanuhura	14.53	20
Komandoo	Komandoo Maldive Island Resort	4.34	130
Kurehdhoo	Kuredu Island Resort	39.13	778
Madhiriguraidhoo	Palm Beach Resort & Spa Maldives	24.79	270
Ookolhufinolhu	Cocoon Maldives	8.45	302
Meemu Atoll			
Hakuraahuraa	Cinnamon Hakuraa Huraa Maldives	5.35	160
Medhufushi	Medhufushi Island Resort	10.34	240
Noonu Atoll			
Fushivelavaru	Velaa Private Island Maldives	5.37	134
Kudafunafaru	Roxy Maldives Resort	18.81	100
Medhafushi	The Sun Siyam Iru Fushi Maldives	20.83	448
Medhufaru	Soneva Jani	51.50	66
Randheli	Cheval Blanc Randheli	9.33	120
Raa Atoll			
Furaveri	Furaveri Island Resort & Spa	23.67	202
Maamin'gili	Loama Resort Maldives at Maamigili	6.88	210
Meedhupparu	Adaaran Select Meedhupparu	17.94	470
Shaviyani Atoll			
Van'garu	Vagaru Island Resort	9.82	159
Thaa Atoll			
Maalefushi	Maalifushi by Como	10.11	152
Vaavu Atoll			
Alimathaa	Alimatha Aquatic Resort	6.88	312
Dhiggiri	Dhiggiri Tourist Resort	2.19	122

ha = hectare.

Source: Ministry of Tourism, 2017.

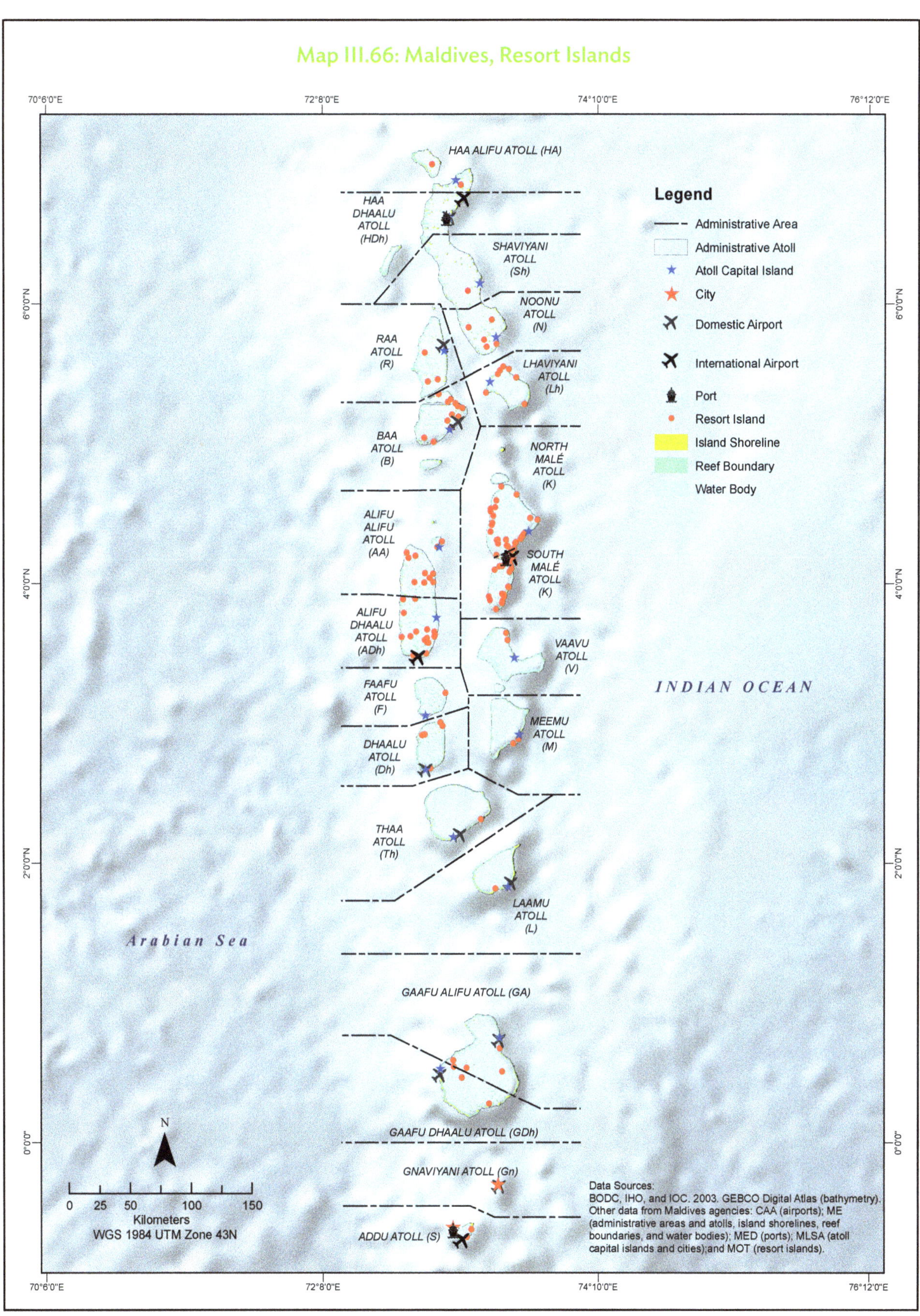

Map III.66: Maldives, Resort Islands

Map III.67: Addu City, Resort Islands

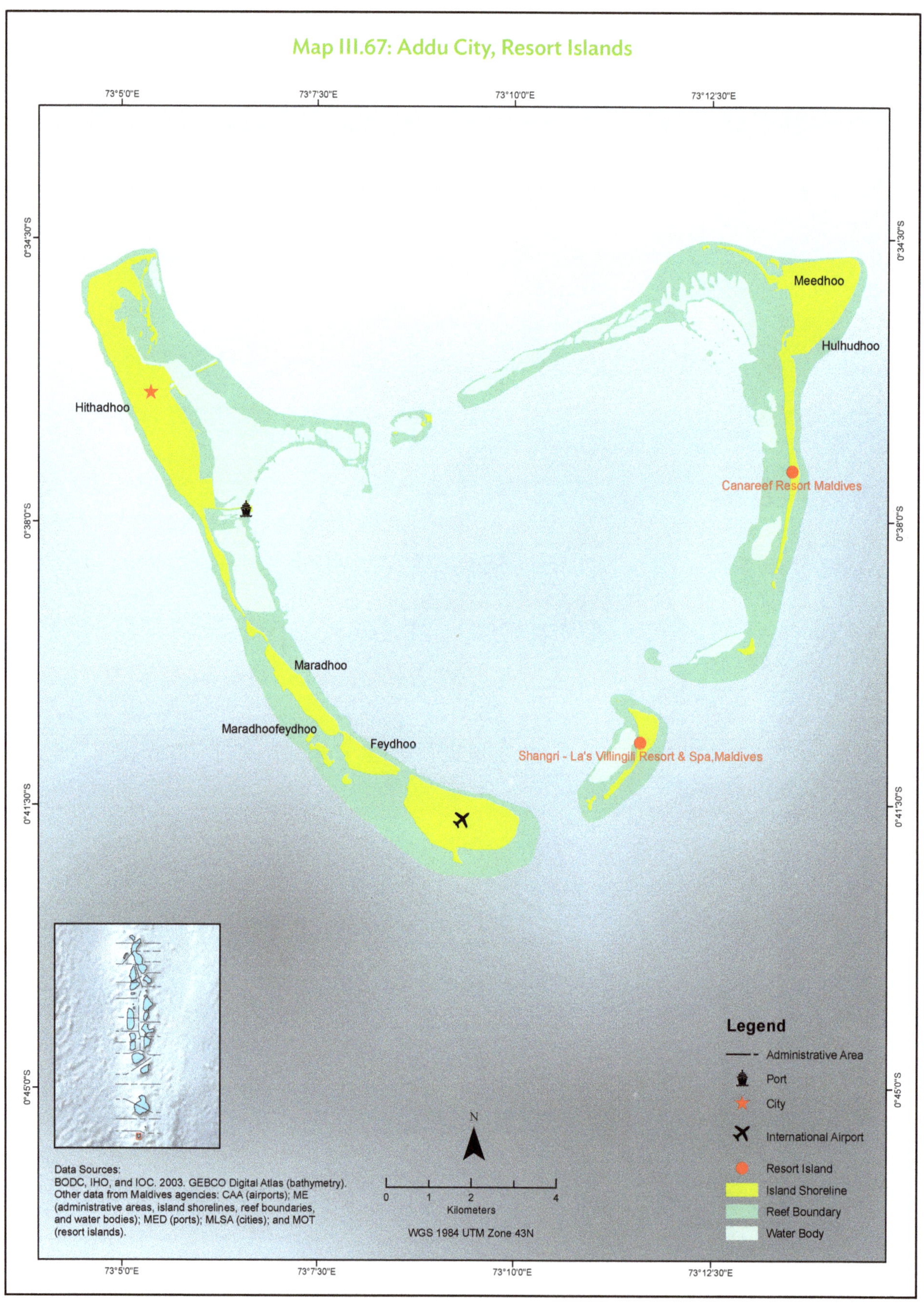

Map III.68: Alifu Alifu Atoll, Resort Islands

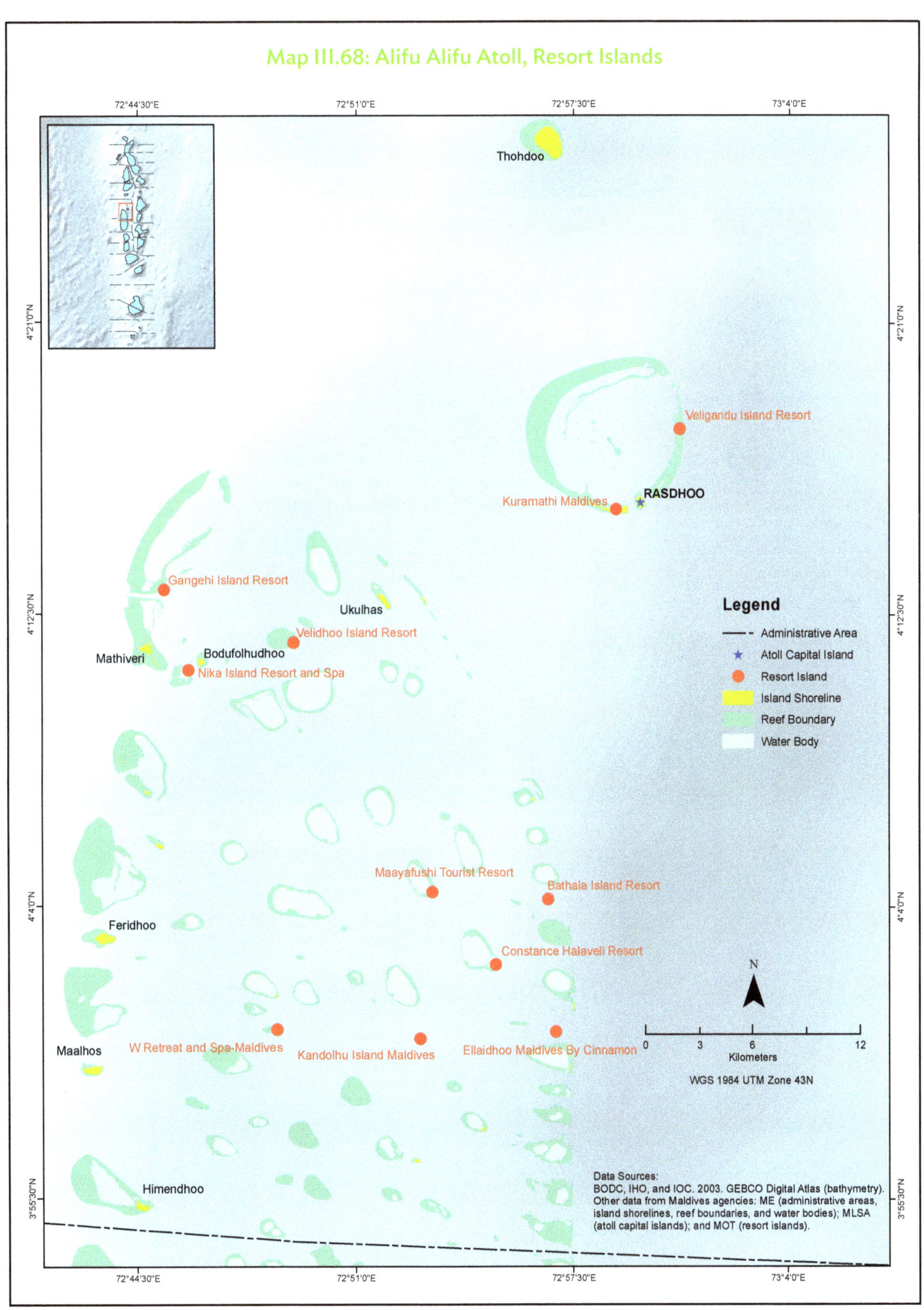

Map III.69: Alifu Dhaalu Atoll, Resort Islands

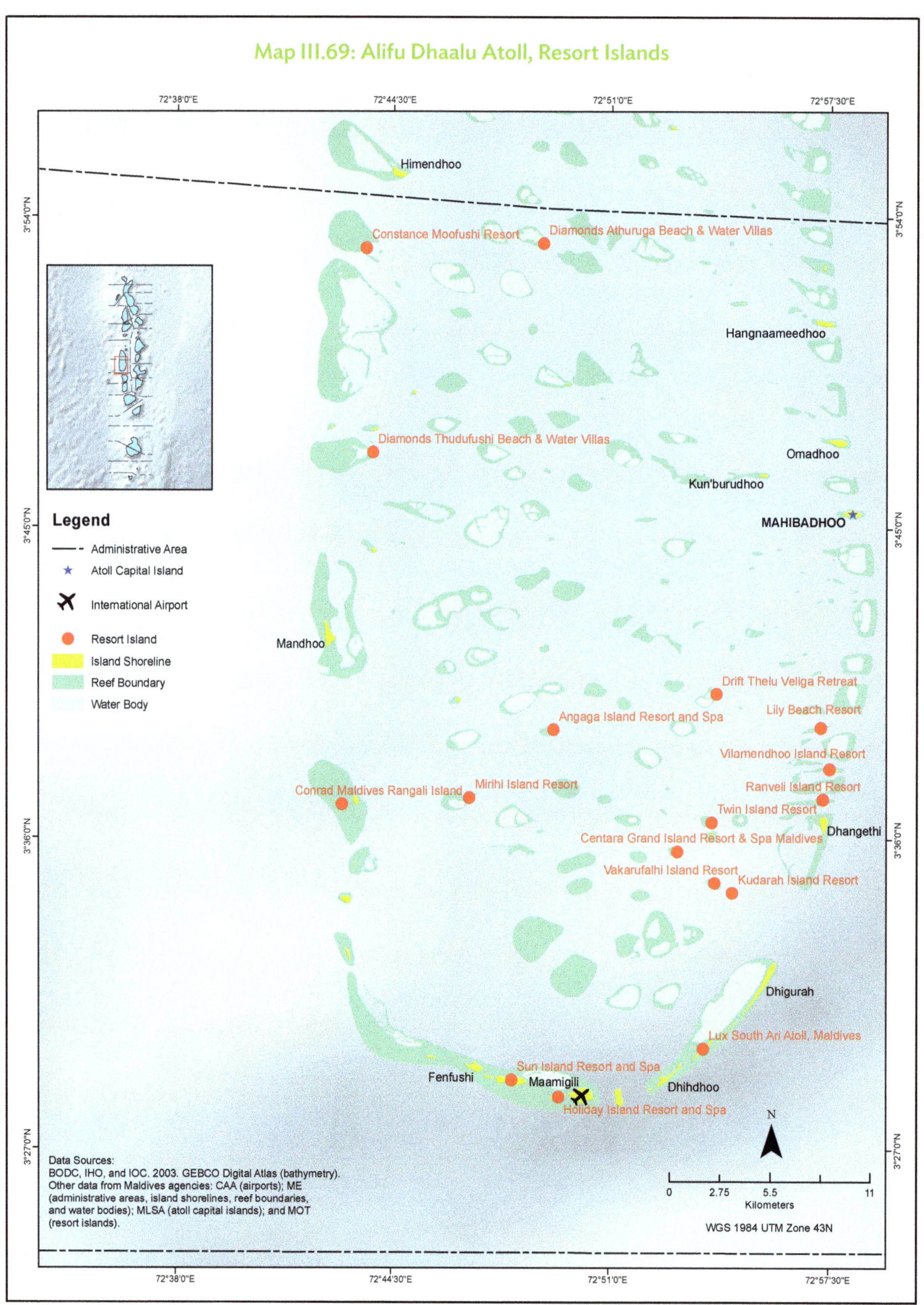

Map III.70: Baa Atoll, Resort Islands

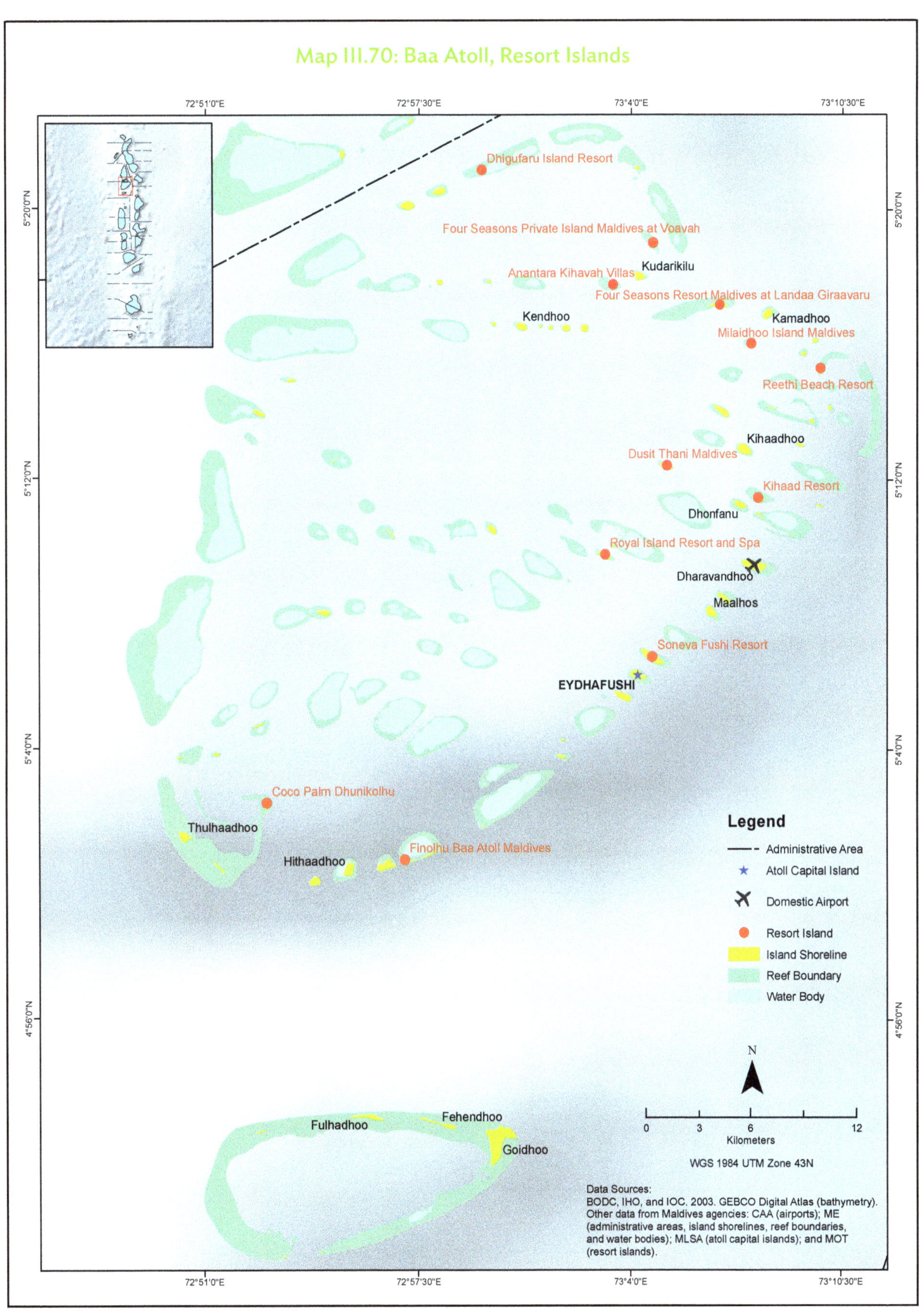

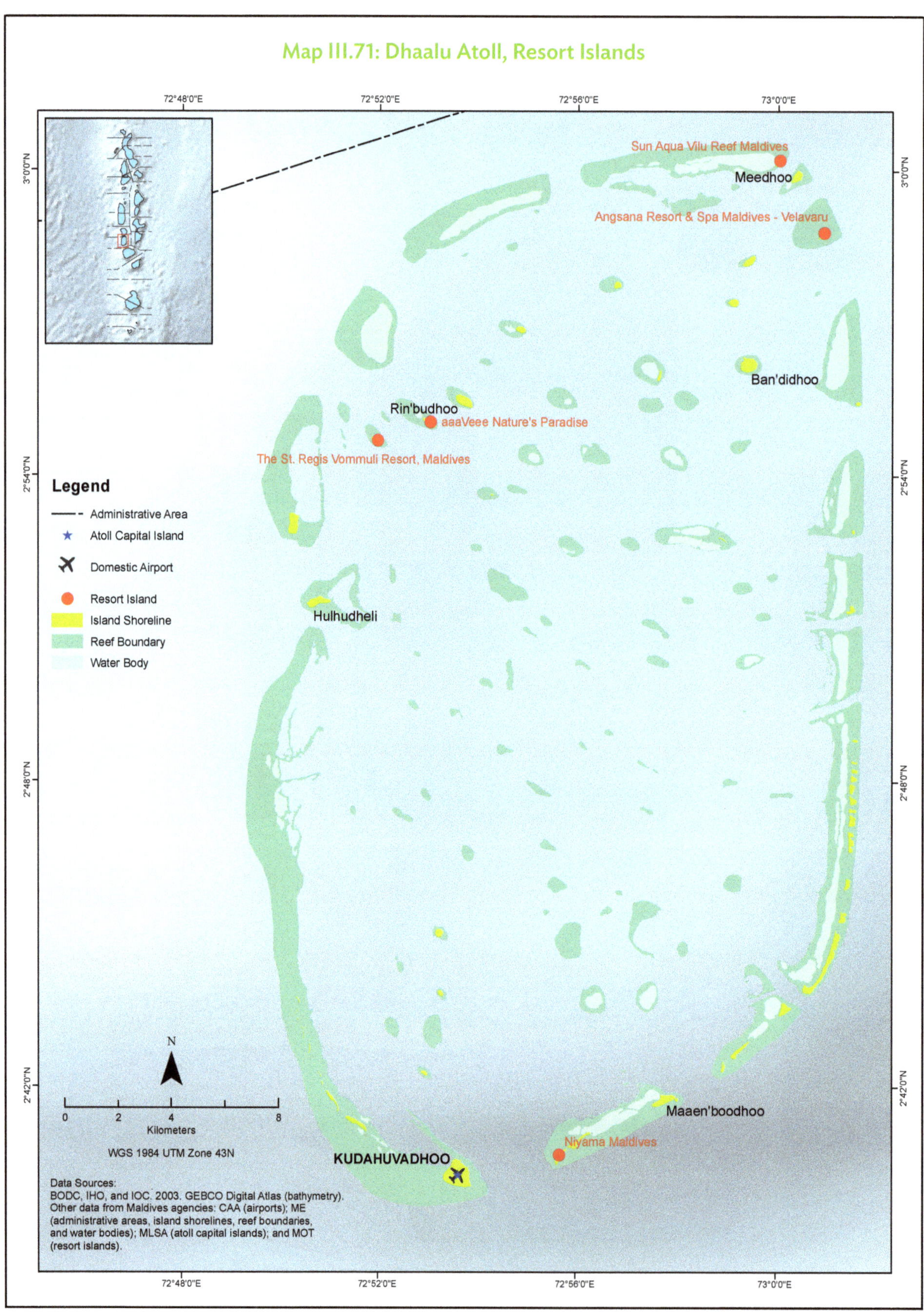
Map III.71: Dhaalu Atoll, Resort Islands
72°48'0"E
72°52'0"E
72°56'0"E
73°0'0"E
3°0'0"N
2°54'0"N
2°48'0"N
2°42'0"N
Sun Aqua Vilu Reef Maldives
Meedhoo
Angsana Resort & Spa Maldives - Velavaru
Ban'didhoo
Rin'budhoo
aaaVeee Nature's Paradise
The St. Regis Vommuli Resort, Maldives
Hulhudheli
Maaen'boodhoo
Niyama Maldives
KUDAHUVADHOO
Legend
Administrative Area
Atoll Capital Island
Domestic Airport
Resort Island
Island Shoreline
Reef Boundary
Water Body
N
0 2 4 8
Kilometers
WGS 1984 UTM Zone 43N
Data Sources:
BODC, IHO, and IOC. 2003. GEBCO Digital Atlas (bathymetry).
Other data from Maldives agencies: CAA (airports); ME (administrative areas, island shorelines, reef boundaries, and water bodies); MLSA (atoll capital islands); and MOT (resort islands).

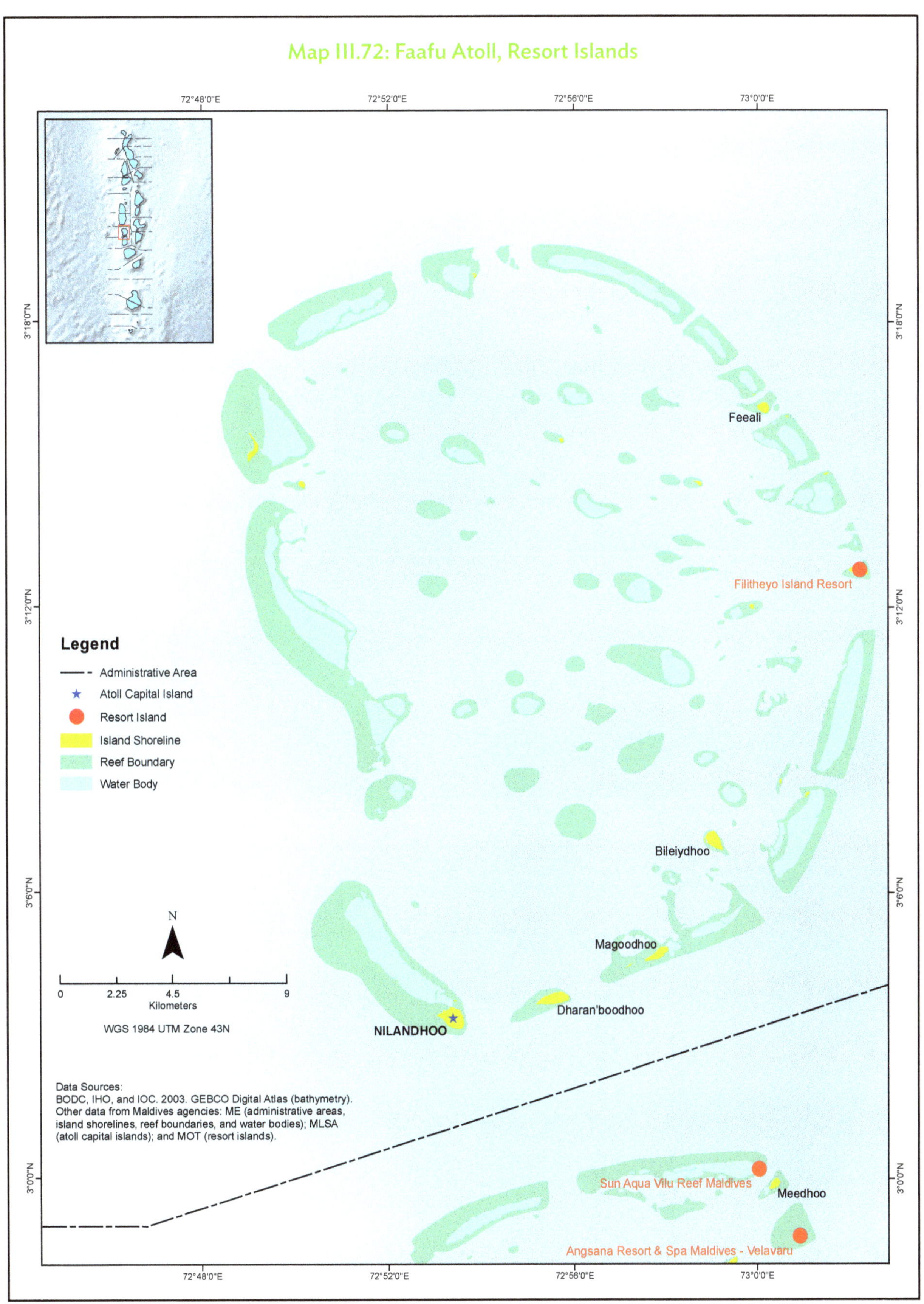

Map III.72: Faafu Atoll, Resort Islands
72°48'0"E
72°52'0"E
72°56'0"E
73°0'0"E
3°18'0"N
3°12'0"N
3°6'0"N
3°0'0"N
Feeali
Filitheyo Island Resort
Bileiydhoo
Magoodhoo
Dharan'boodhoo
NILANDHOO
Sun Aqua Vilu Reef Maldives
Meedhoo
Angsana Resort & Spa Maldives - Velavaru
Legend
Administrative Area
Atoll Capital Island
Resort Island
Island Shoreline
Reef Boundary
Water Body
N
0
2.25
4.5
9
Kilometers
WGS 1984 UTM Zone 43N
Data Sources:
BODC, IHO, and IOC. 2003. GEBCO Digital Atlas (bathymetry).
Other data from Maldives agencies: ME (administrative areas, island shorelines, reef boundaries, and water bodies); MLSA (atoll capital islands); and MOT (resort islands).

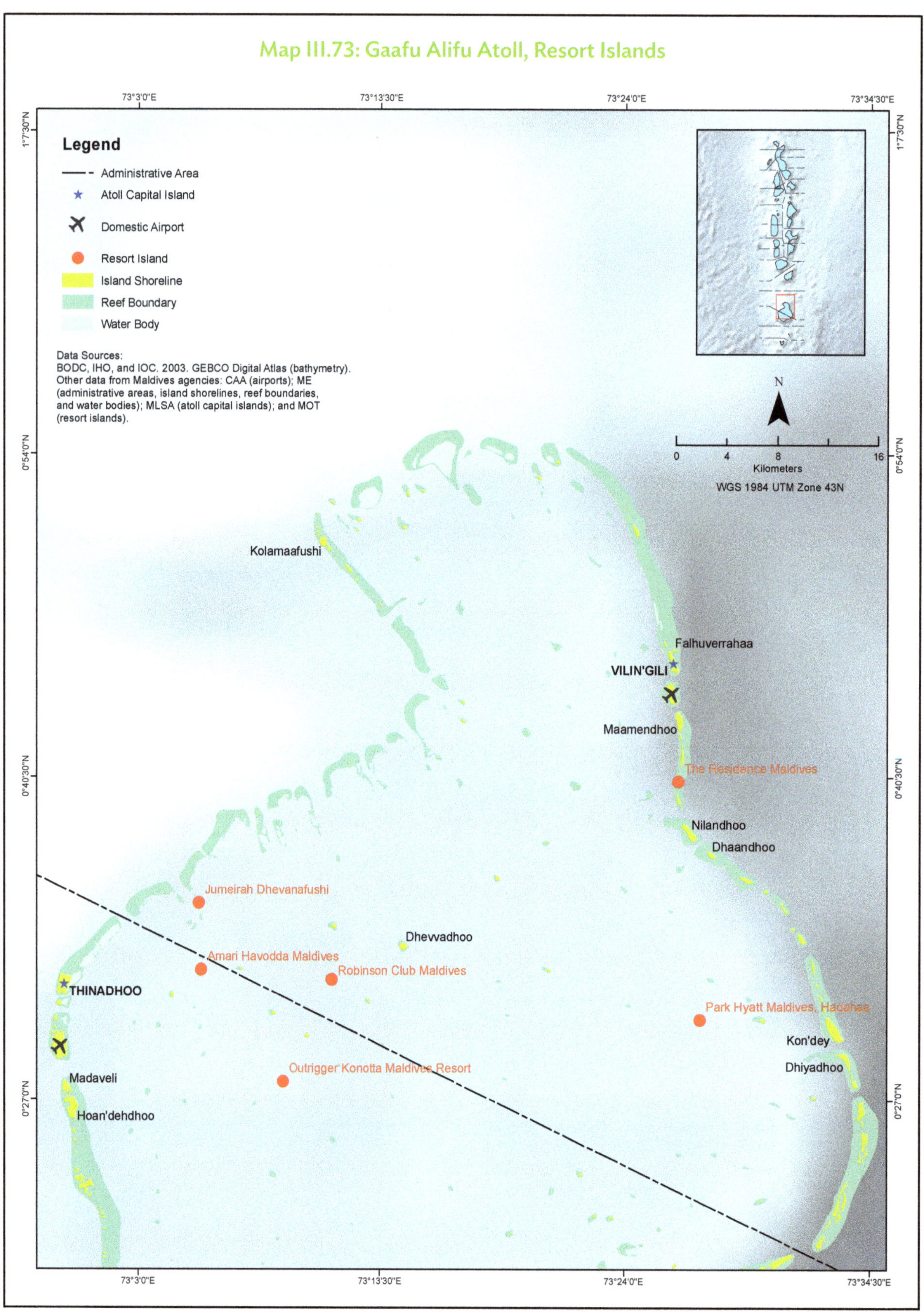
Map III.73: Gaafu Alifu Atoll, Resort Islands
73°3'0"E
73°13'30"E
73°24'0"E
73°34'30"E
1°7'30"N
0°54'0"N
0°40'30"N
0°27'0"N
Legend
Administrative Area
Atoll Capital Island
Domestic Airport
Resort Island
Island Shoreline
Reef Boundary
Water Body
Data Sources:
BODC, IHO, and IOC. 2003. GEBCO Digital Atlas (bathymetry).
Other data from Maldives agencies: CAA (airports); ME (administrative areas, island shorelines, reef boundaries, and water bodies); MLSA (atoll capital islands); and MOT (resort islands).
N
0 4 8 16
Kilometers
WGS 1984 UTM Zone 43N
Kolamaafushi
Falhuverrahaa
VILIN'GILI
Maamendhoo
The Residence Maldives
Nilandhoo
Dhaandhoo
Jumeirah Dhevanafushi
Dhevvadhoo
Amari Havodda Maldives
Robinson Club Maldives
THINADHOO
Park Hyatt Maldives, Hadahaa
Kon'dey
Dhiyadhoo
Outrigger Konotta Maldives Resort
Madaveli
Hoan'dehdhoo

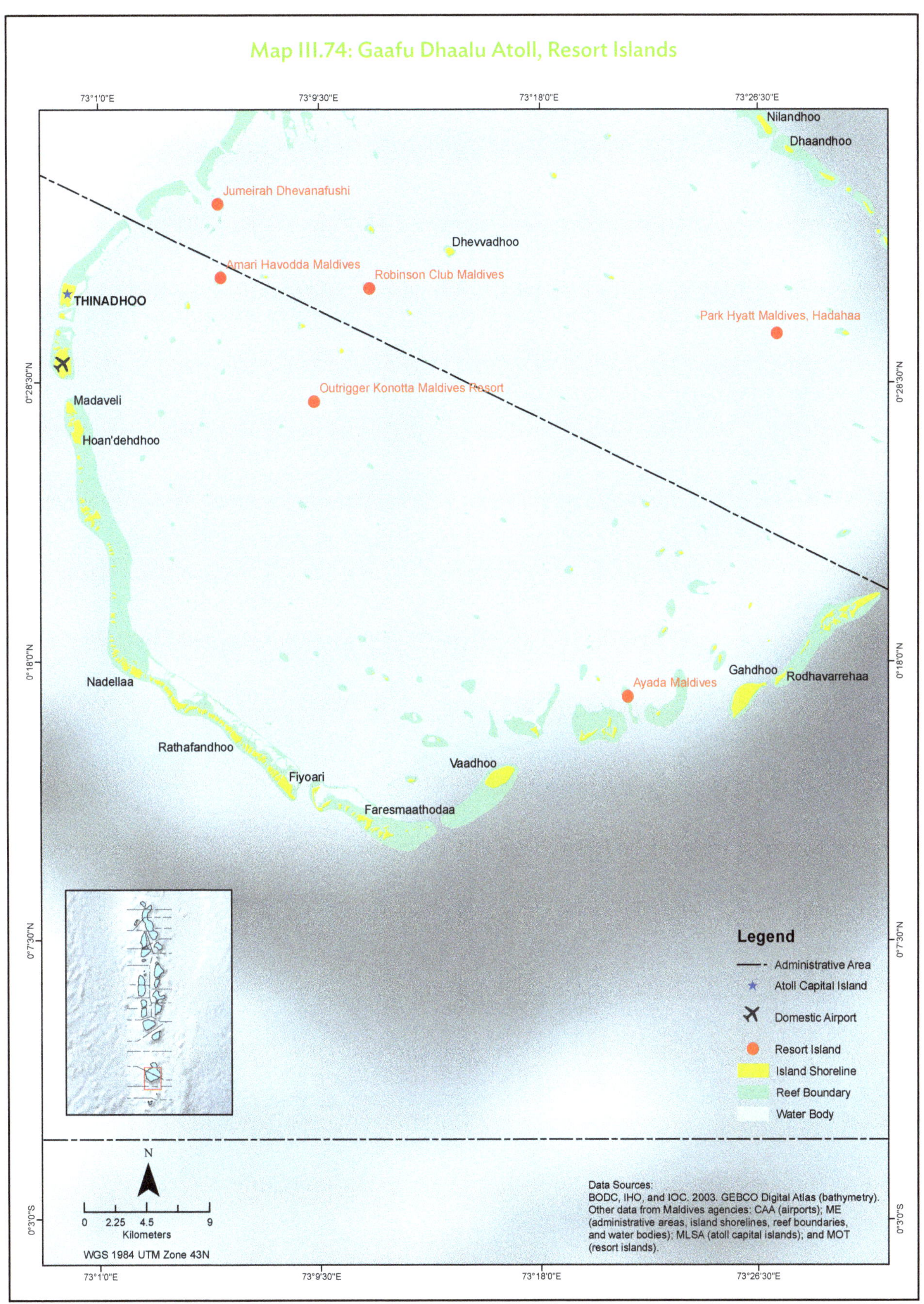
Map III.74: Gaafu Dhaalu Atoll, Resort Islands
73°1'0"E
73°9'30"E
73°18'0"E
73°26'30"E
0°28'30"N
0°18'0"N
0°7'30"N
0°3'0"S
Nilandhoo
Dhaandhoo
Jumeirah Dhevanafushi
Dhevvadhoo
Amari Havodda Maldives
Robinson Club Maldives
THINADHOO
Park Hyatt Maldives, Hadahaa
Outrigger Konotta Maldives Resort
Madaveli
Hoan'dehdhoo
Nadellaa
Gahdhoo
Rodhavarrehaa
Ayada Maldives
Rathafandhoo
Vaadhoo
Fiyoari
Faresmaathodaa
Legend
Administrative Area
Atoll Capital Island
Domestic Airport
Resort Island
Island Shoreline
Reef Boundary
Water Body
N
0 2.25 4.5 9
Kilometers
WGS 1984 UTM Zone 43N
Data Sources:
BODC, IHO, and IOC. 2003. GEBCO Digital Atlas (bathymetry).
Other data from Maldives agencies: CAA (airports); ME (administrative areas, island shorelines, reef boundaries, and water bodies); MLSA (atoll capital islands); and MOT (resort islands).

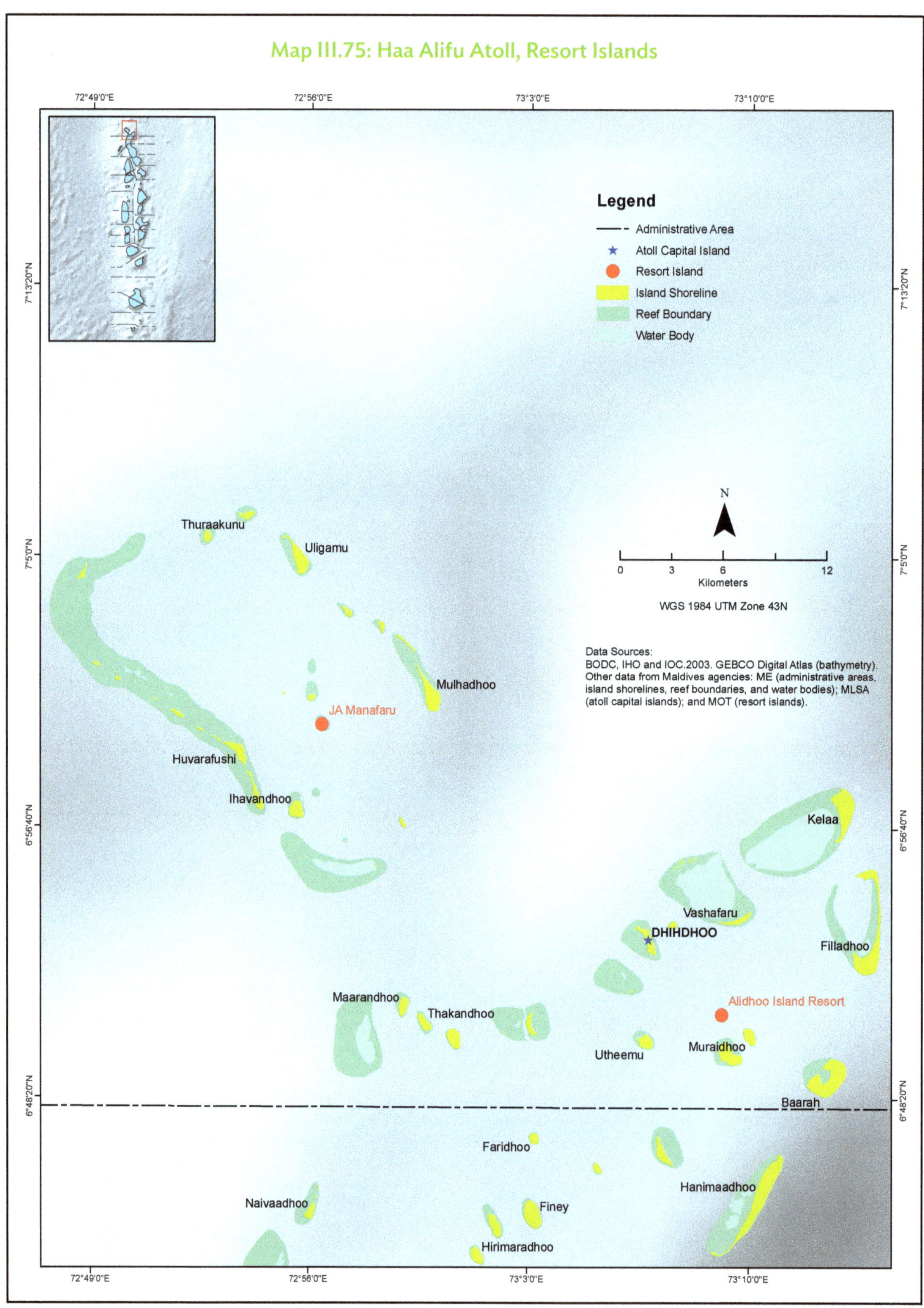
Map III.75: Haa Alifu Atoll, Resort Islands
72°49'0"E
72°56'0"E
73°3'0"E
73°10'0"E
7°13'20"N
7°5'0"N
6°56'40"N
6°48'20"N
Legend
Administrative Area
Atoll Capital Island
Resort Island
Island Shoreline
Reef Boundary
Water Body
N
0 3 6 12
Kilometers
WGS 1984 UTM Zone 43N
Data Sources:
BODC, IHO and IOC.2003. GEBCO Digital Atlas (bathymetry).
Other data from Maldives agencies: ME (administrative areas, island shorelines, reef boundaries, and water bodies); MLSA (atoll capital islands); and MOT (resort islands).
Thuraakunu
Uligamu
Mulhadhoo
JA Manafaru
Huvarafushi
Ihavandhoo
Kelaa
Vashafaru
DHIHDHOO
Filladhoo
Maarandhoo
Thakandhoo
Alidhoo Island Resort
Utheemu
Muraidhoo
Baarah
Faridhoo
Hanimaadhoo
Naivaadhoo
Finey
Hirimaradhoo

Map III.76: Laamu Atoll, Resort Islands

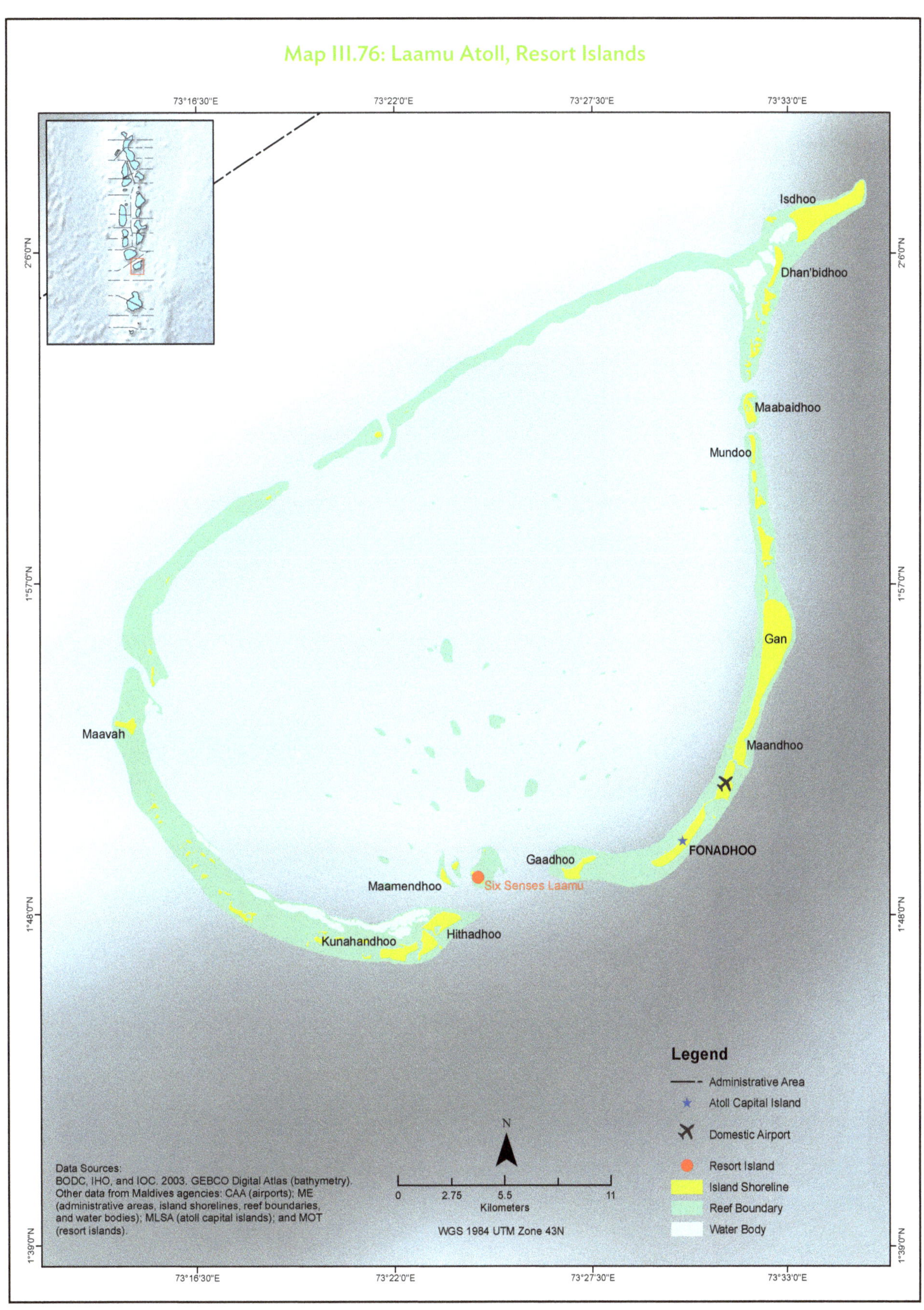

Map III.77: Lhaviyani Atoll, Resort Islands

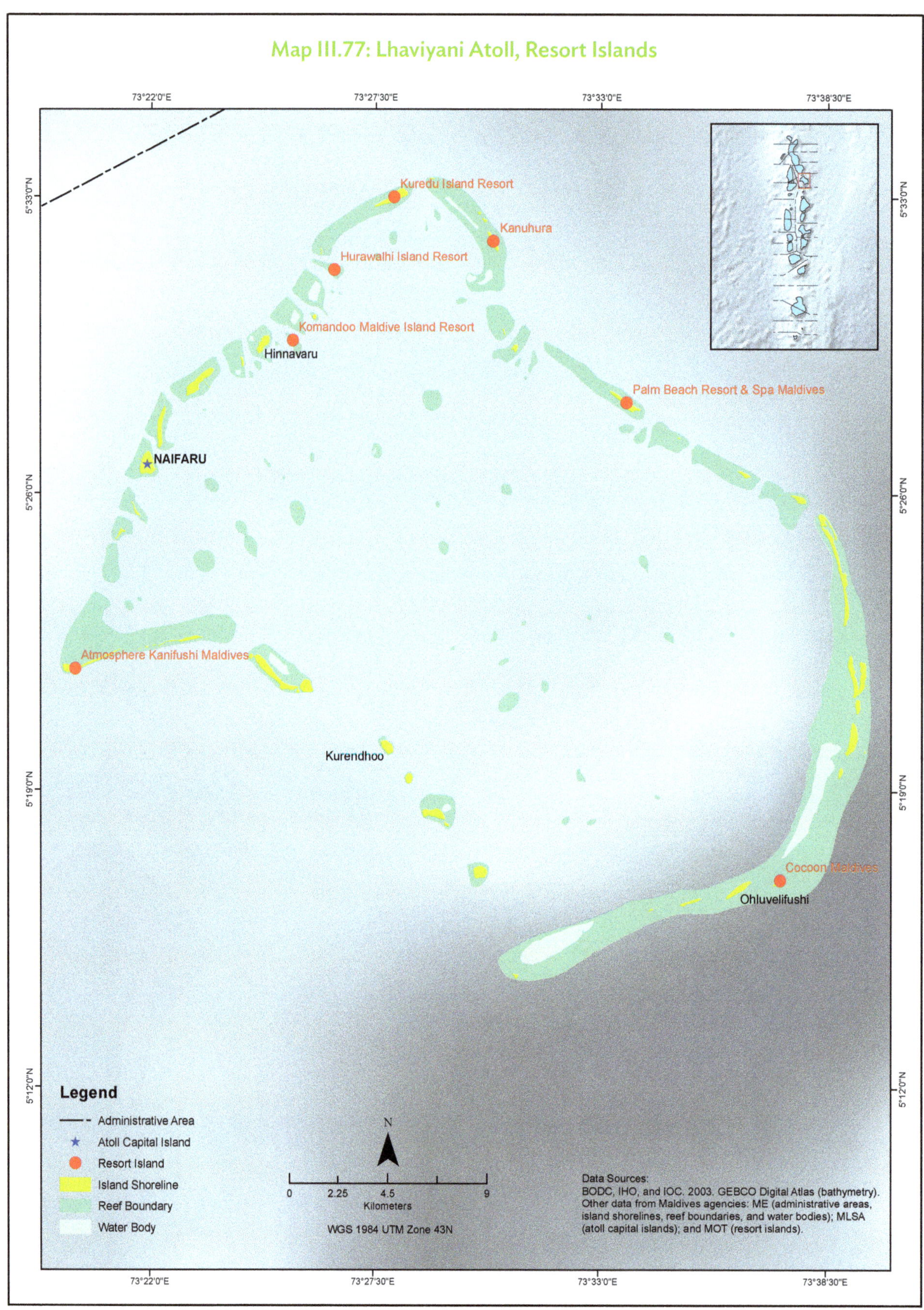

Map III.78: Meemu Atoll, Resort Islands

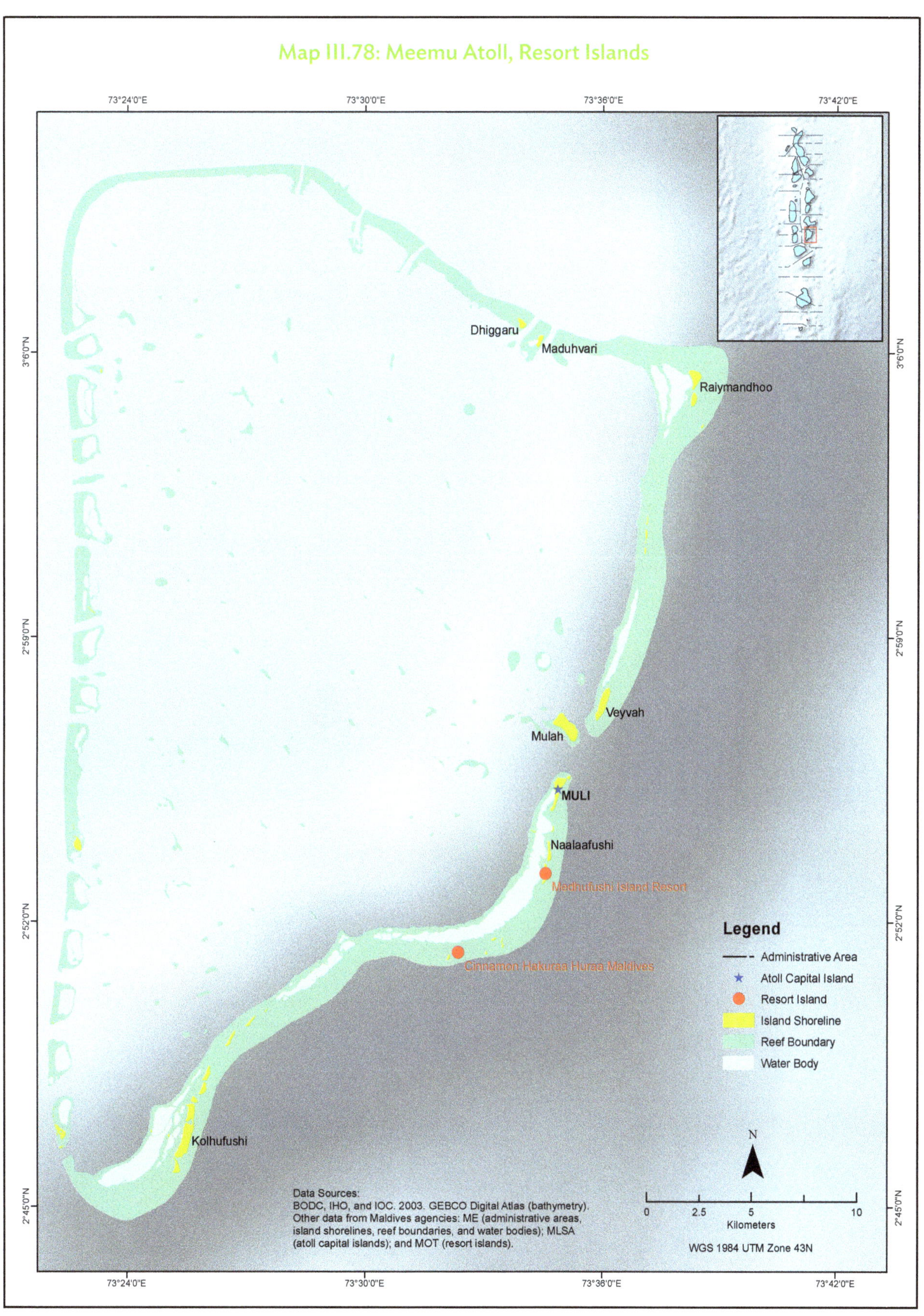

Map III.79: Noonu Atoll, Resort Islands

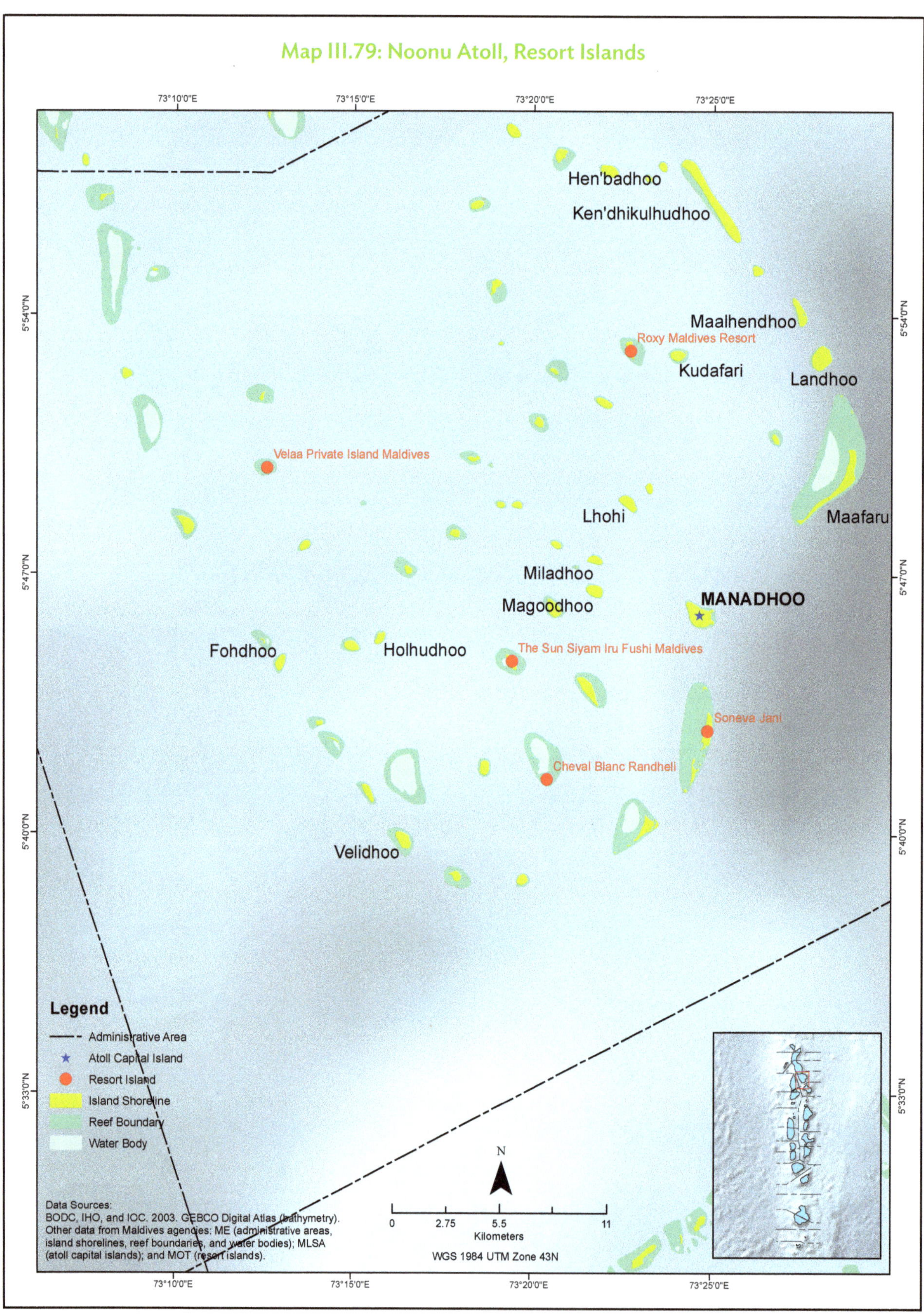

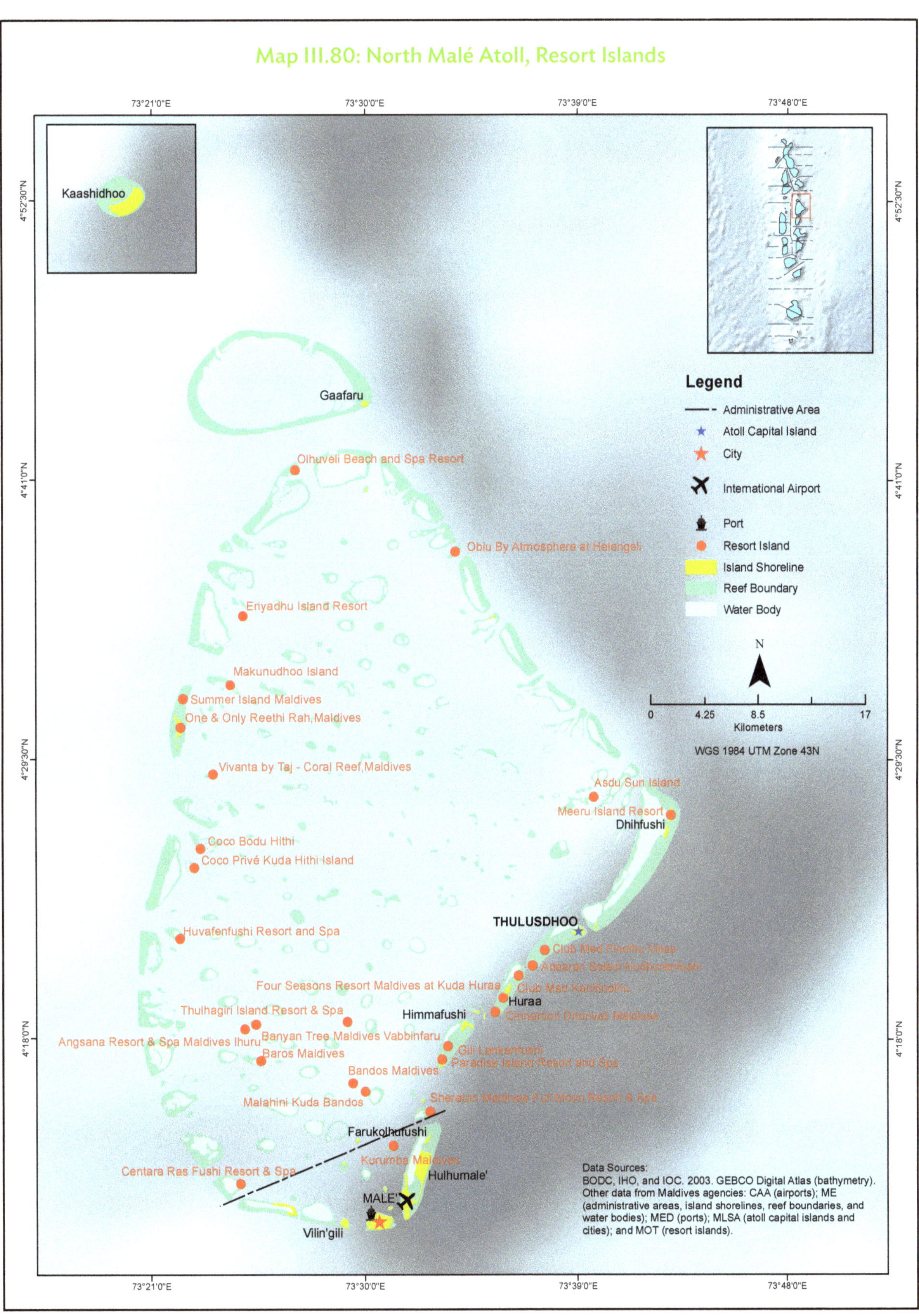

Map III.80: North Malé Atoll, Resort Islands

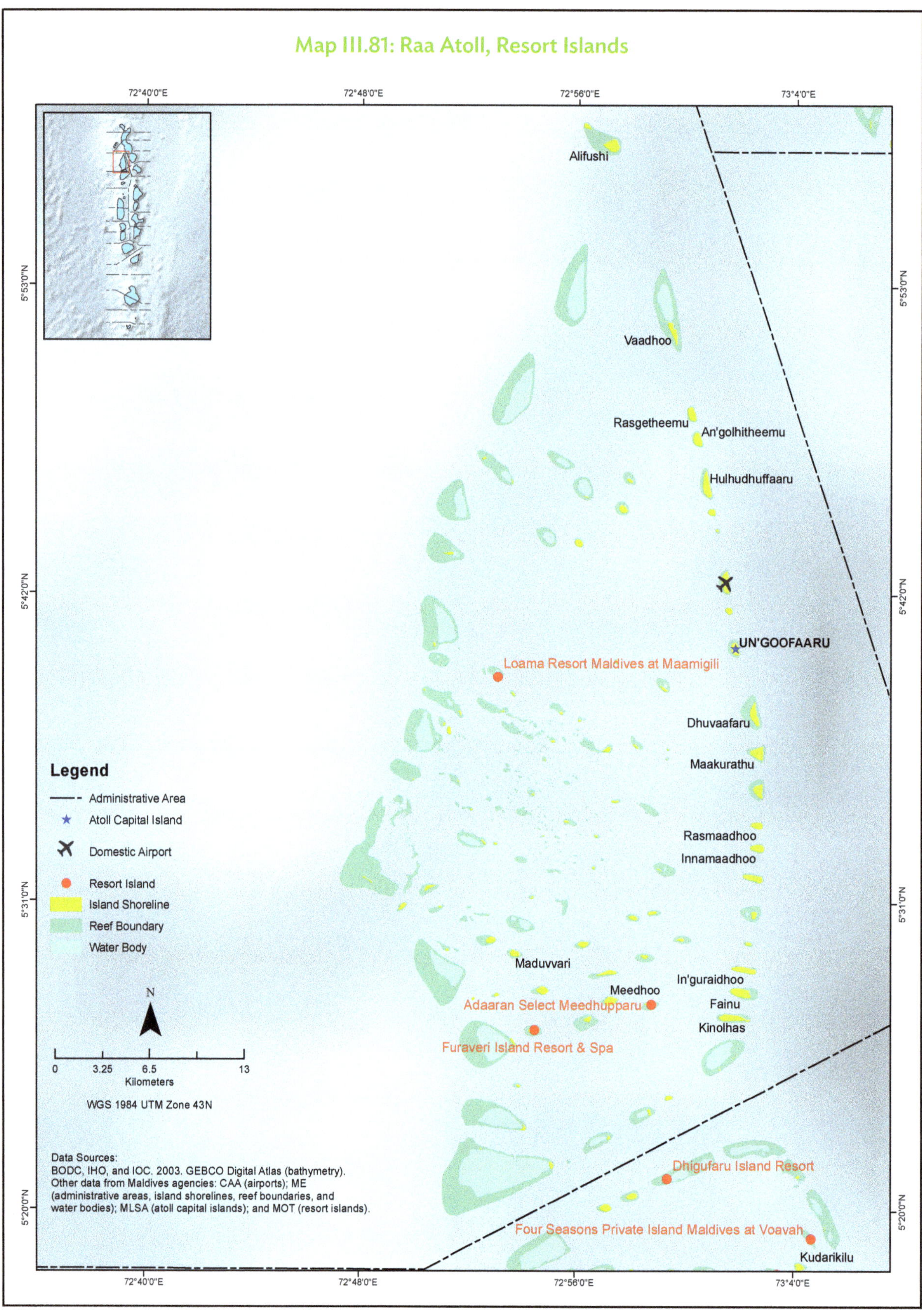
Map III.81: Raa Atoll, Resort Islands
72°40'0"E
72°48'0"E
72°56'0"E
73°4'0"E
5°53'0"N
5°42'0"N
5°31'0"N
5°20'0"N
Alifushi
Vaadhoo
Rasgetheemu
An'golhitheemu
Hulhudhuffaaru
UN'GOOFAARU
Loama Resort Maldives at Maamigili
Dhuvaafaru
Maakurathu
Rasmaadhoo
Innamaadhoo
Maduvvari
In'guraidhoo
Meedhoo
Adaaran Select Meedhupparu
Fainu
Kinolhas
Furaveri Island Resort & Spa
Dhigufaru Island Resort
Four Seasons Private Island Maldives at Voavah
Kudarikilu
Legend
Administrative Area
Atoll Capital Island
Domestic Airport
Resort Island
Island Shoreline
Reef Boundary
Water Body
N
0
3.25
6.5
13
Kilometers
WGS 1984 UTM Zone 43N
Data Sources:
BODC, IHO, and IOC. 2003. GEBCO Digital Atlas (bathymetry).
Other data from Maldives agencies: CAA (airports); ME (administrative areas, island shorelines, reef boundaries, and water bodies); MLSA (atoll capital islands); and MOT (resort islands).

Map III.82: Shaviyani Atoll, Resort Islands

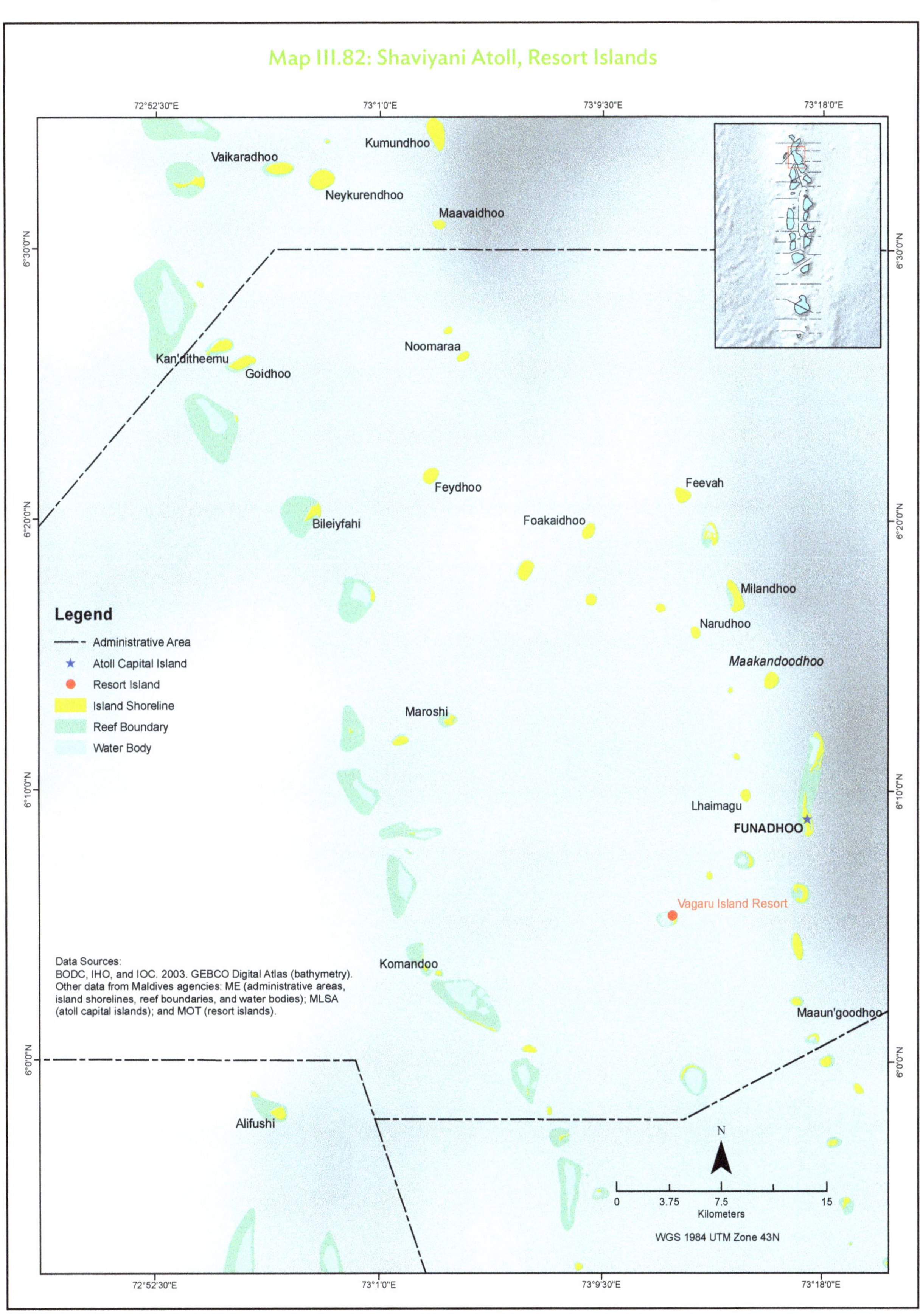

Map III.83: South Malé Atoll, Resort Islands

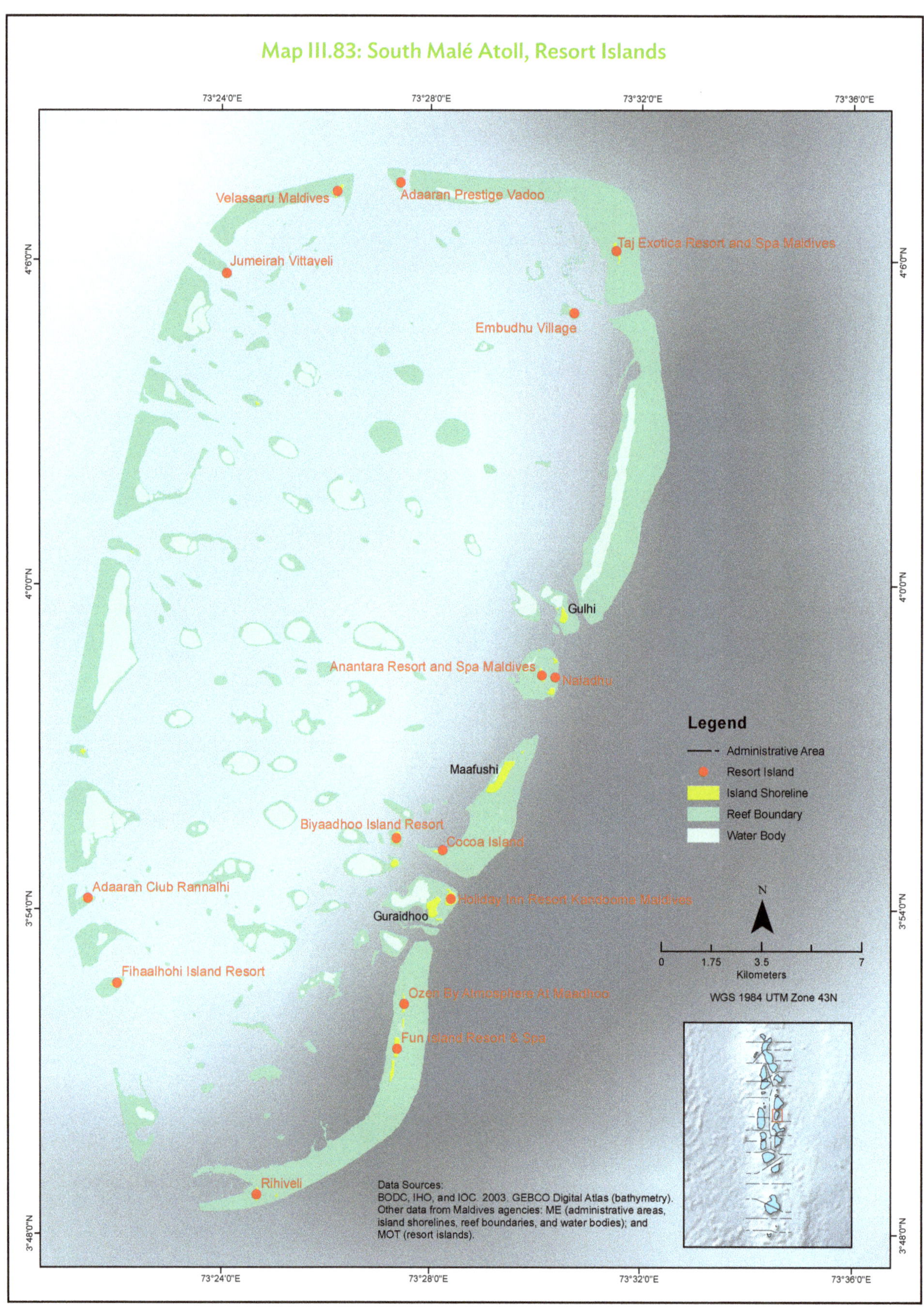

Map III.84: Thaa Atoll, Resort Islands

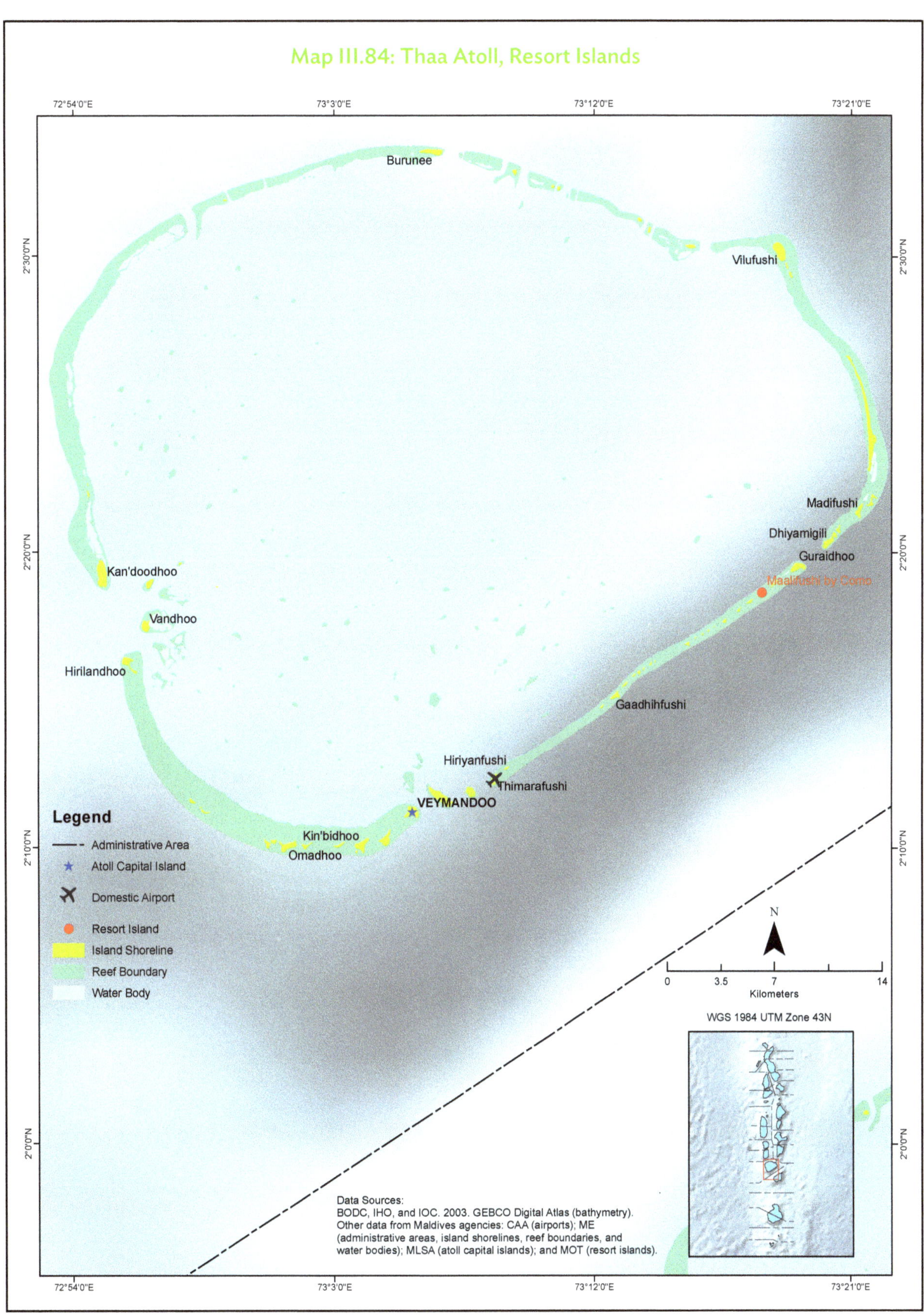

Map III.85: Vaavu Atoll, Resort Islands

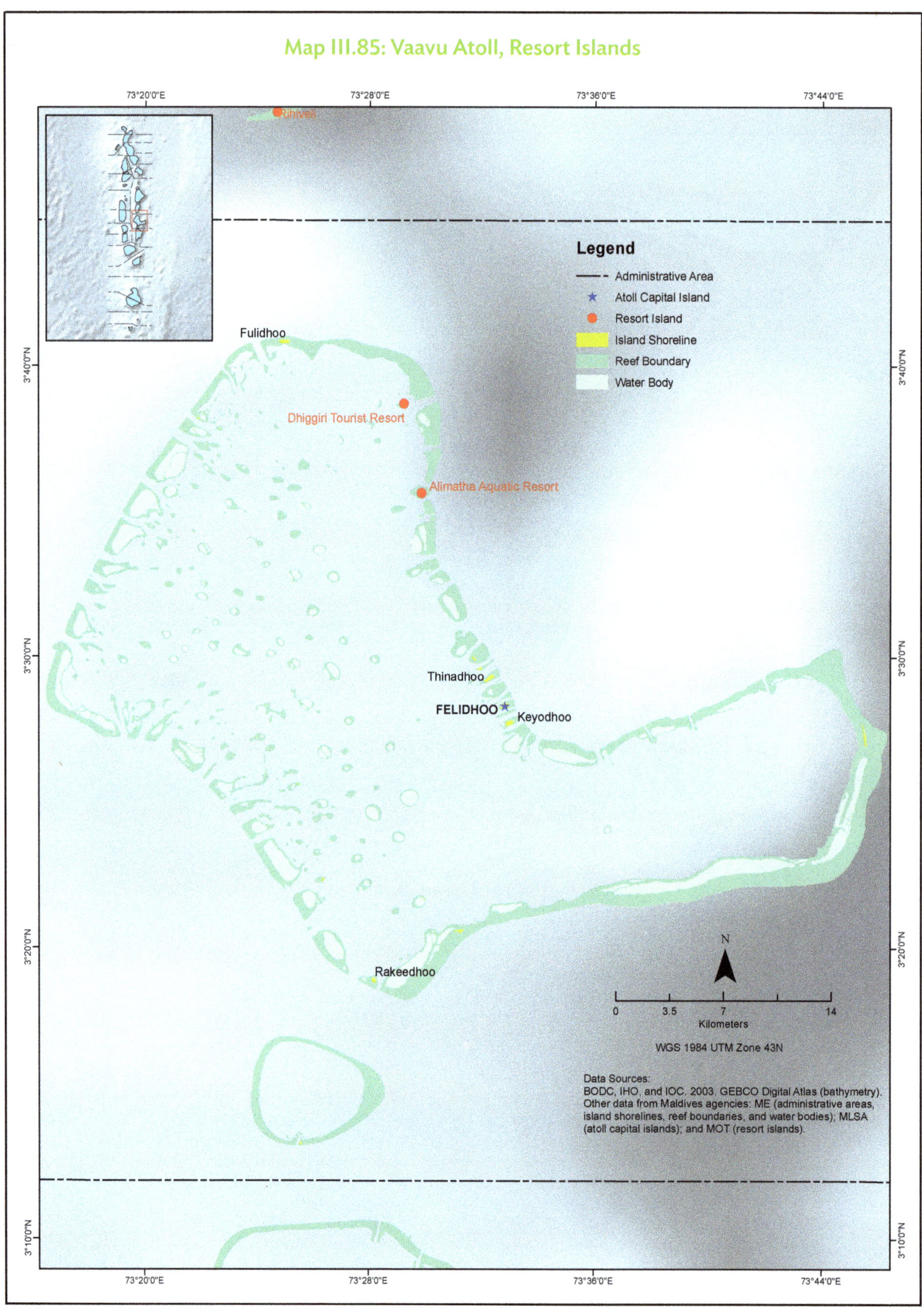

Transportation

In Maldives, traveling from one end of the island to the other does not take too long, and there are various modes of transportation. People can choose to travel by air or sea.

There are three ports for traveling and transporting goods: (i) Kulhudhuffushi Regional Port in Haa Dhaalu, (ii) Hithadhoo Regional Port in Addu City, and (iii) Malé Commercial Harbor in Malé City. People can also board planes through the international and domestic airports. One alternative mode of transport is via floatplane, which can be accessed from floating platforms scattered across the islands.

Means of transport. A seaplane docked at a port in Maldives is a usual sight and transport mode (photo by Ibahim Asad).

Table III.7: Airports in Maldives

Atoll	Island Name	Aerodrome
Domestic Airports		
Baa	Dharavandhoo	Dharavandhoo Airport
Gnaviyani	Fuvahmulah	Fuvahmulah Airport
Dhaalu	Kudahuvadhoo	Dhaalu Airport
Thaa	Thimarafushi	Thimarafushi Airport
Laamu	Kahdhoo	Kadhdhoo Airport
Gaafu Dhaalu	Kaadehdhoo	Kaadedhdhoo Airport
Alifu Dhaalu	Maamin'gili	Villa Airport Maamigili
Gaafu Alifu	Kooddoo	Koodoo Airport
Raa	Ifuru	Ifuru Airport
Haa Dhaalu	Kulhudhuffushi	Kulhudhuffushi Airport
International Airports		
Noonu	Maafaru	Maafaru International Airport
Haa Dhaalu	Hanimaadhoo	Hanimaadhoo International Airport
Kaafu	Hulhulé	Velana International Airport
Seenu	Gan	Gan International Airport
Future Airports		
Gaafu Dhaalu	Faresmaathodaa	
Gaafu Dhaalu	Maavaarulu	
Haa Alifu	Hoarafushi	
Lhaviyani	Madivaru	
Meemu	Muli	
Shaviyani	Funadhoo	

Source: Civil Aviation Authority, 2019.

Map III.86: Maldives, Transportation

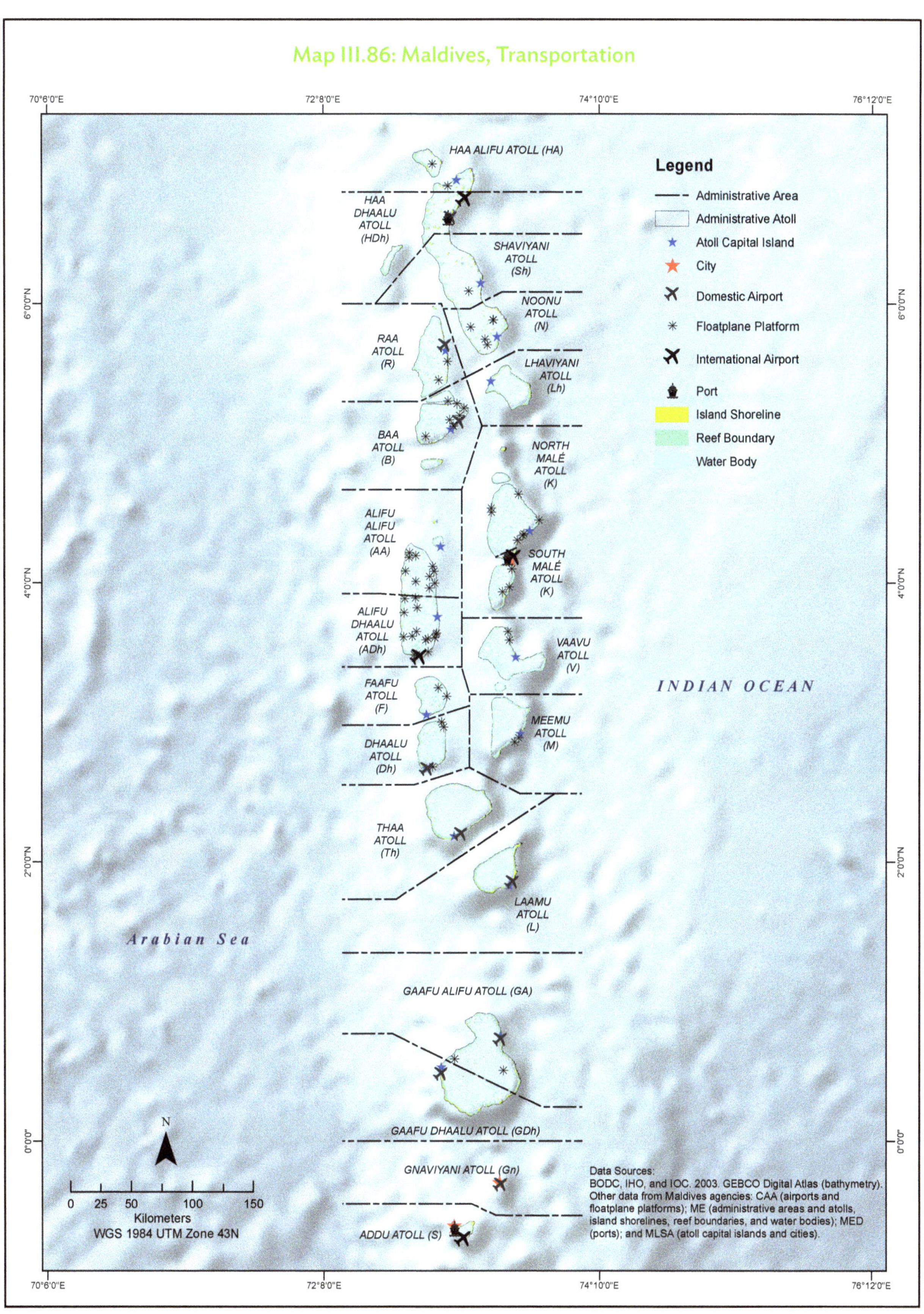

Map III.87: Addu City, Transportation

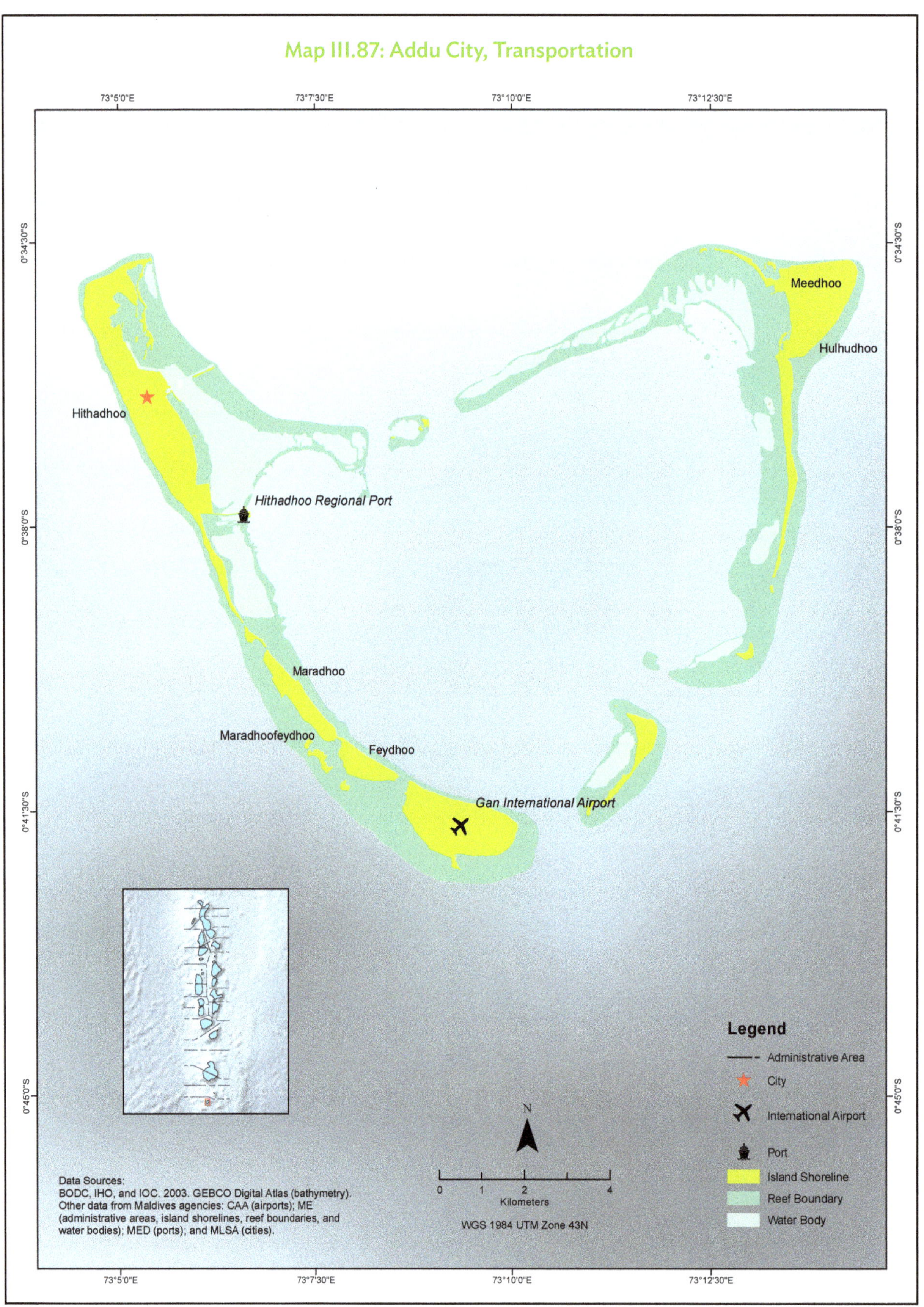

Data Sources:
BODC, IHO, and IOC. 2003. GEBCO Digital Atlas (bathymetry).
Other data from Maldives agencies: CAA (airports); ME (administrative areas, island shorelines, reef boundaries, and water bodies); MED (ports); and MLSA (cities).

Map III.88: Alifu Alifu Atoll, Transportation

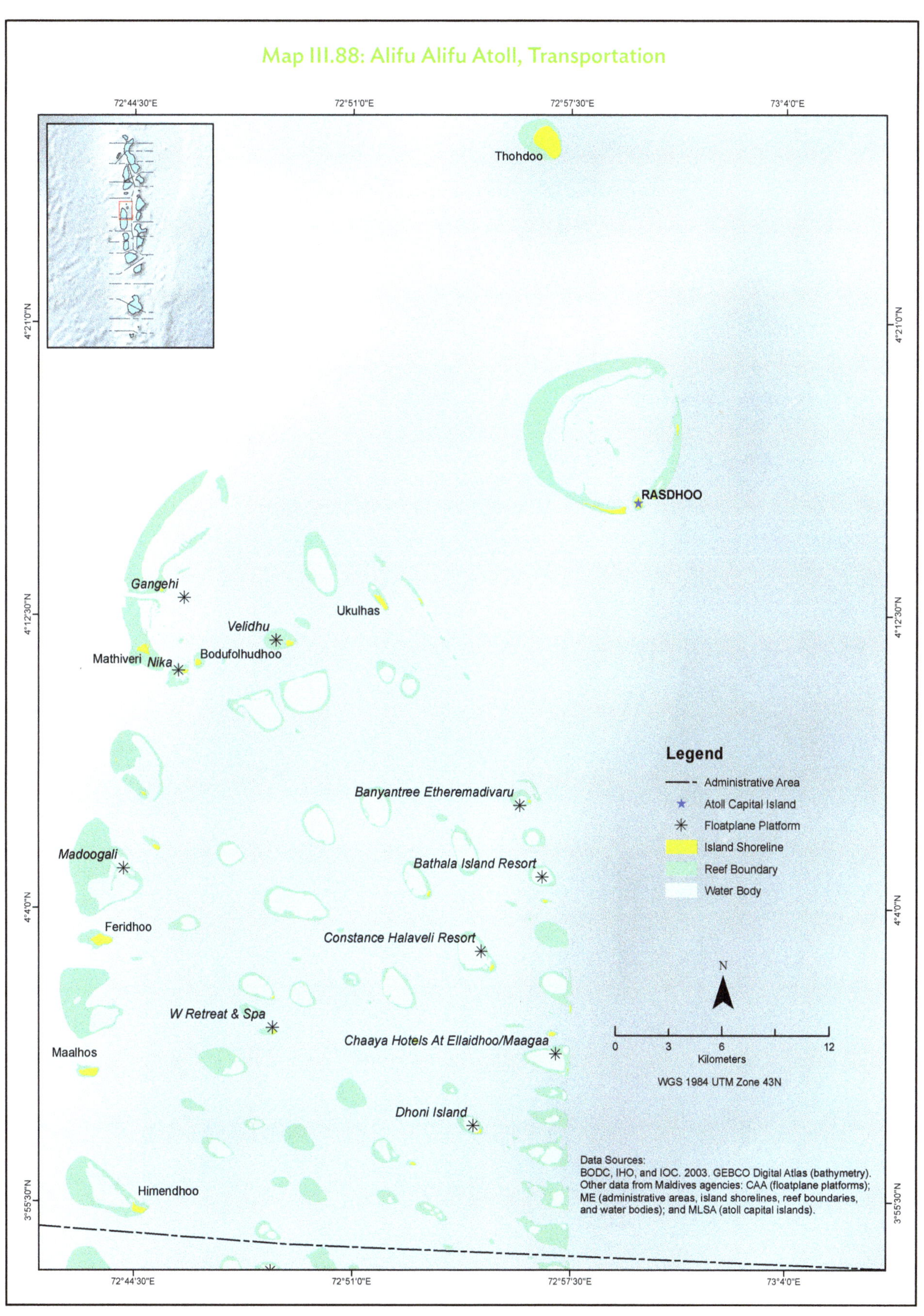

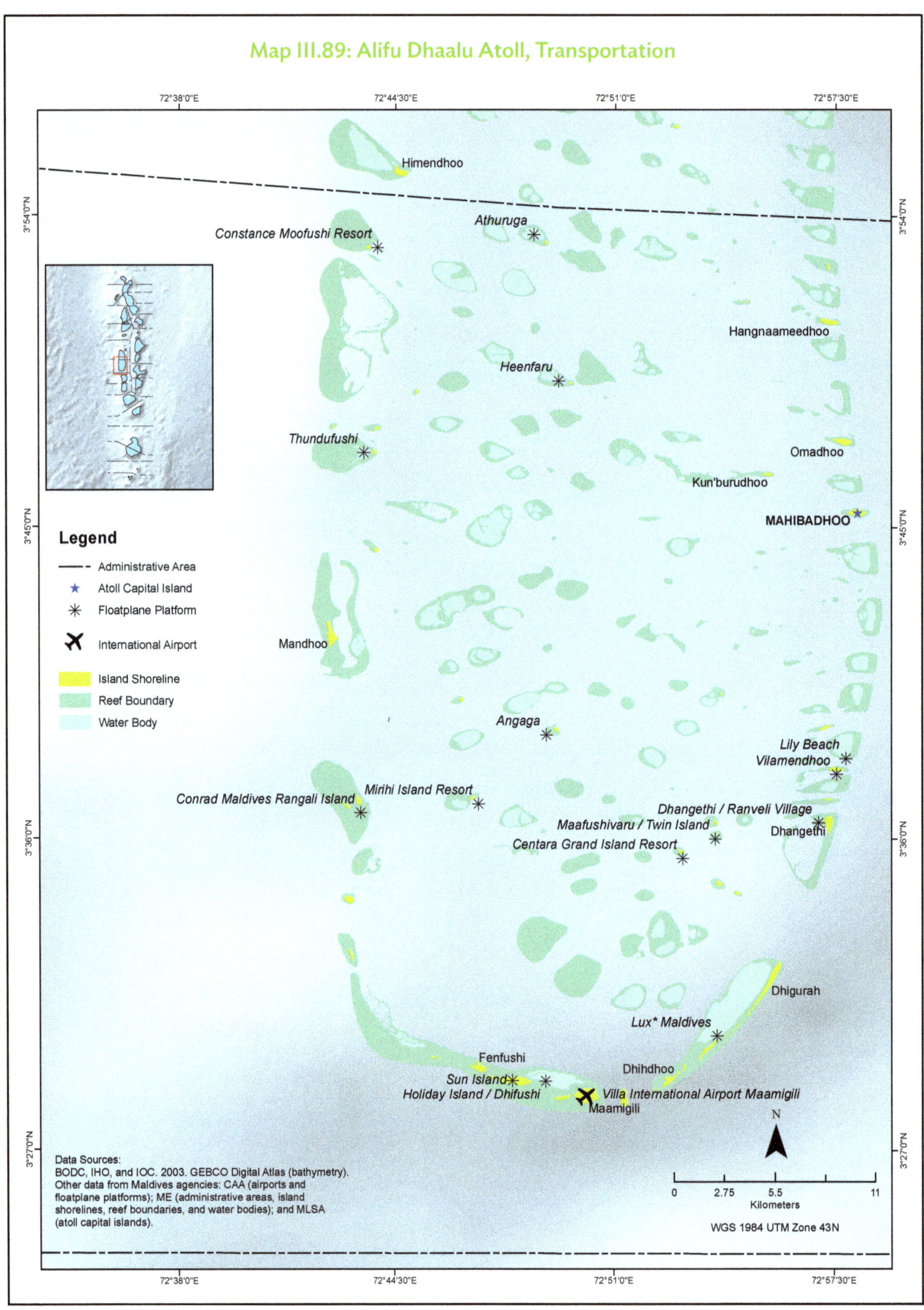

Map III.89: Alifu Dhaalu Atoll, Transportation
72°38'0"E
72°44'30"E
72°51'0"E
72°57'30"E
3°54'0"N
3°45'0"N
3°36'0"N
3°27'0"N
Himendhoo
Athuruga
Constance Moofushi Resort
Hangnaameedhoo
Heenfaru
Thundufushi
Omadhoo
Kun'burudhoo
MAHIBADHOO
Legend
Administrative Area
Atoll Capital Island
Floatplane Platform
International Airport
Island Shoreline
Reef Boundary
Water Body
Mandhoo
Angaga
Lily Beach
Vilamendhoo
Mirihi Island Resort
Conrad Maldives Rangali Island
Dhangethi / Ranveli Village
Maafushivaru / Twin Island
Dhangethi
Centara Grand Island Resort
Dhigurah
Lux* Maldives
Fenfushi
Dhihdhoo
Sun Island
Holiday Island / Dhifushi
Villa International Airport Maamigili
Maamigili
N
0
2.75
5.5
11
Kilometers
WGS 1984 UTM Zone 43N
Data Sources:
BODC, IHO, and IOC. 2003. GEBCO Digital Atlas (bathymetry).
Other data from Maldives agencies: CAA (airports and floatplane platforms); ME (administrative areas, island shorelines, reef boundaries, and water bodies); and MLSA (atoll capital islands).

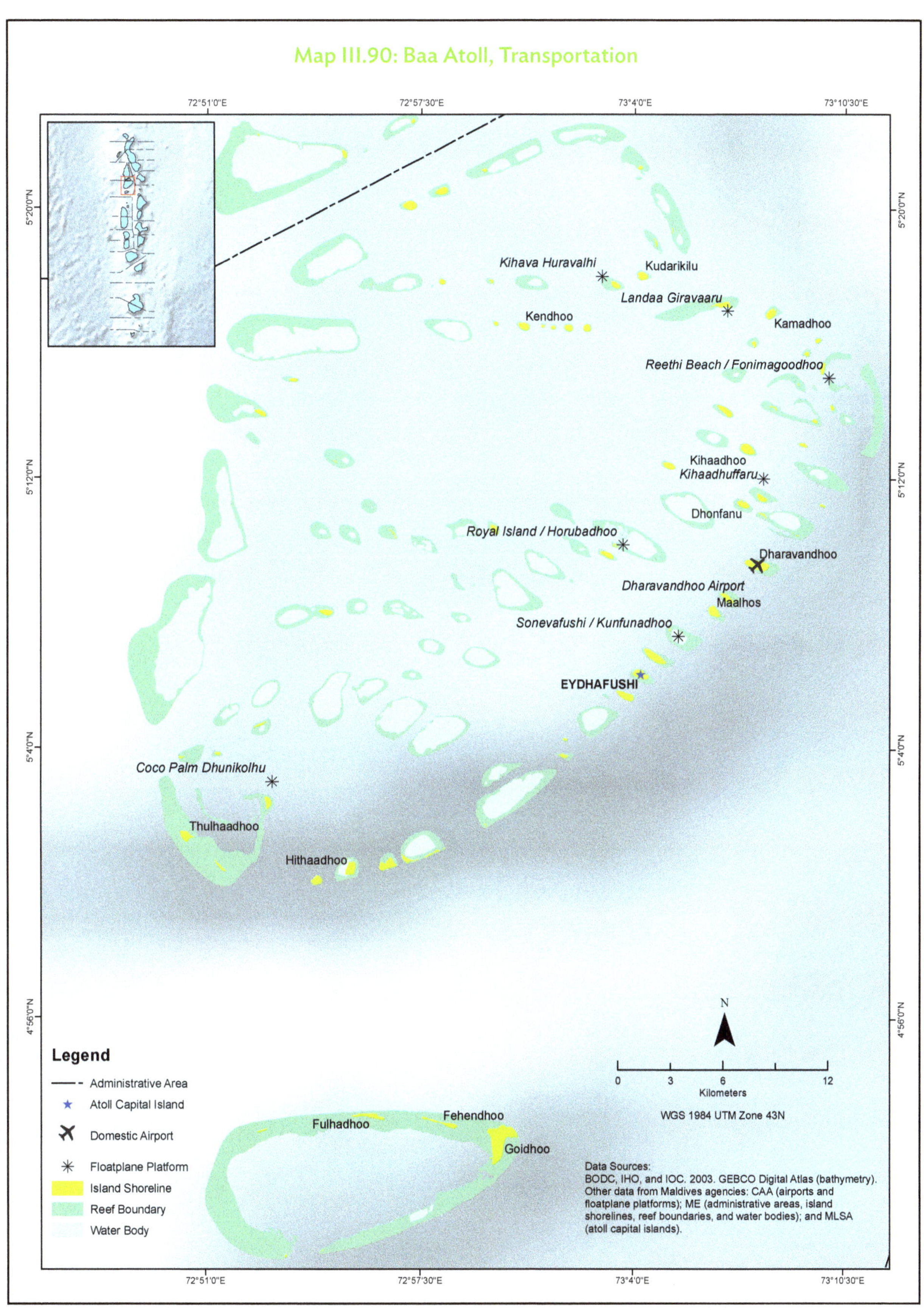

Map III.90: Baa Atoll, Transportation
72°51'0"E
72°57'30"E
73°4'0"E
73°10'30"E
5°20'0"N
5°12'0"N
5°4'0"N
4°56'0"N
Kihava Huravalhi
Kudarikilu
Landaa Giravaaru
Kendhoo
Kamadhoo
Reethi Beach / Fonimagoodhoo
Kihaadhoo
Kihaadhuffaru
Dhonfanu
Royal Island / Horubadhoo
Dharavandhoo
Dharavandhoo Airport
Maalhos
Sonevafushi / Kunfunadhoo
EYDHAFUSHI
Coco Palm Dhunikolhu
Thulhaadhoo
Hithaadhoo
Fulhadhoo
Fehendhoo
Goidhoo
N
0 3 6 12
Kilometers
WGS 1984 UTM Zone 43N
Legend
Administrative Area
Atoll Capital Island
Domestic Airport
Floatplane Platform
Island Shoreline
Reef Boundary
Water Body
Data Sources:
BODC, IHO, and IOC. 2003. GEBCO Digital Atlas (bathymetry).
Other data from Maldives agencies: CAA (airports and floatplane platforms); ME (administrative areas, island shorelines, reef boundaries, and water bodies); and MLSA (atoll capital islands).

Map III.91: Dhaalu Atoll, Transportation

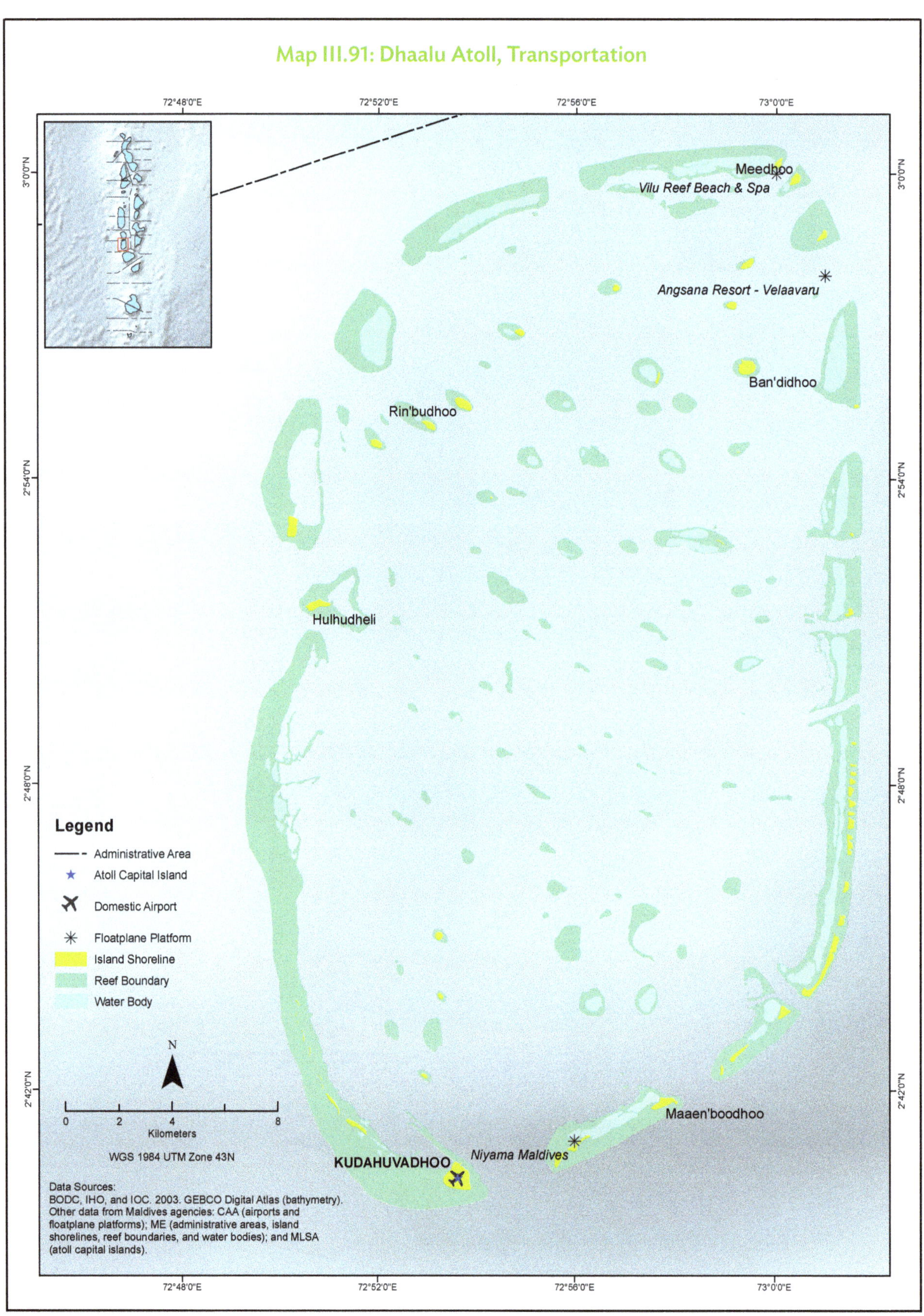

Map III.92: Faafu Atoll, Transportation

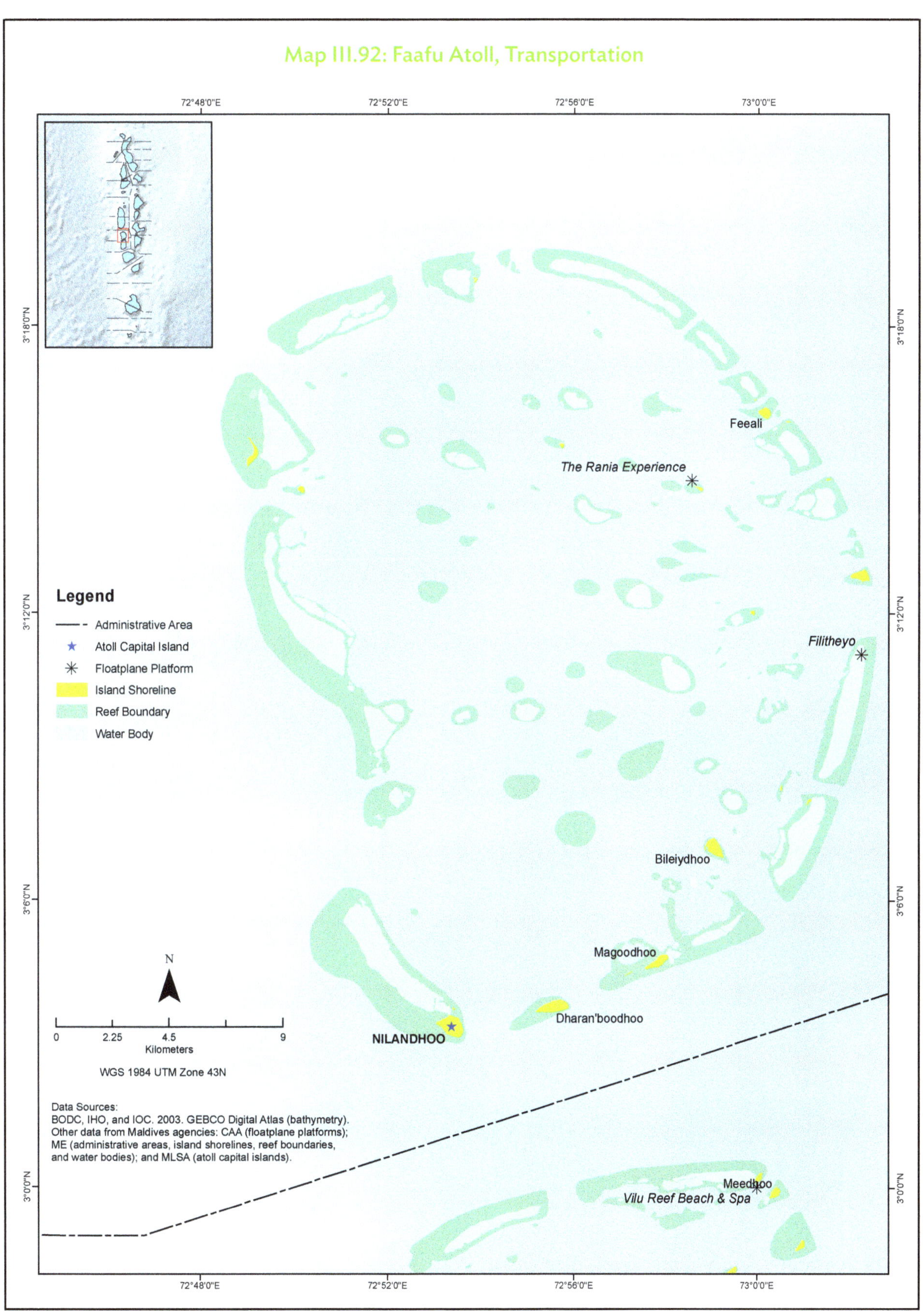

Map III.93: Gaafu Alifu Atoll, Transportation

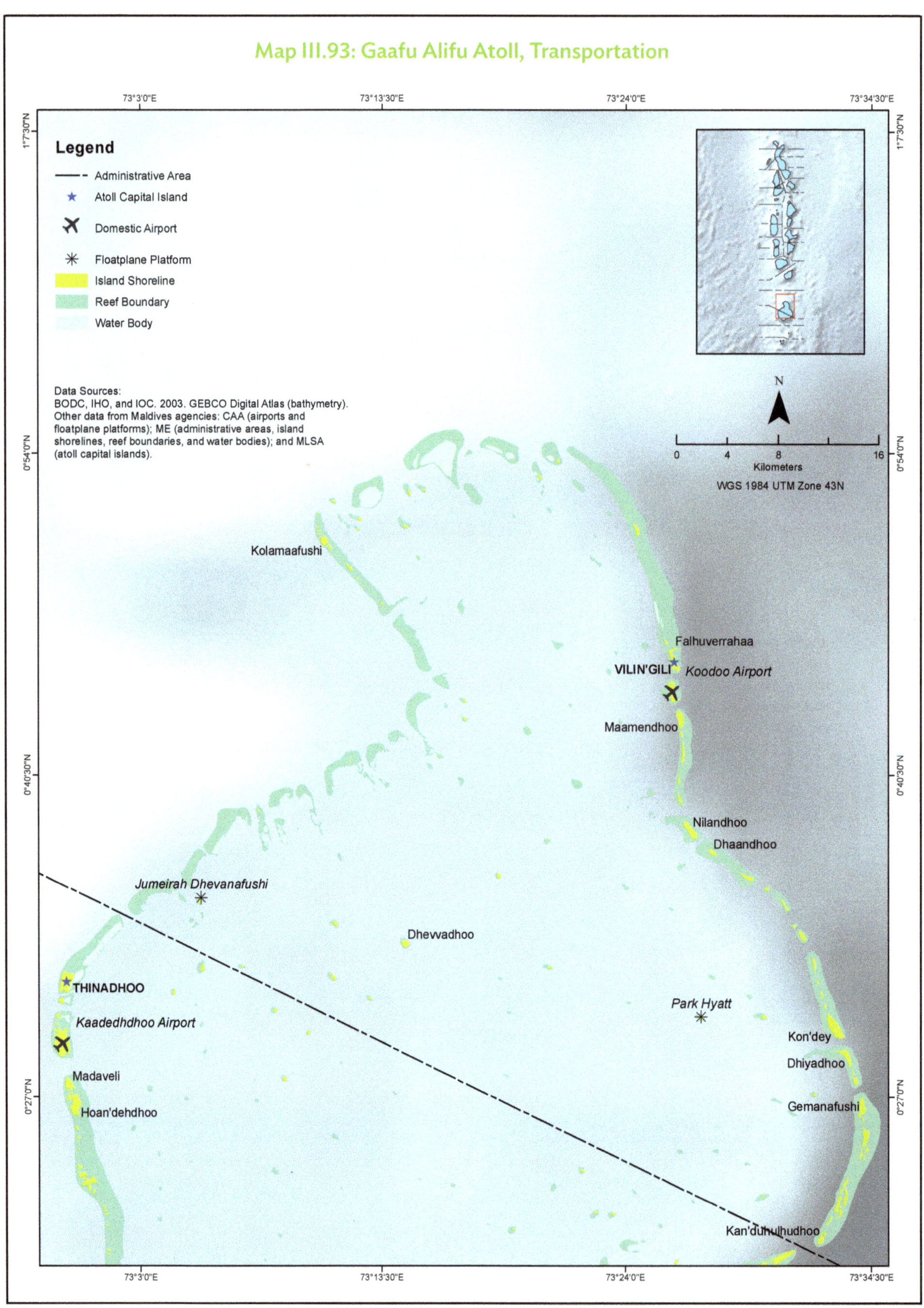

Map III.94: Gaafu Dhaalu Atoll, Transportation

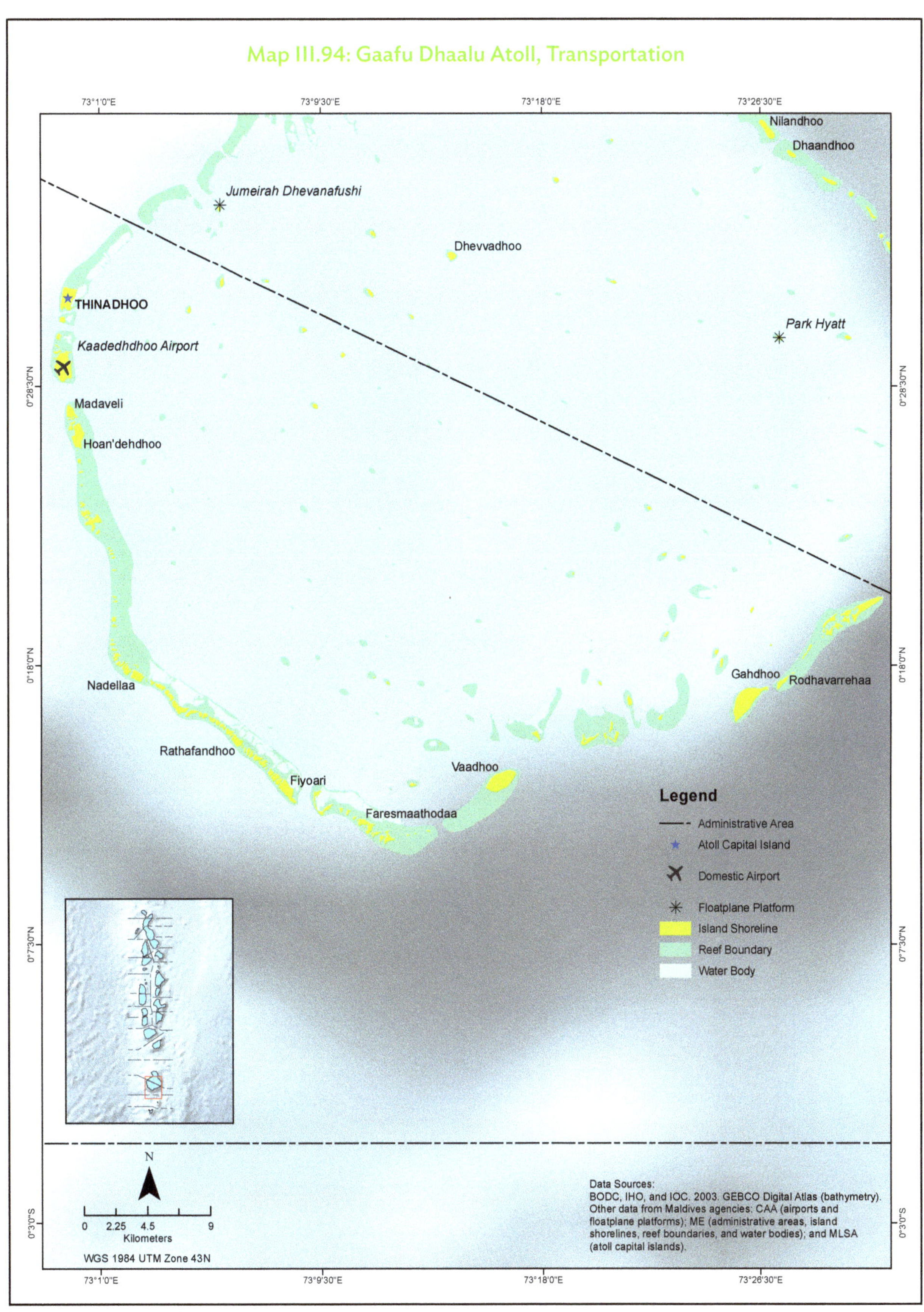

Map III.95: Gnaviyani Atoll, Transportation

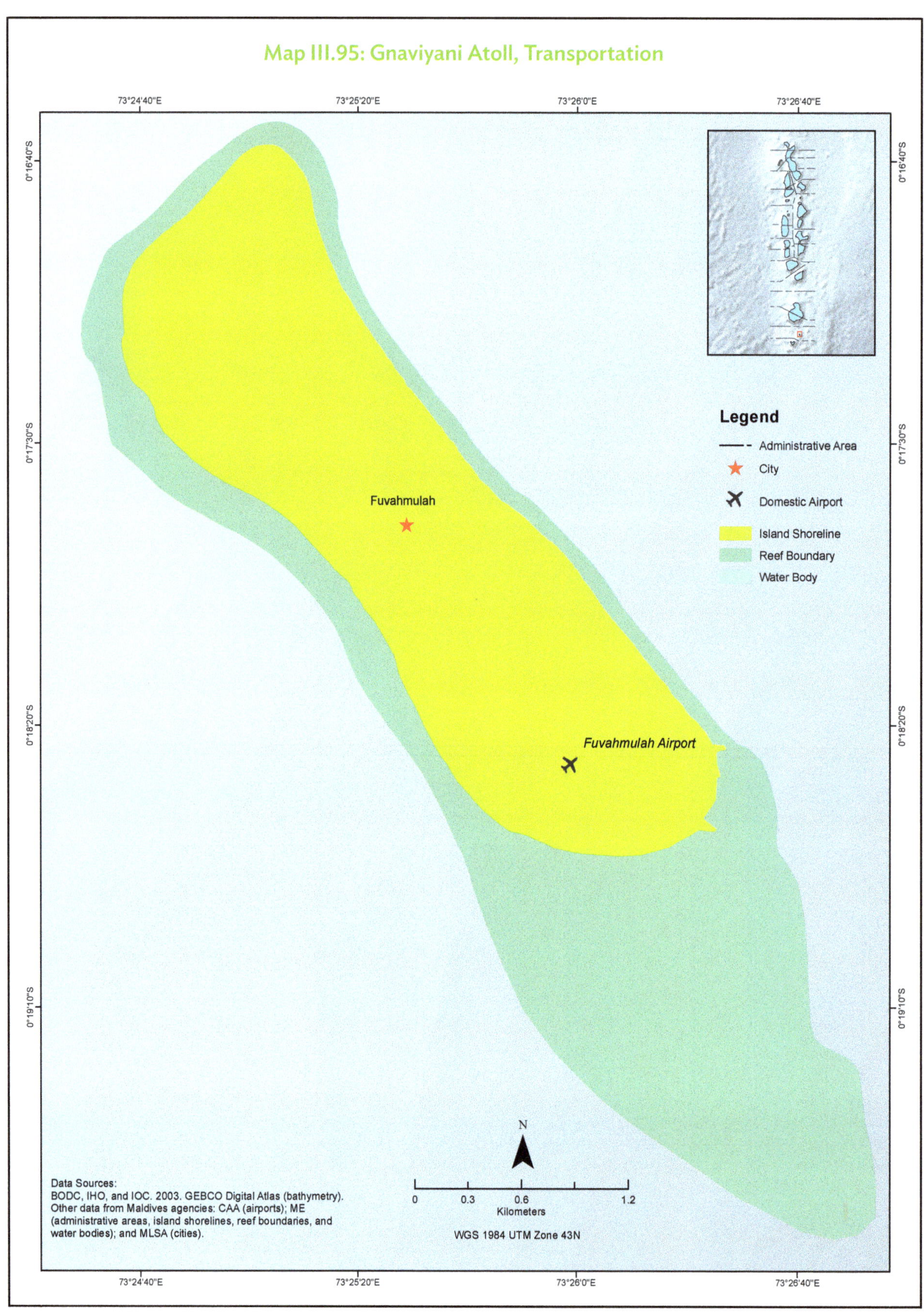

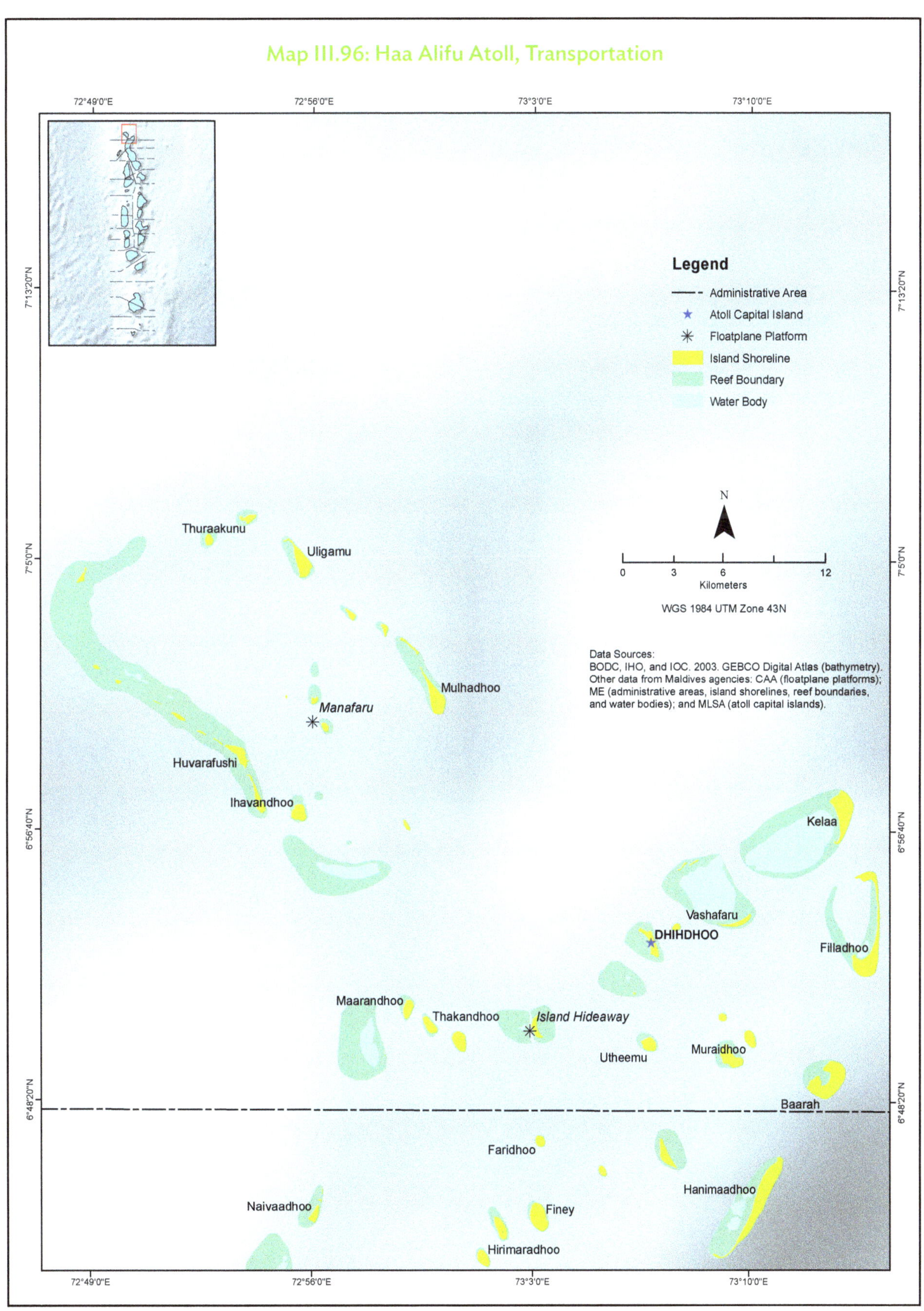
Map III.96: Haa Alifu Atoll, Transportation
72°49'0"E
72°56'0"E
73°3'0"E
73°10'0"E
7°13'20"N
7°5'0"N
6°56'40"N
6°48'20"N
Legend
Administrative Area
Atoll Capital Island
Floatplane Platform
Island Shoreline
Reef Boundary
Water Body
N
0 3 6 12
Kilometers
WGS 1984 UTM Zone 43N
Data Sources:
BODC, IHO, and IOC. 2003. GEBCO Digital Atlas (bathymetry).
Other data from Maldives agencies: CAA (floatplane platforms);
ME (administrative areas, island shorelines, reef boundaries,
and water bodies); and MLSA (atoll capital islands).
Thuraakunu
Uligamu
Mulhadhoo
Manafaru
Huvarafushi
Ihavandhoo
Kelaa
Vashafaru
DHIHDHOO
Filladhoo
Maarandhoo
Thakandhoo
Island Hideaway
Utheemu
Muraidhoo
Baarah
Faridhoo
Hanimaadhoo
Naivaadhoo
Finey
Hirimaradhoo

Map III.97: Haa Dhaalu Atoll, Transportation

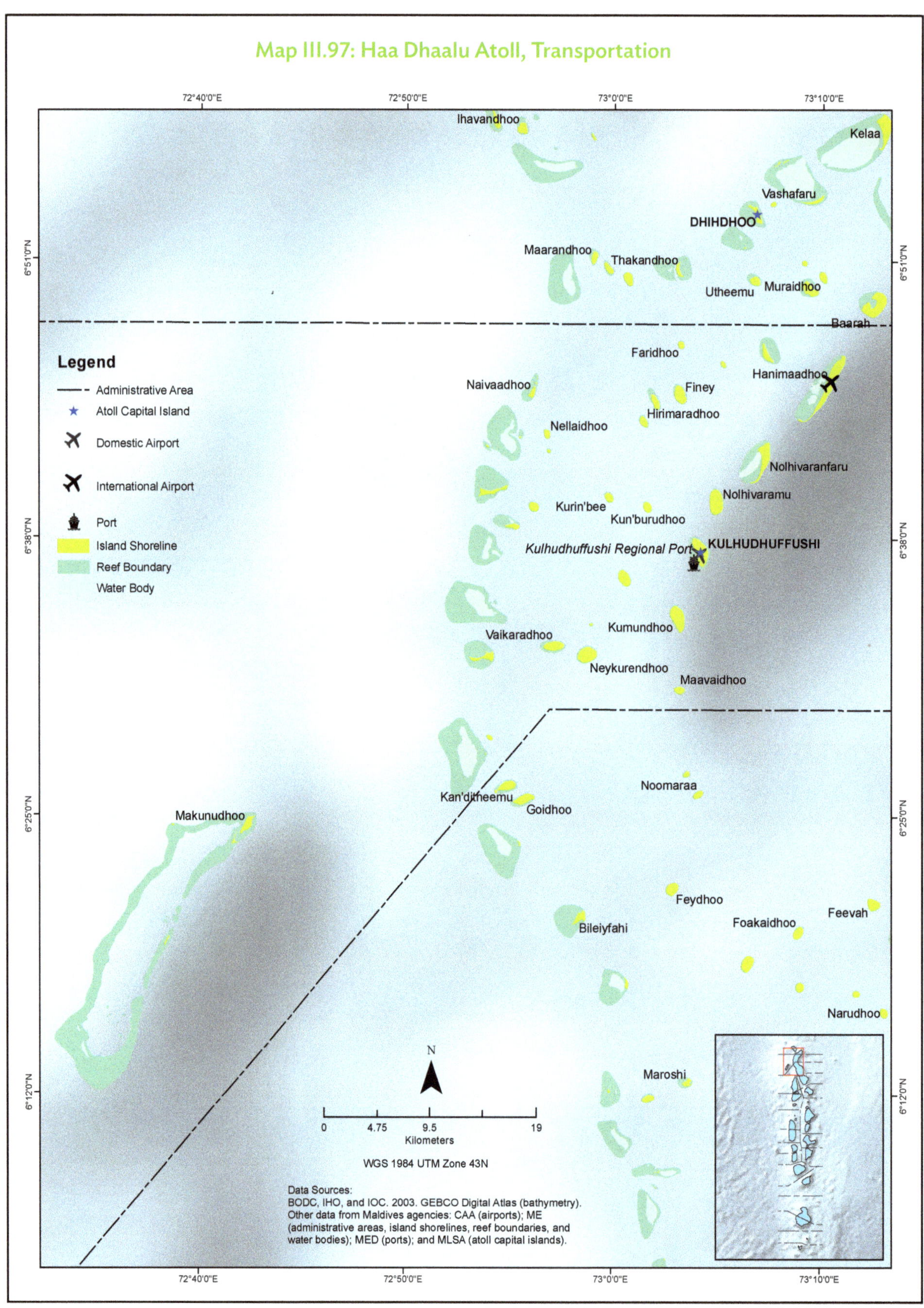

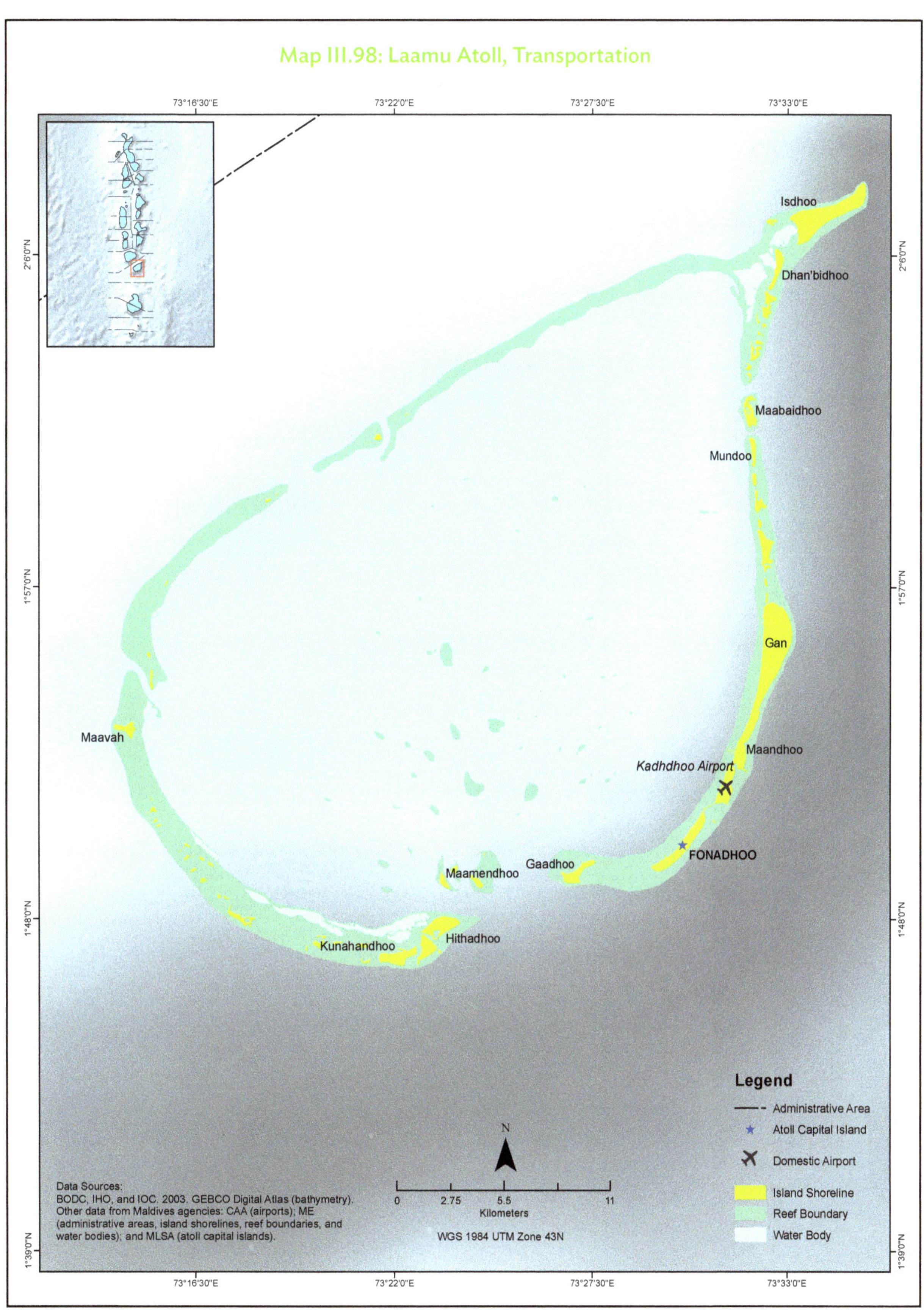
Map III.98: Laamu Atoll, Transportation
73°16'30"E
73°22'0"E
73°27'30"E
73°33'0"E
2°6'0"N
1°57'0"N
1°48'0"N
1°39'0"N
Isdhoo
Dhan'bidhoo
Maabaidhoo
Mundoo
Gan
Maavah
Maandhoo
Kadhdhoo Airport
FONADHOO
Gaadhoo
Maamendhoo
Hithadhoo
Kunahandhoo
Legend
Administrative Area
Atoll Capital Island
Domestic Airport
Island Shoreline
Reef Boundary
Water Body
N
0 2.75 5.5 11
Kilometers
WGS 1984 UTM Zone 43N
Data Sources:
BODC, IHO, and IOC. 2003. GEBCO Digital Atlas (bathymetry).
Other data from Maldives agencies: CAA (airports); ME (administrative areas, island shorelines, reef boundaries, and water bodies); and MLSA (atoll capital islands).

Map III.99: Lhaviyani Atoll, Transportation

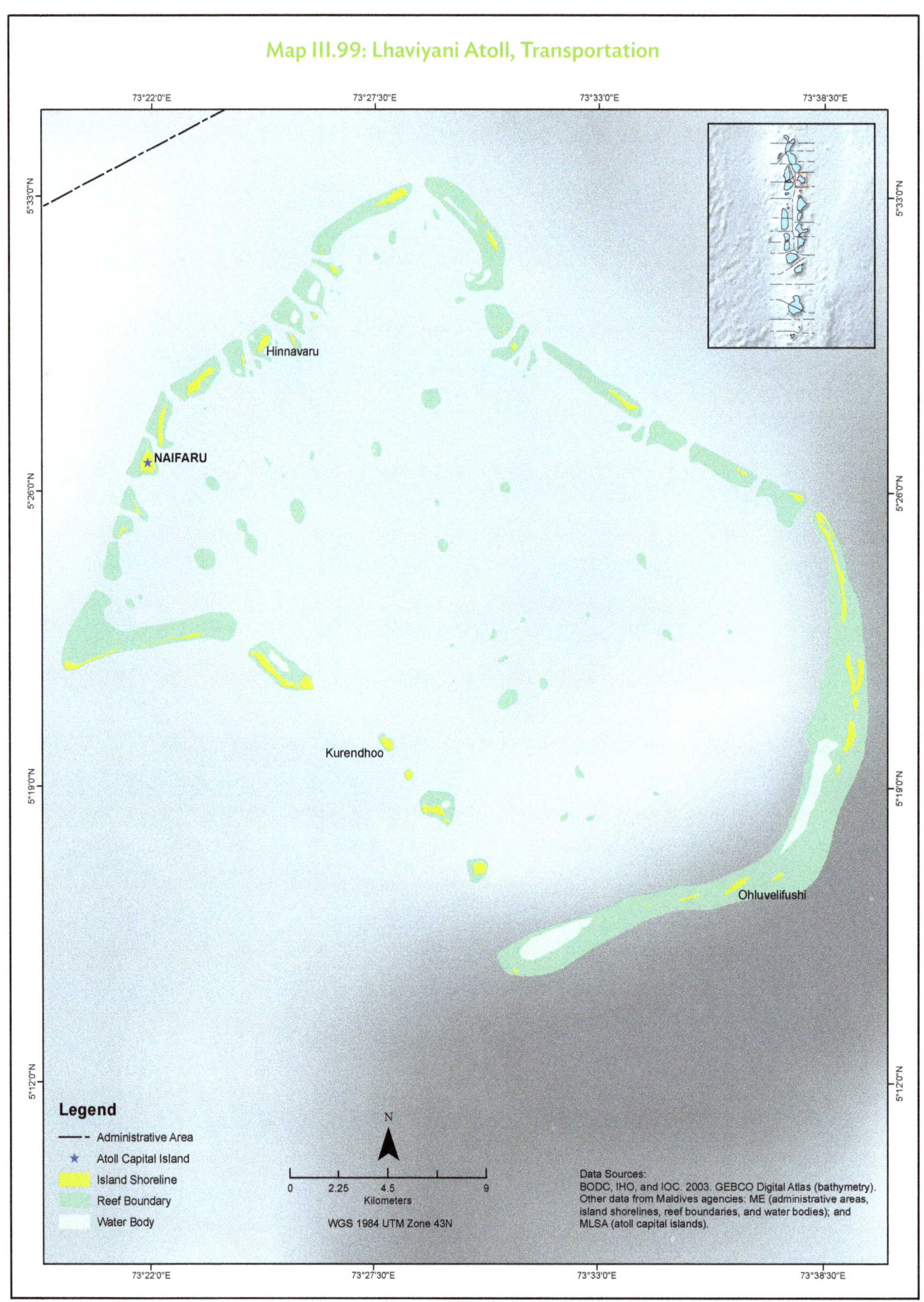

Map III.100: Meemu Atoll, Transportation

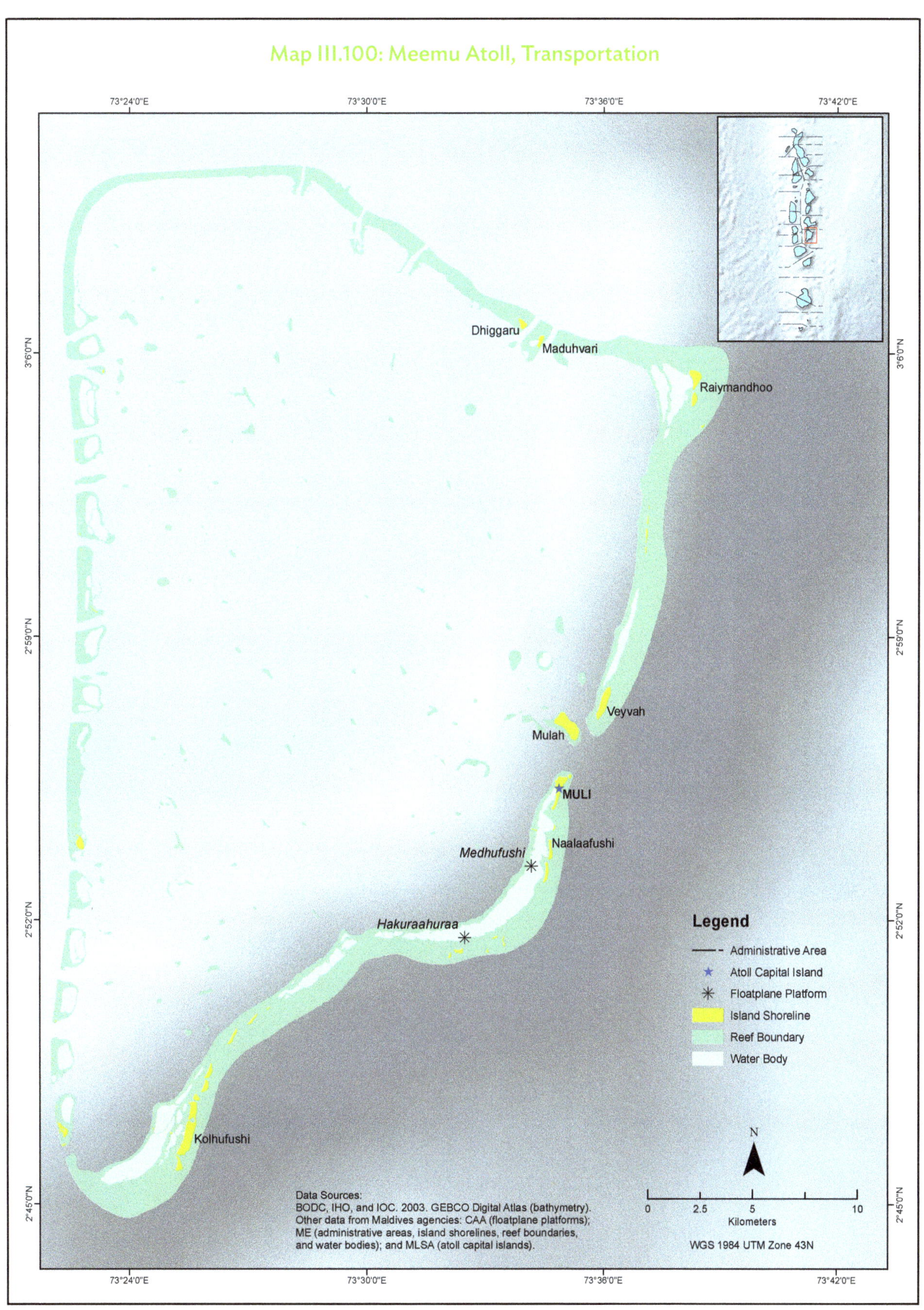

Map III.101: Noonu Atoll, Transportation

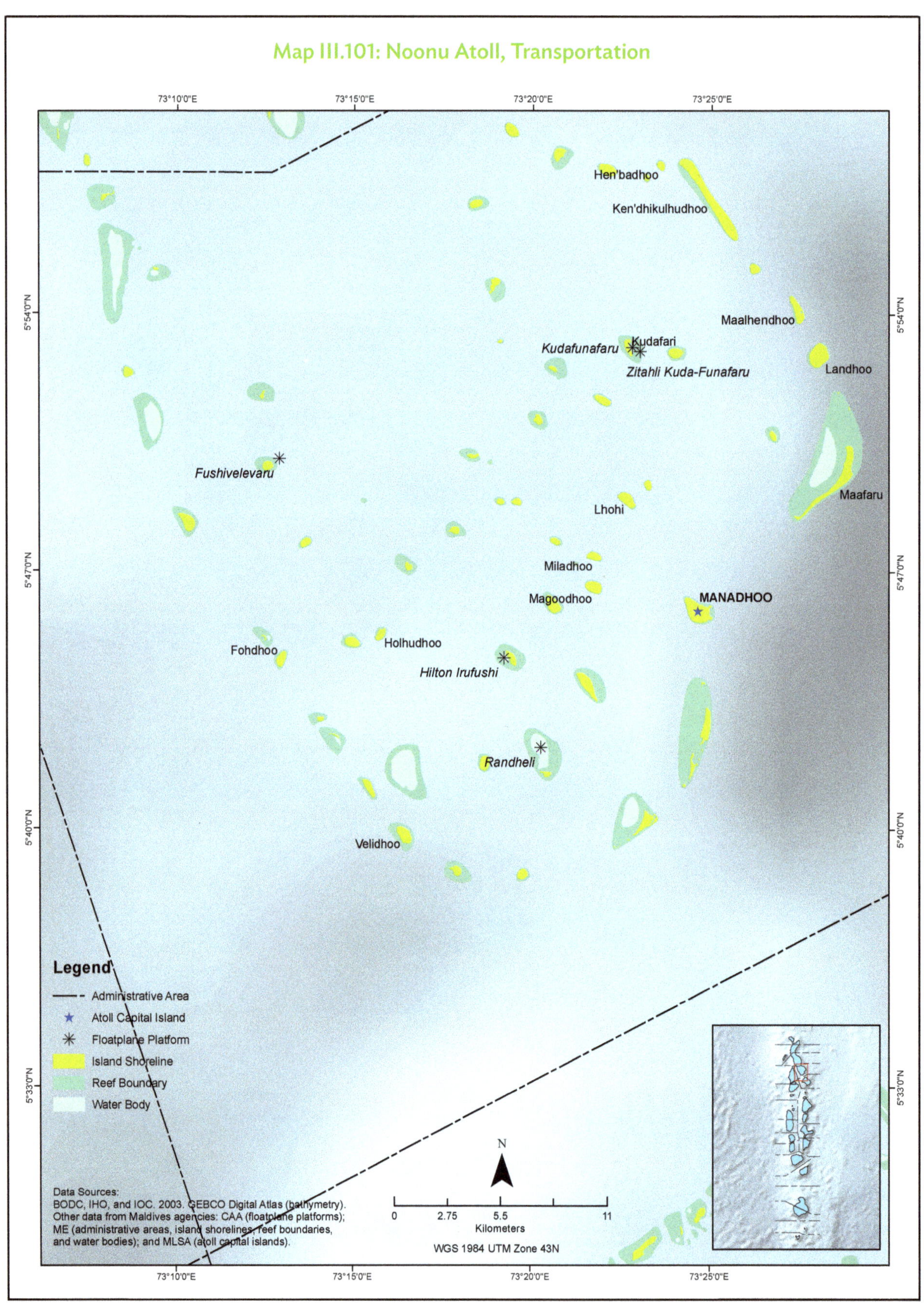

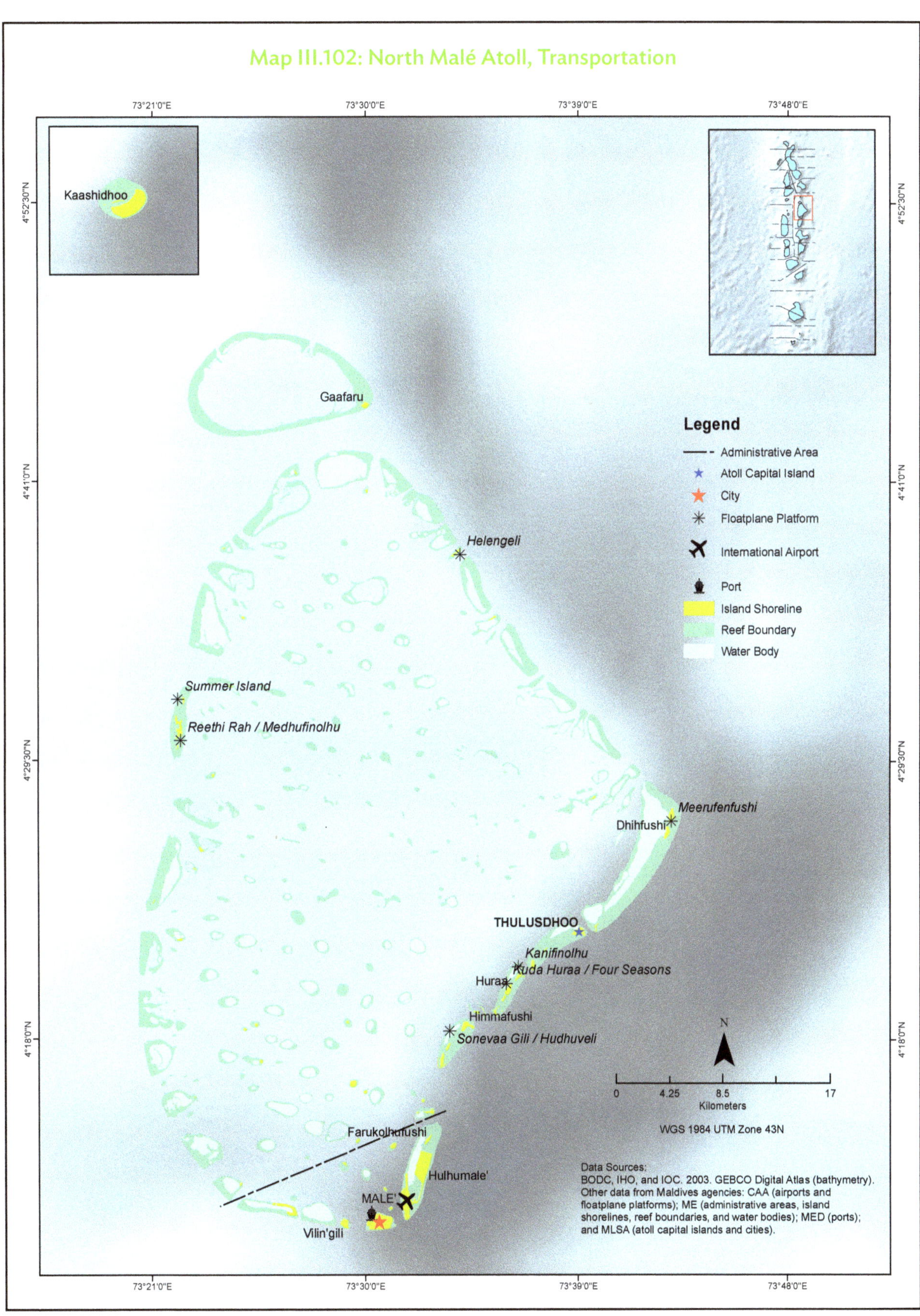

Map III.102: North Malé Atoll, Transportation
73°21'0"E
73°30'0"E
73°39'0"E
73°48'0"E
4°52'30"N
4°41'0"N
4°29'30"N
4°18'0"N
Kaashidhoo
Gaafaru
Helengeli
Summer Island
Reethi Rah / Medhufinolhu
Meerufenfushi
Dhihfushi
THULUSDHOO
Kanifinolhu
Kuda Huraa / Four Seasons
Huraa
Himmafushi
Sonevaa Gili / Hudhuveli
Farukolhufushi
Hulhumale'
MALE'
Vilin'gili
Legend
Administrative Area
Atoll Capital Island
City
Floatplane Platform
International Airport
Port
Island Shoreline
Reef Boundary
Water Body
N
0
4.25
8.5
17
Kilometers
WGS 1984 UTM Zone 43N
Data Sources;
BODC, IHO, and IOC. 2003. GEBCO Digital Atlas (bathymetry).
Other data from Maldives agencies: CAA (airports and floatplane platforms); ME (administrative areas, island shorelines, reef boundaries, and water bodies); MED (ports); and MLSA (atoll capital islands and cities).

Map III.103: Raa Atoll, Transportation

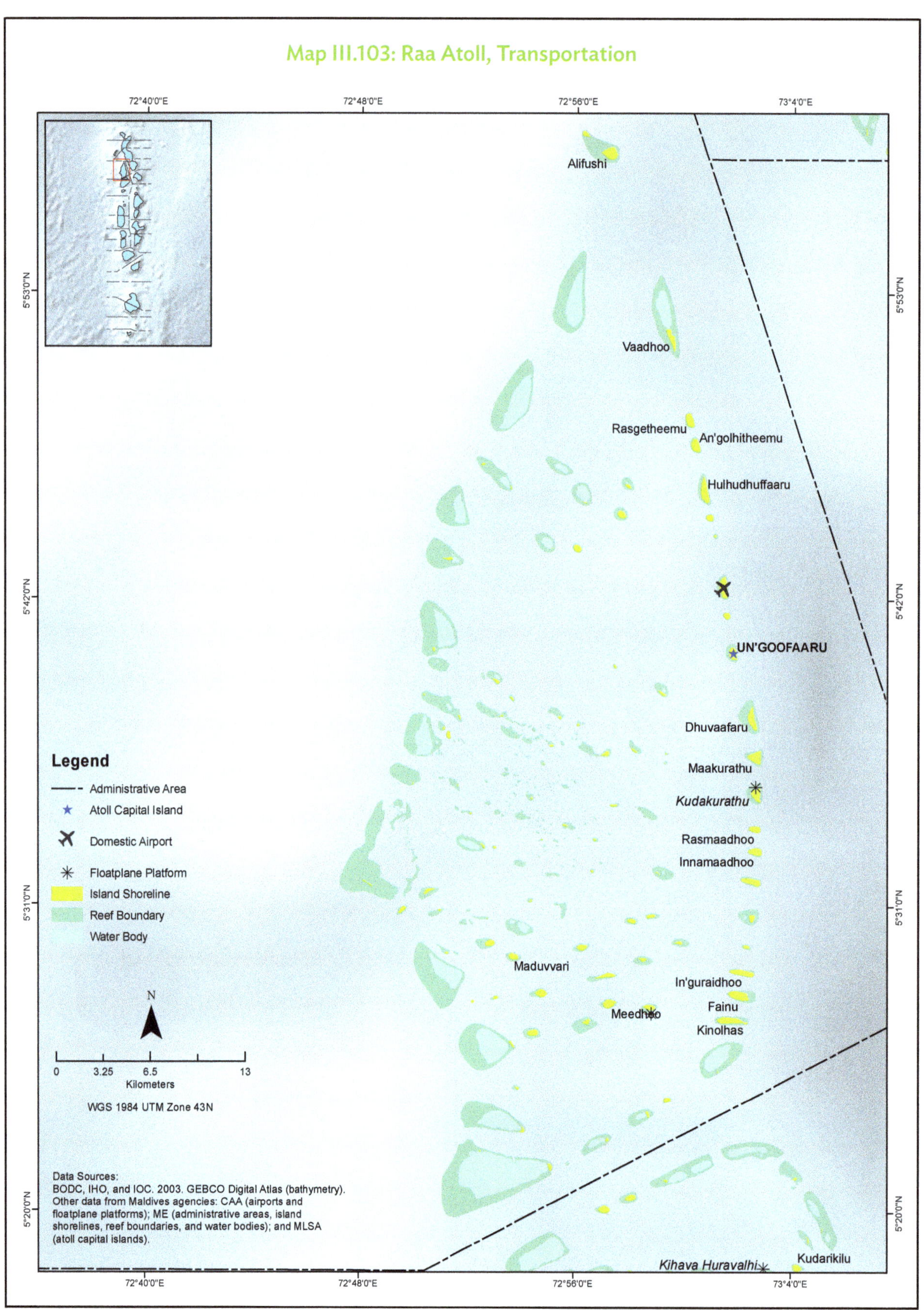

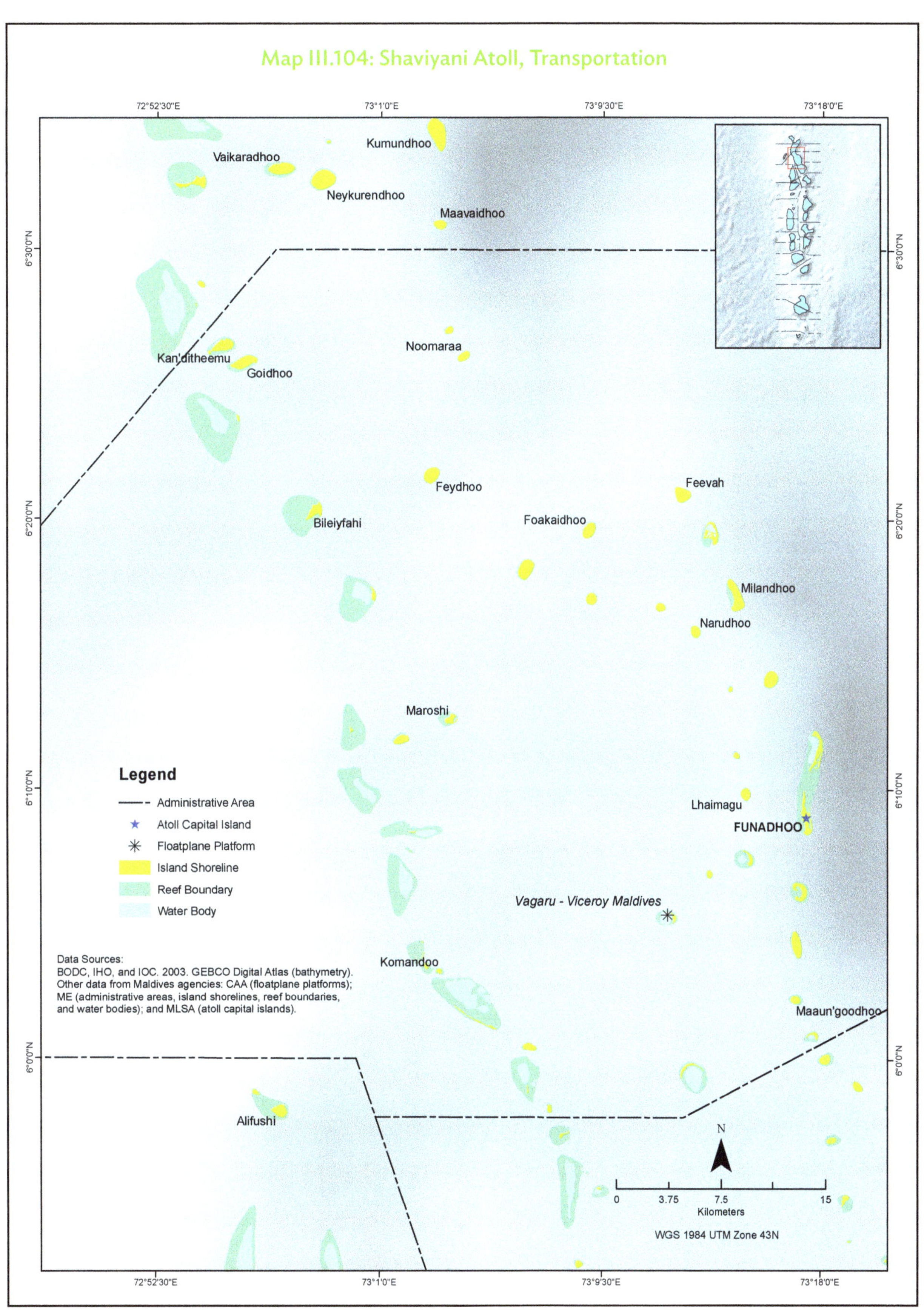

Map III.104: Shaviyani Atoll, Transportation
72°52'30"E
73°1'0"E
73°9'30"E
73°18'0"E
6°30'0"N
6°20'0"N
6°10'0"N
6°0'0"N
Vaikaradhoo
Kumundhoo
Neykurendhoo
Maavaidhoo
Noomaraa
Kan'ditheemu
Goidhoo
Feydhoo
Feevah
Bileiyfahi
Foakaidhoo
Milandhoo
Narudhoo
Maroshi
Lhaimagu
FUNADHOO
Vagaru - Viceroy Maldives
Komandoo
Maaun'goodhoo
Alifushi
Legend
Administrative Area
Atoll Capital Island
Floatplane Platform
Island Shoreline
Reef Boundary
Water Body
Data Sources:
BODC, IHO, and IOC. 2003. GEBCO Digital Atlas (bathymetry).
Other data from Maldives agencies: CAA (floatplane platforms);
ME (administrative areas, island shorelines, reef boundaries, and water bodies); and MLSA (atoll capital islands).
N
0
3.75
7.5
15
Kilometers
WGS 1984 UTM Zone 43N

Map III.105: South Malé Atoll, Transportation

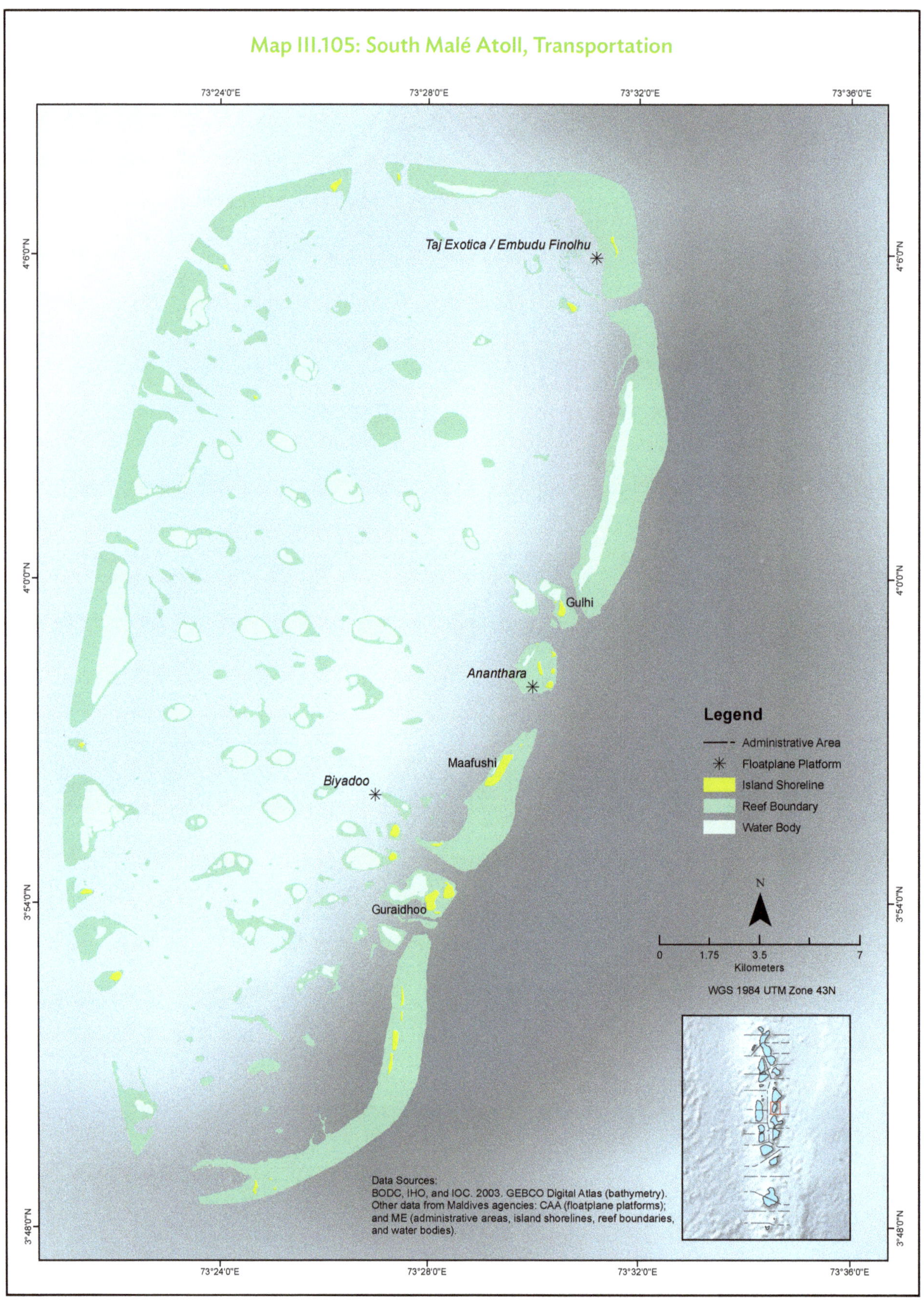

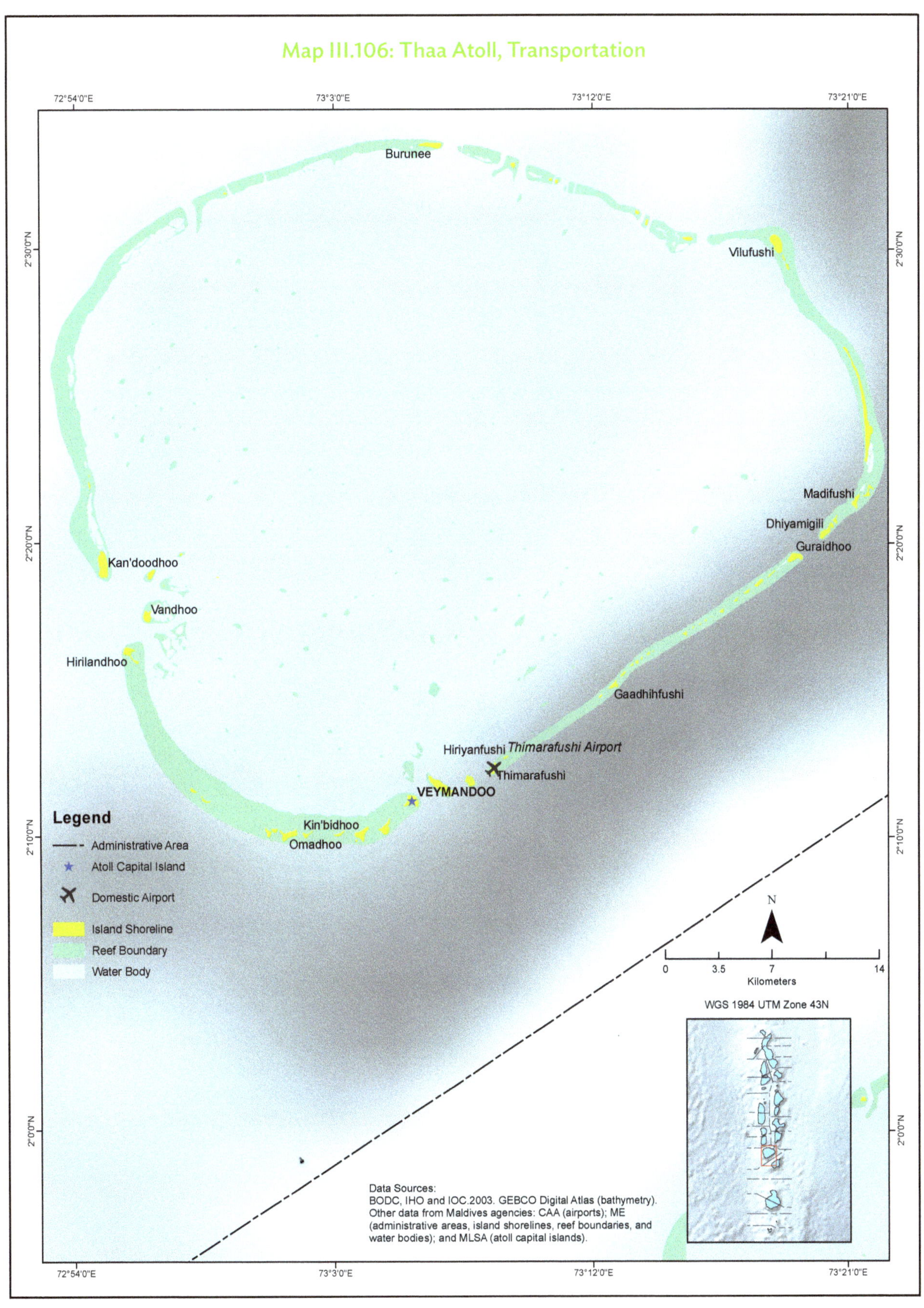
Map III.106: Thaa Atoll, Transportation
72°54'0"E
73°3'0"E
73°12'0"E
73°21'0"E
2°30'0"N
2°20'0"N
2°10'0"N
2°0'0"N
Burunee
Vilufushi
Madifushi
Dhiyamigili
Guraidhoo
Kan'doodhoo
Vandhoo
Hirilandhoo
Gaadhihfushi
Hiriyanfushi
Thimarafushi Airport
Thimarafushi
VEYMANDOO
Kin'bidhoo
Omadhoo
Legend
Administrative Area
Atoll Capital Island
Domestic Airport
Island Shoreline
Reef Boundary
Water Body
N
0
3.5
7
14
Kilometers
WGS 1984 UTM Zone 43N
Data Sources:
BODC, IHO and IOC.2003. GEBCO Digital Atlas (bathymetry).
Other data from Maldives agencies: CAA (airports); ME (administrative areas, island shorelines, reef boundaries, and water bodies); and MLSA (atoll capital islands).

Map III.107: Vaavu Atoll, Transportation

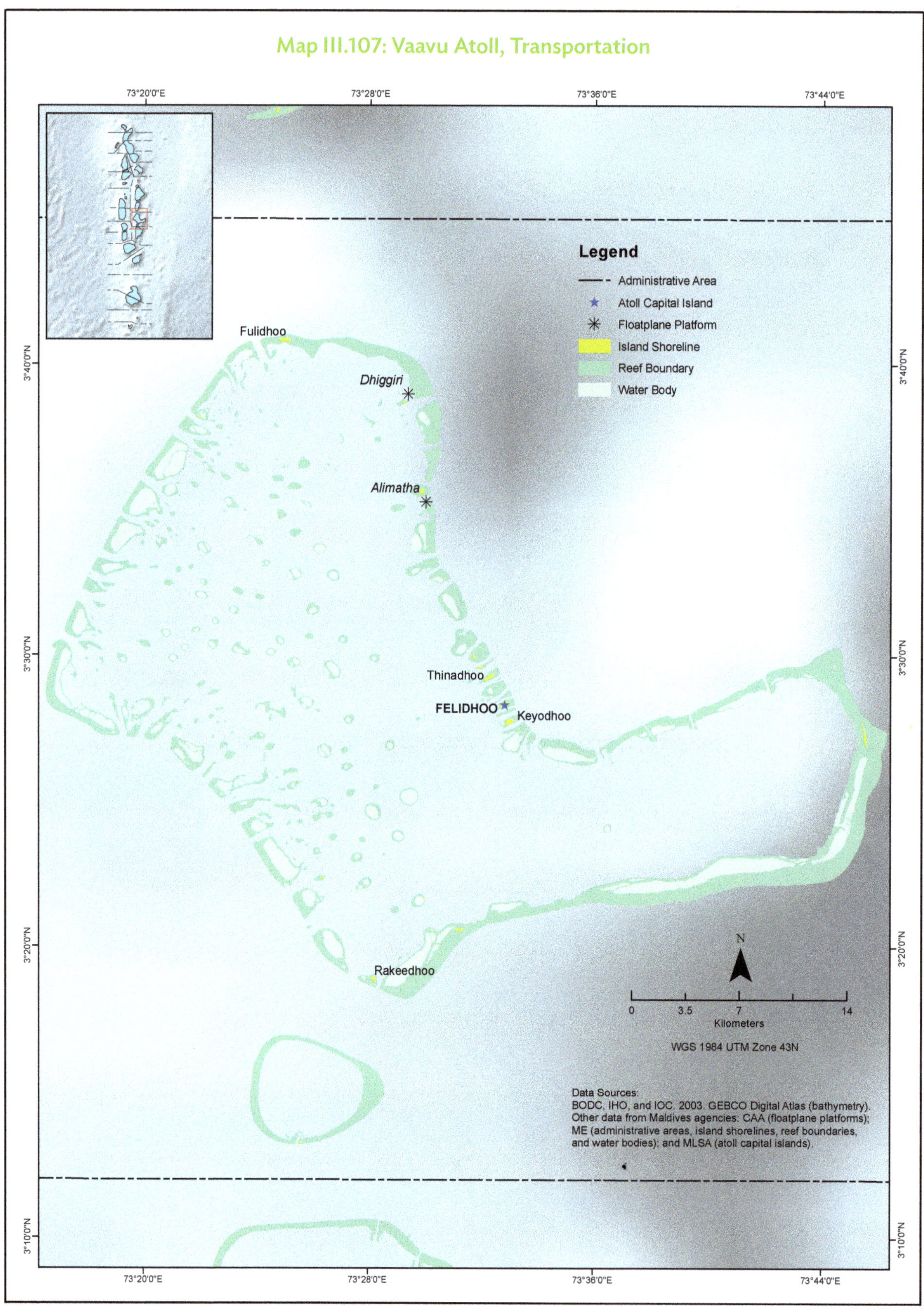

Harbor Facilities

For an archipelagic country like Maldives, harbors are important nodes in the transportation system. People can mobilize through these harbors. However, not all the inhabited islands have harbor facilities.

Boats known locally as *dhoni* and ferries are other modes of transportation in Maldives. They dock in harbor facilities installed along the coasts of 170 islands, 17 of which are atoll capitals. Out of 199 harbors in the country, Raa and Shaviyani atolls have the most with 15 harbors each.

Table III.8: Harbors in Maldives

Atoll	Number of Harbors
Raa Atoll	15
Shaviyani Atoll	15
Laamu Atoll	14
Haa Dhaalu Atoll	14
Haa Alifu Atoll	13
Baa Atoll	13
Malé City	13
Thaa Atoll	11
Gaafu Alifu Atoll	10
Alifu Dhaalu Atoll	10
Noonu Atoll	10
Gaafu Dhaalu Atoll	9
Alifu Alifu Atoll	8
Kaafu Atoll	8
Meemu Atoll	7
Dhaalu Atoll	6
Faafu Atoll	6
Lhaviyani Atoll	6
Addu City	6
Vaavu Atoll	4
Gnaviyani Atoll	1

Source: Ministry of National Planning and Infrastructure, 2017.

Common port scene. Small motorboats are a usual sight at ports in Maldives (photo by Department of Foreign Affairs and Trade).

Traditional boats. Traditional boats called *dhoni* docked at a port in Maldives (photo by Azwar Thaufeeq).

Map III.108: Maldives, Harbor Facilities

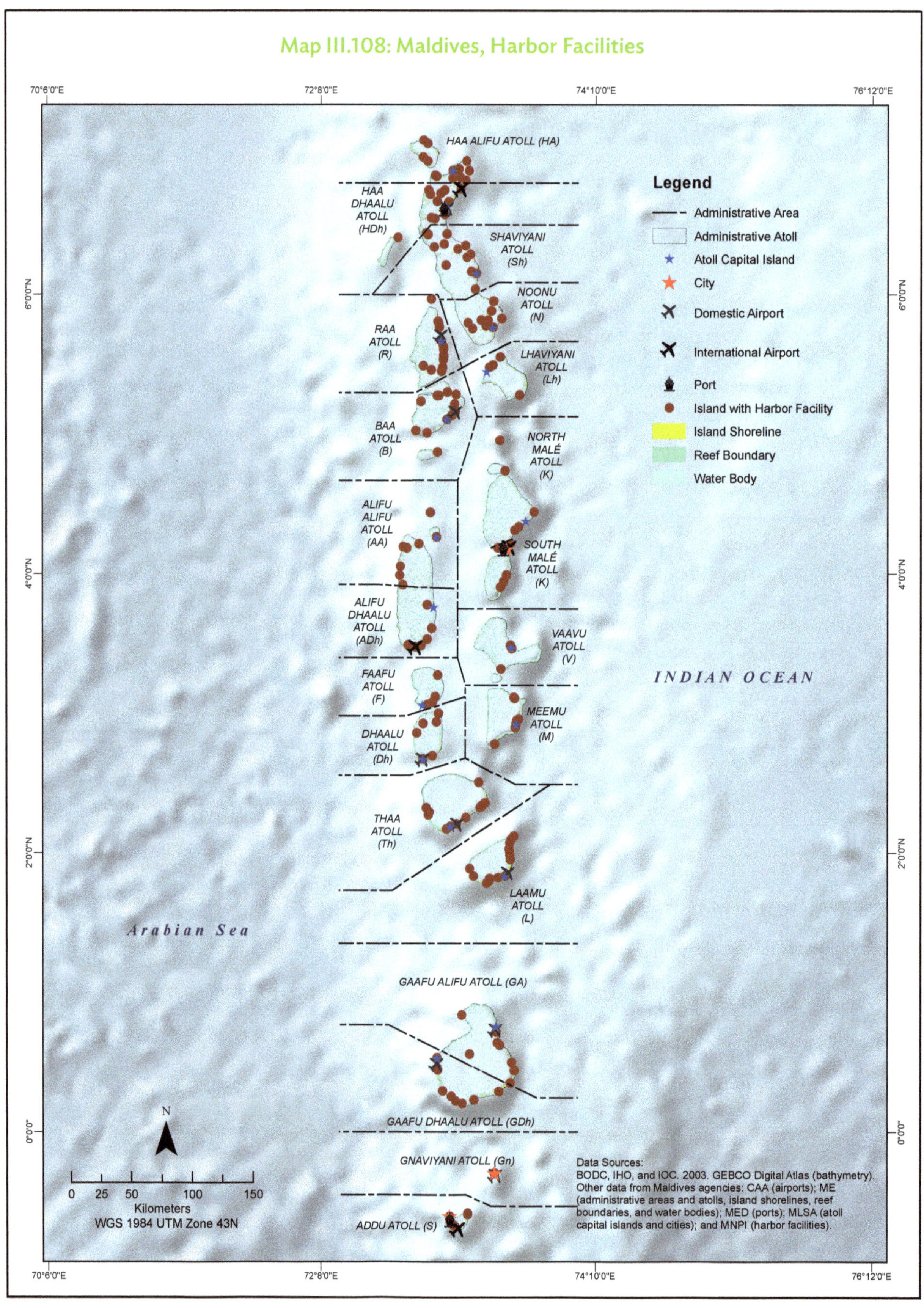

Map III.109: Addu City, Harbor Facilities

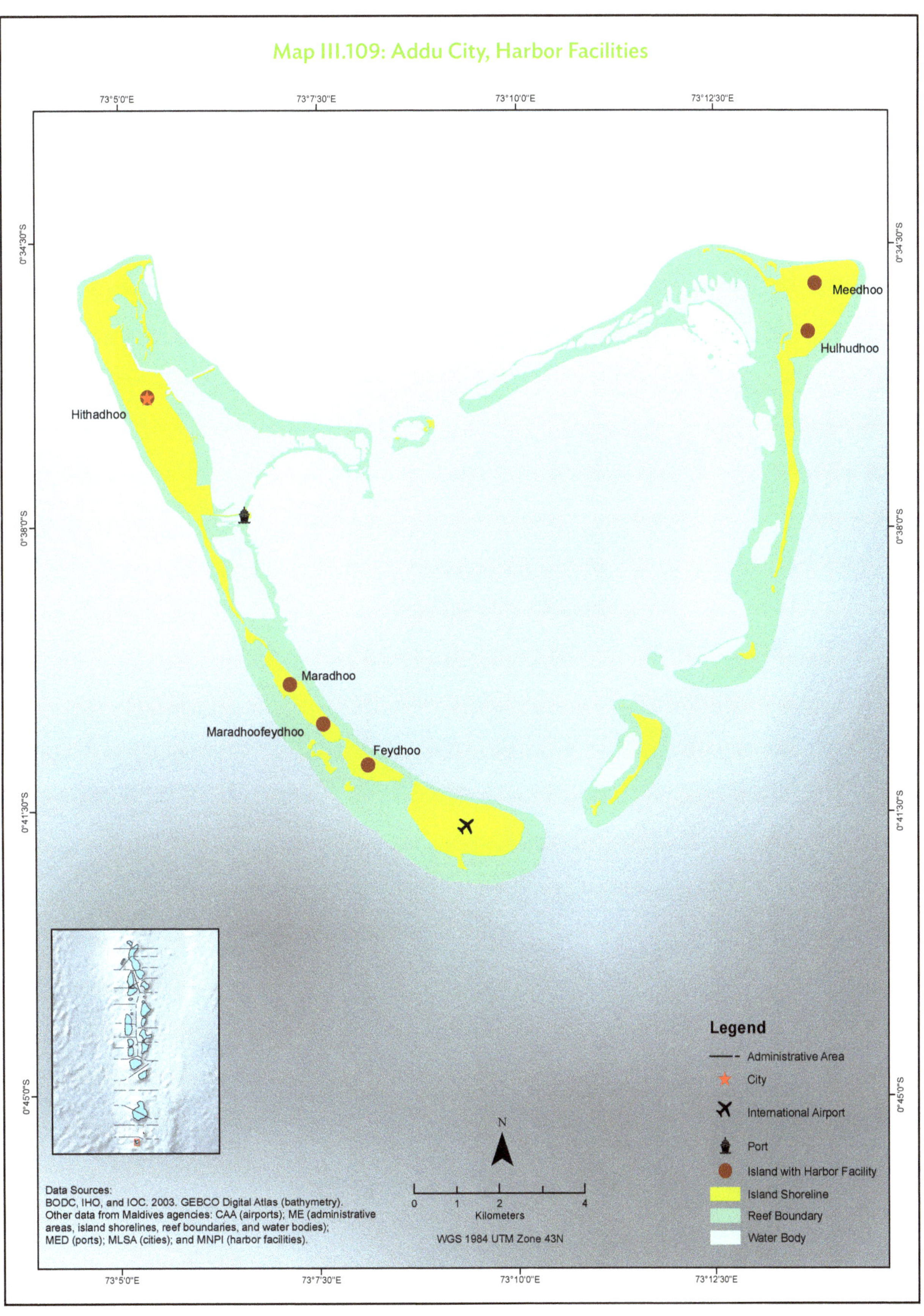

Data Sources:
BODC, IHO, and IOC. 2003. GEBCO Digital Atlas (bathymetry).
Other data from Maldives agencies: CAA (airports); ME (administrative areas, island shorelines, reef boundaries, and water bodies); MED (ports); MLSA (cities); and MNPI (harbor facilities).

Map III.110: Alifu Alifu Atoll, Harbor Facilities

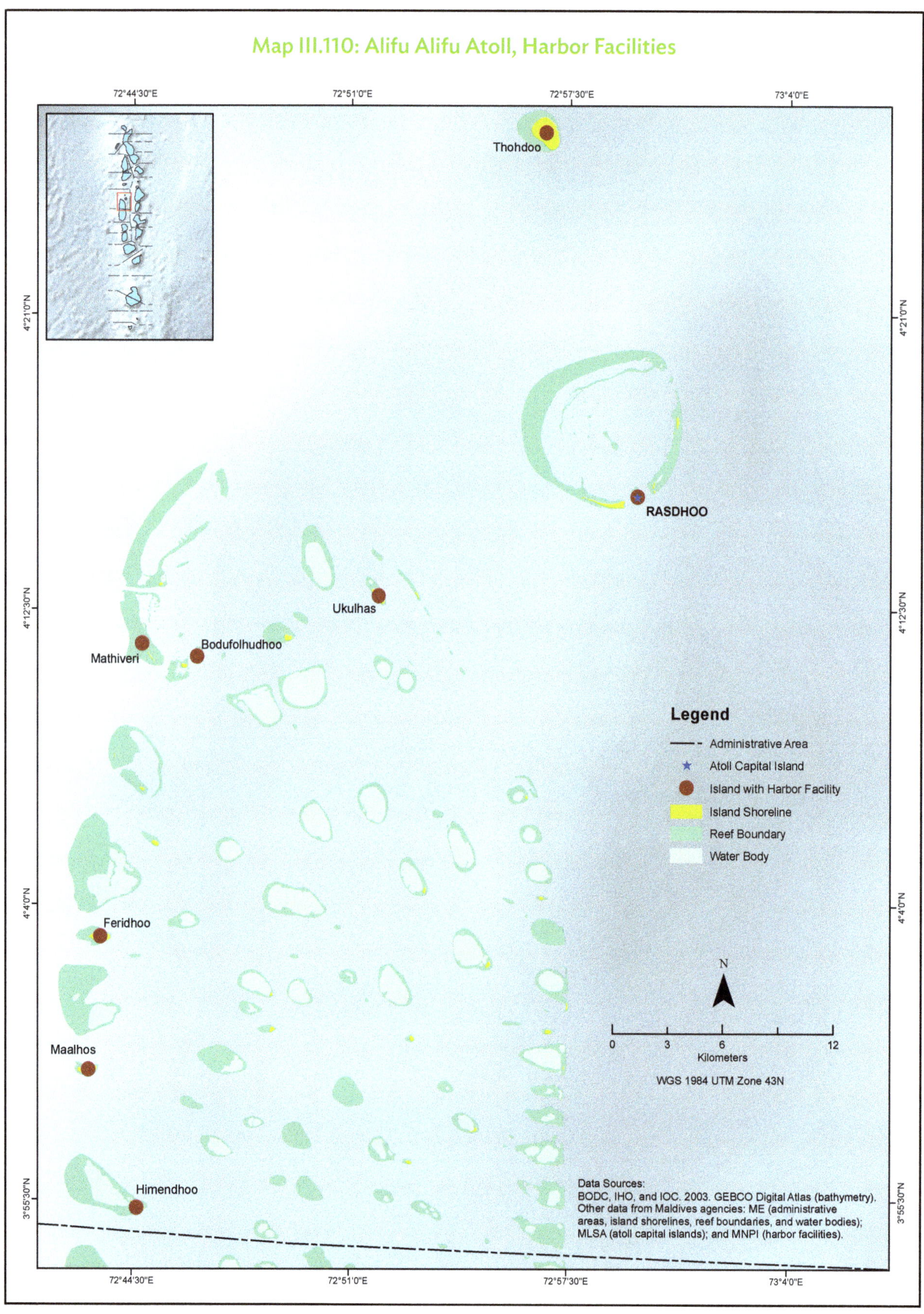

Map III.111: Alifu Dhaalu Atoll, Harbor Facilities

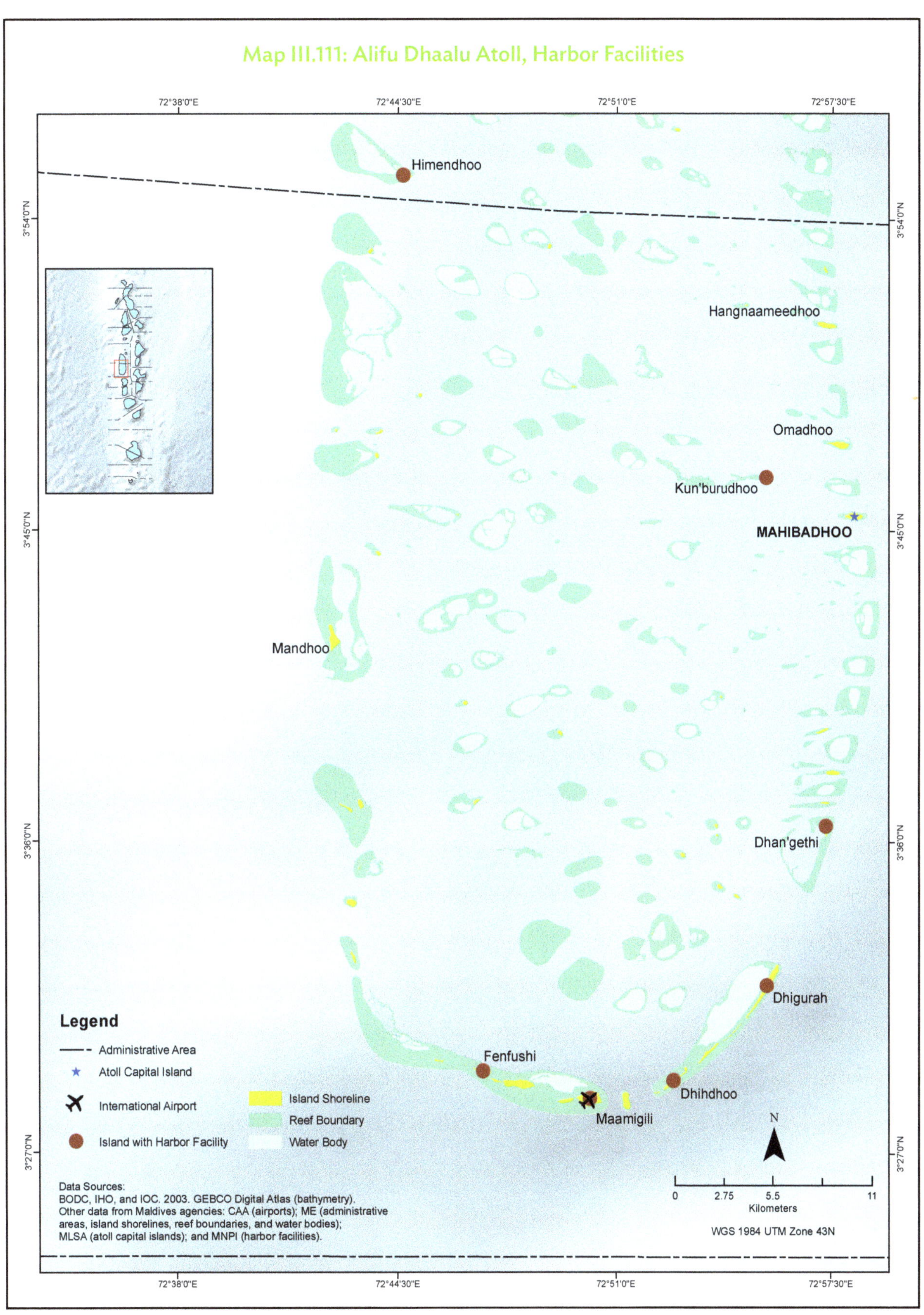

Map III.112: Baa Atoll, Harbor Facilities

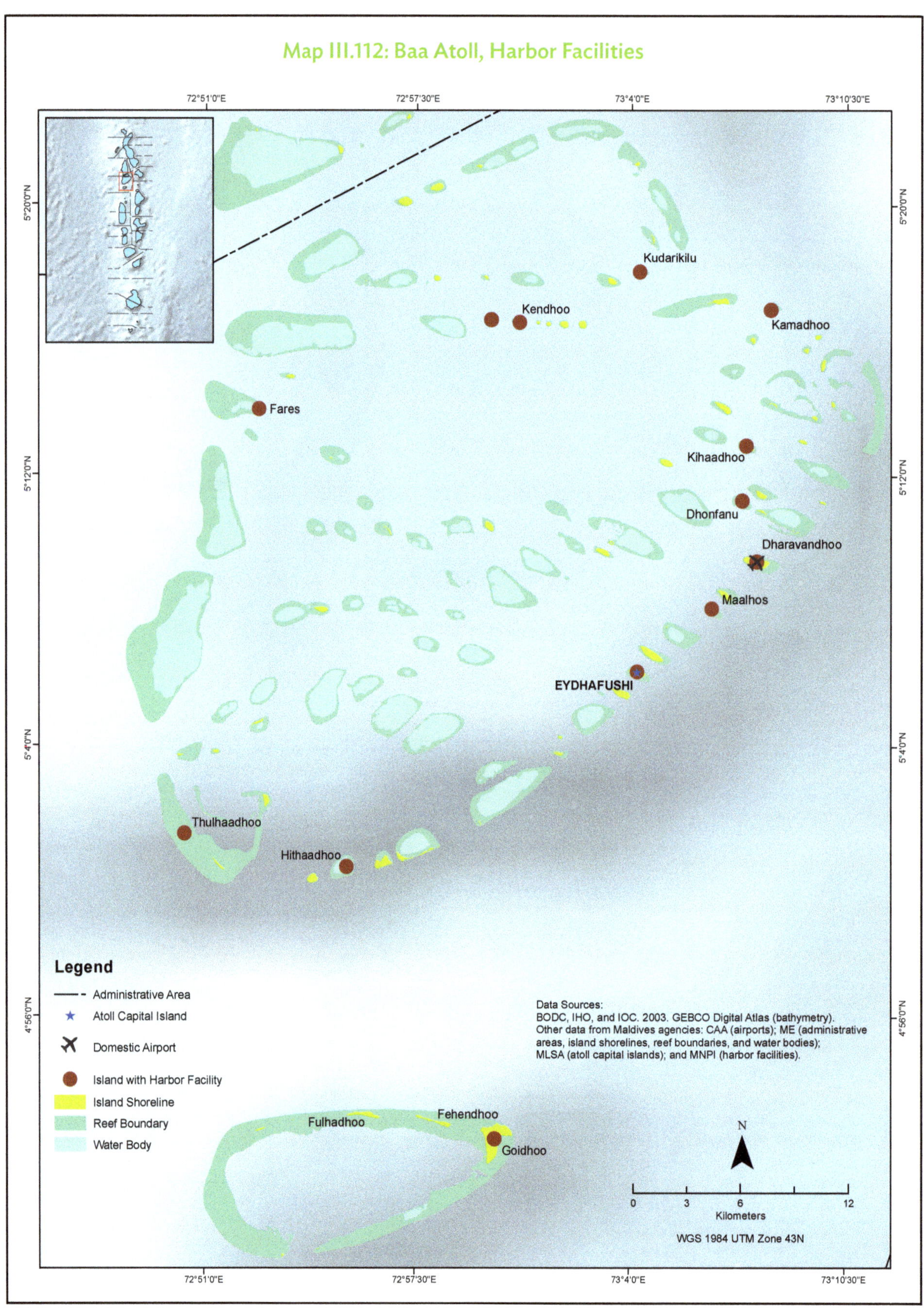

Map III.113: Dhaalu Atoll, Harbor Facilities

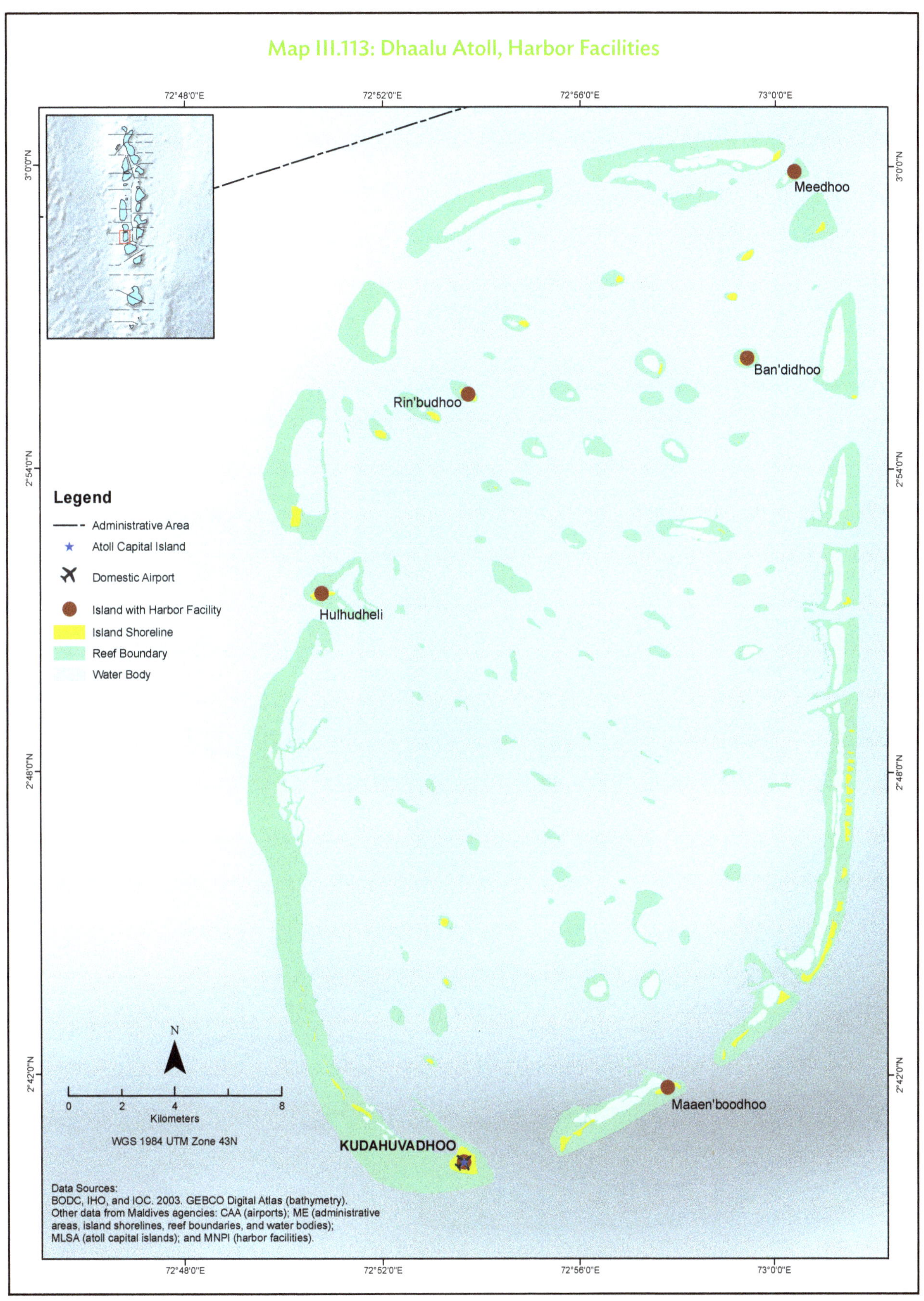

Map III.114: Faafu Atoll, Harbor Facilities

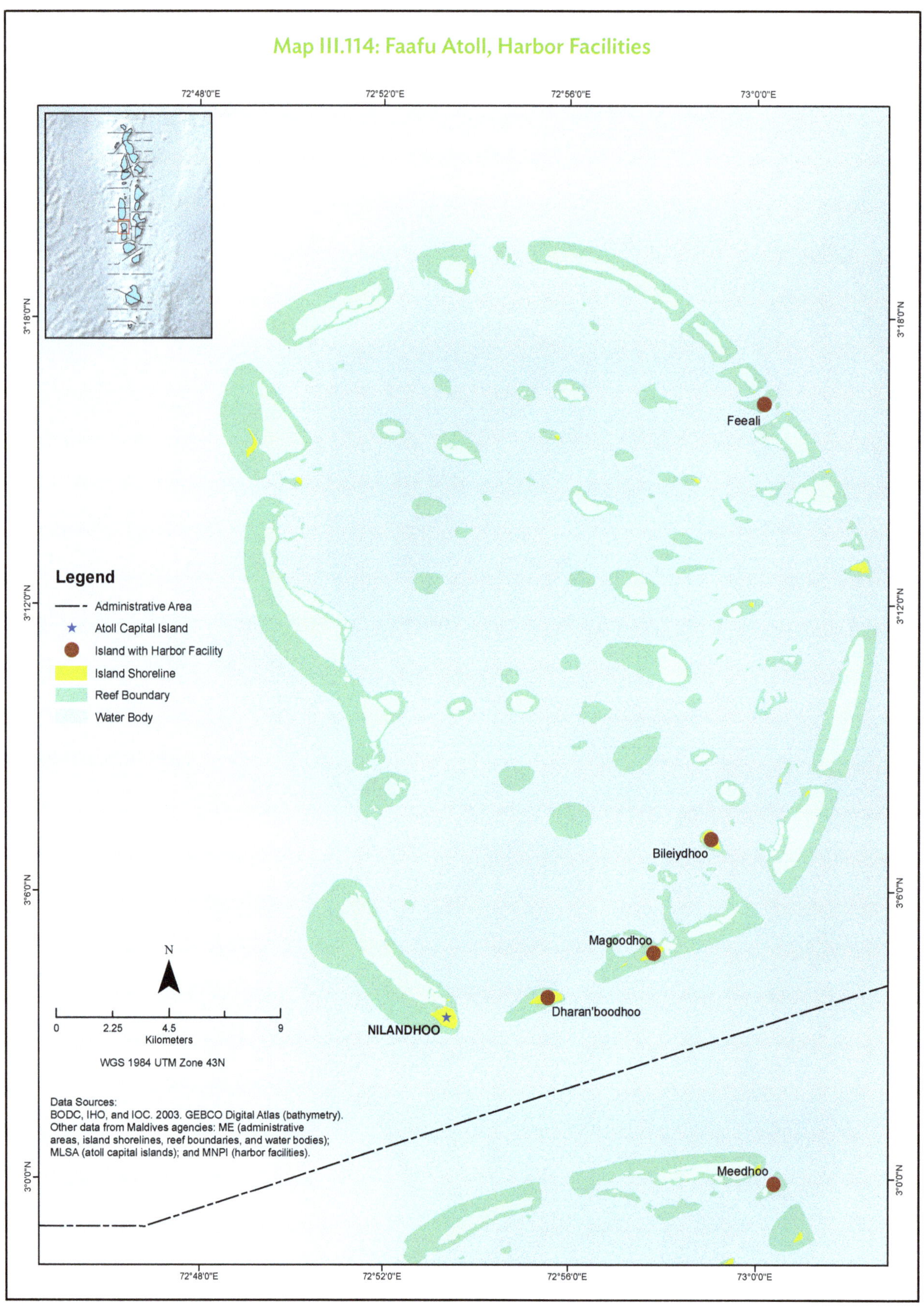

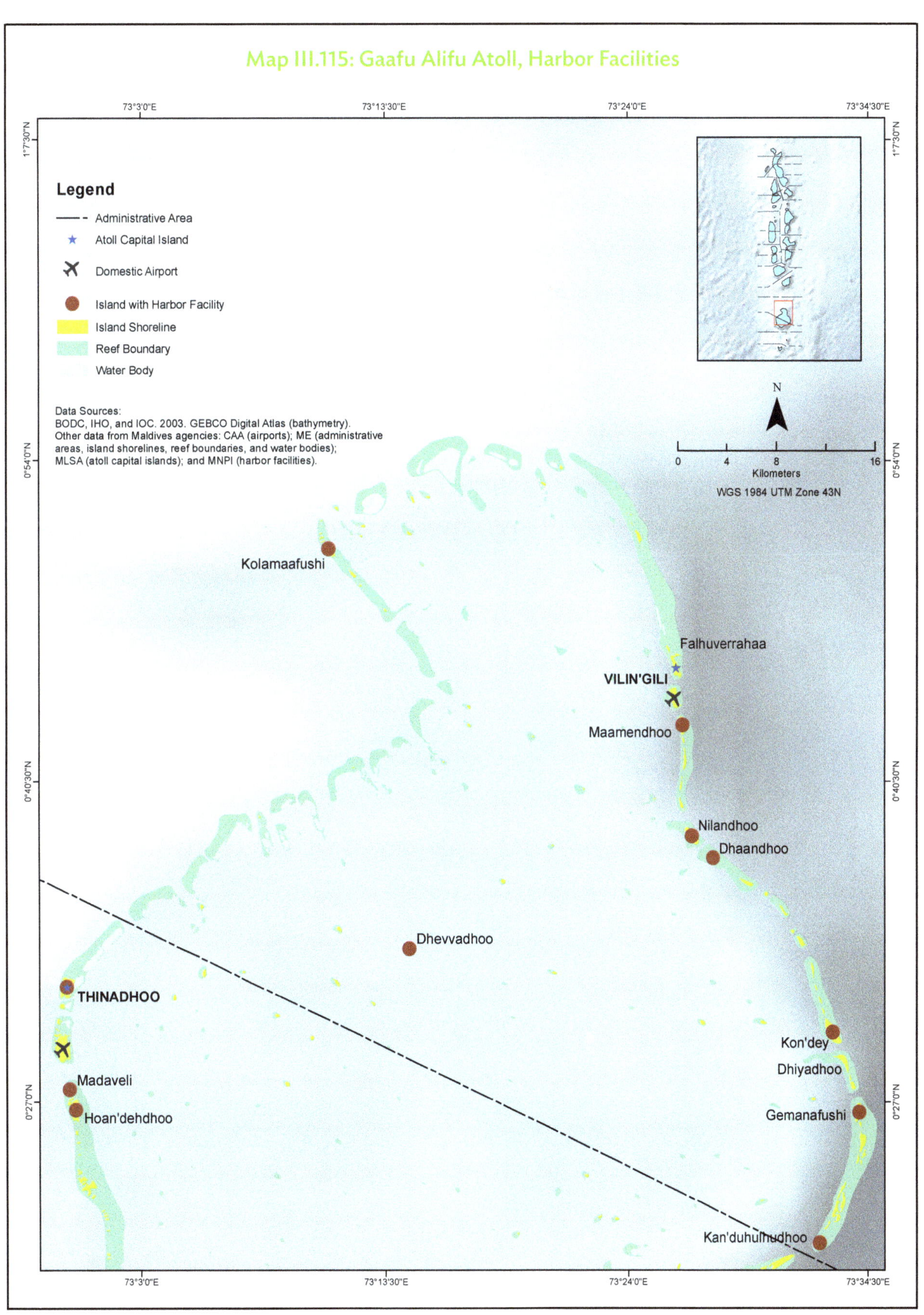

Map III.115: Gaafu Alifu Atoll, Harbor Facilities

Map III.116: Gaafu Dhaalu Atoll, Harbor Facilities

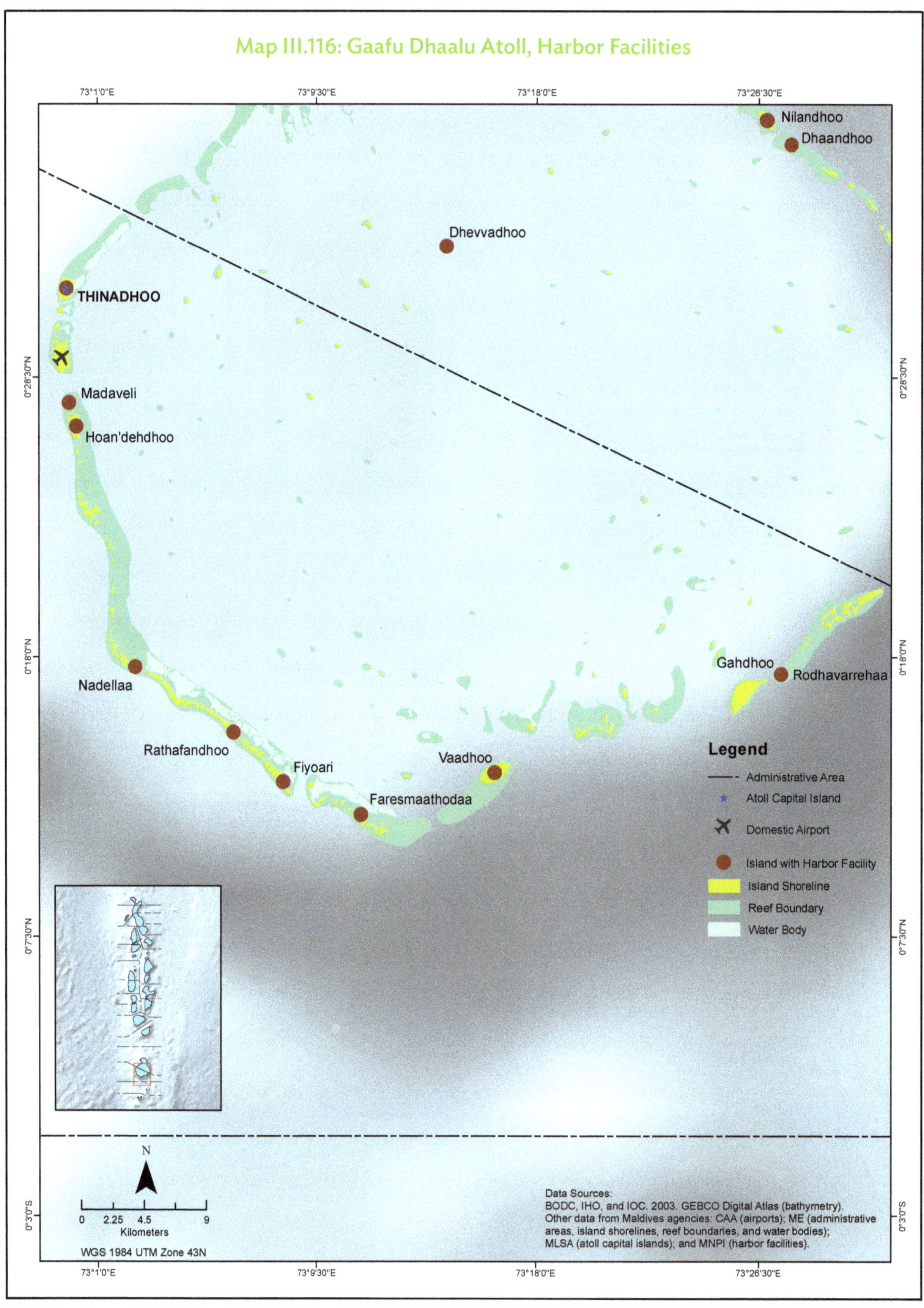

Map III.117: Gnaviyani Atoll, Harbor Facilities

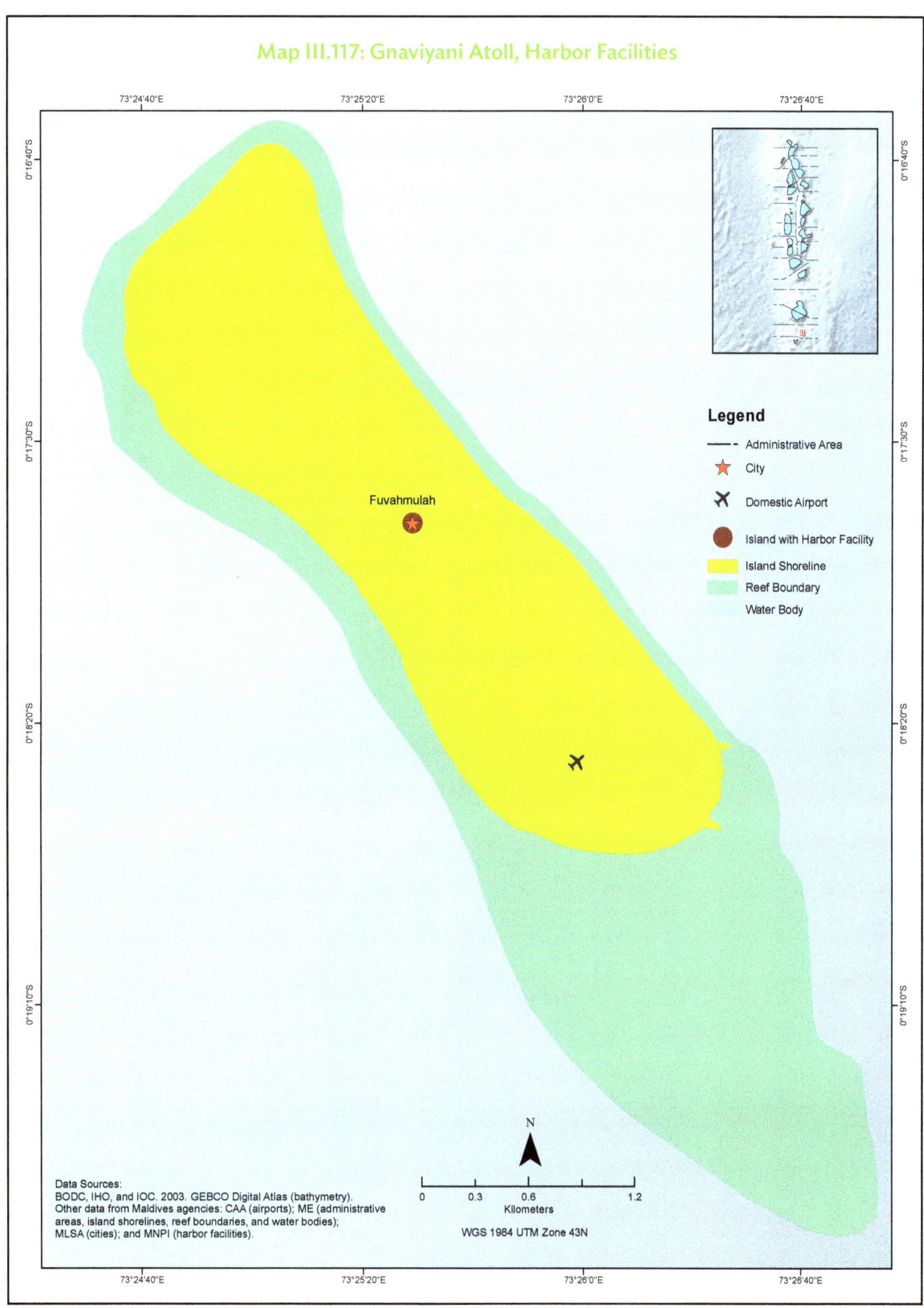

Data Sources:
BODC, IHO, and IOC. 2003. GEBCO Digital Atlas (bathymetry).
Other data from Maldives agencies: CAA (airports); ME (administrative areas, island shorelines, reef boundaries, and water bodies); MLSA (cities); and MNPI (harbor facilities).

Map III.118: Haa Alifu Atoll, Harbor Facilities

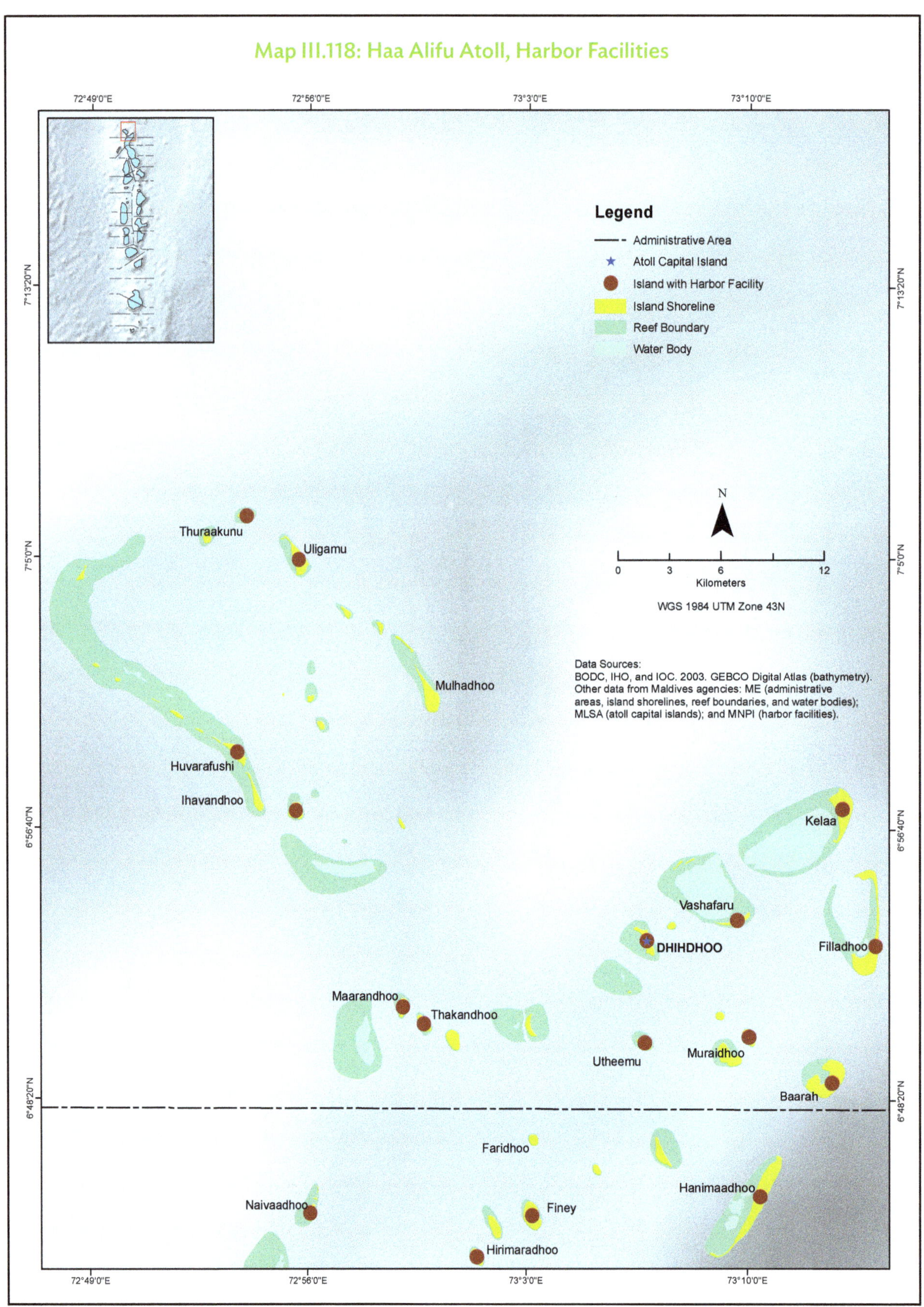

Map III.119: Haa Dhaalu Atoll, Harbor Facilities

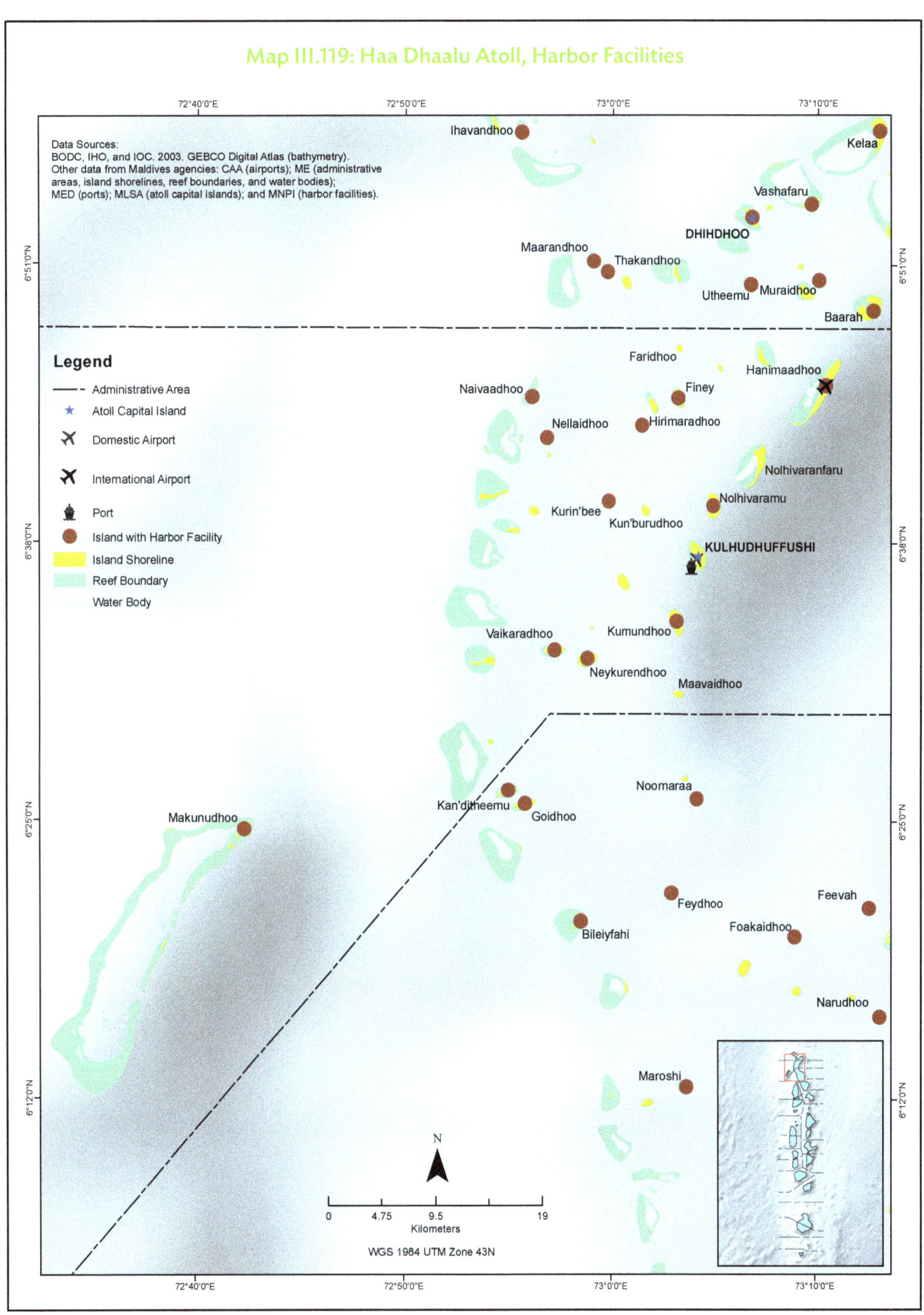

Map III.120: Laamu Atoll, Harbor Facilities

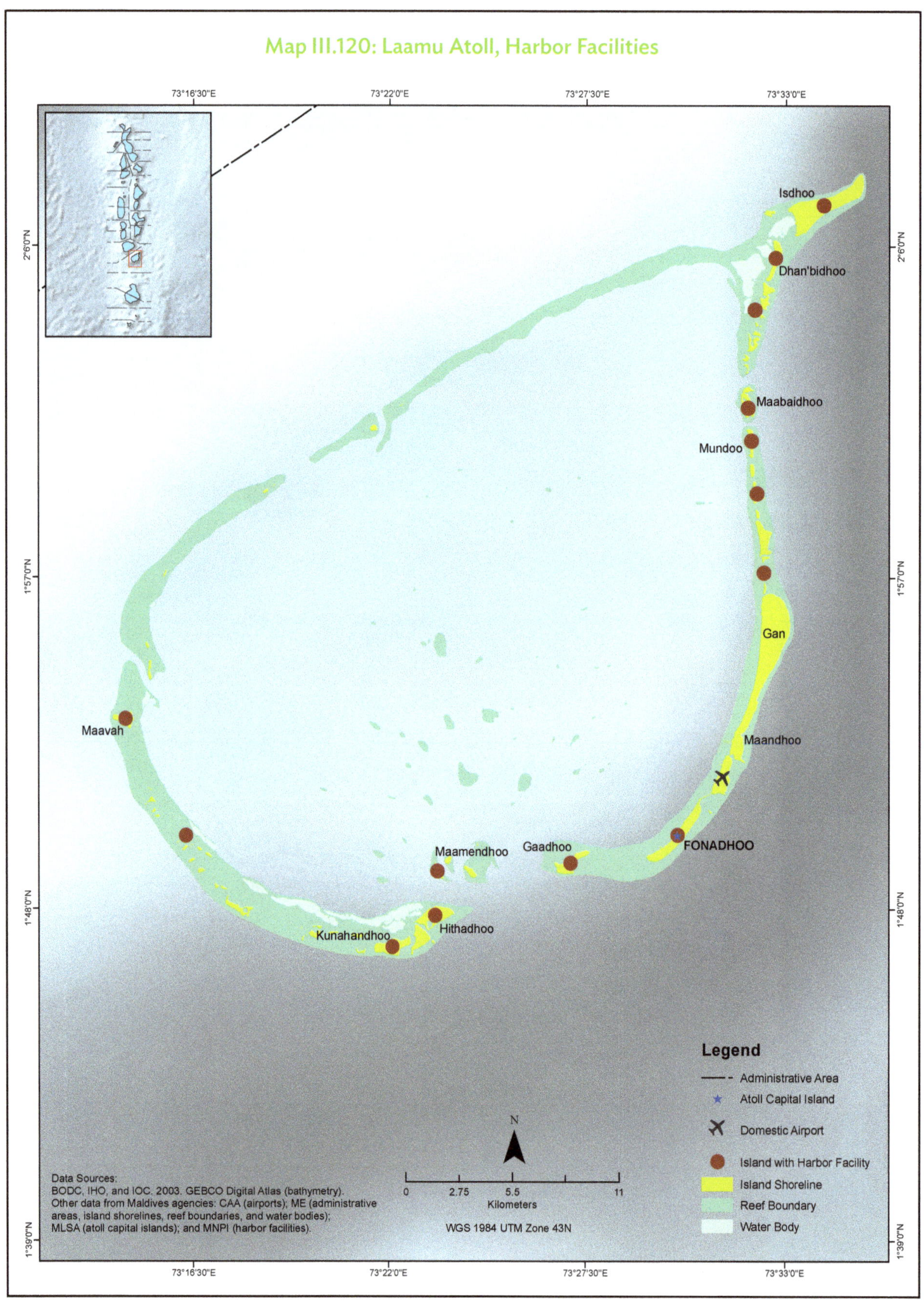

Map III.121: Lhaviyani Atoll, Harbor Facilities

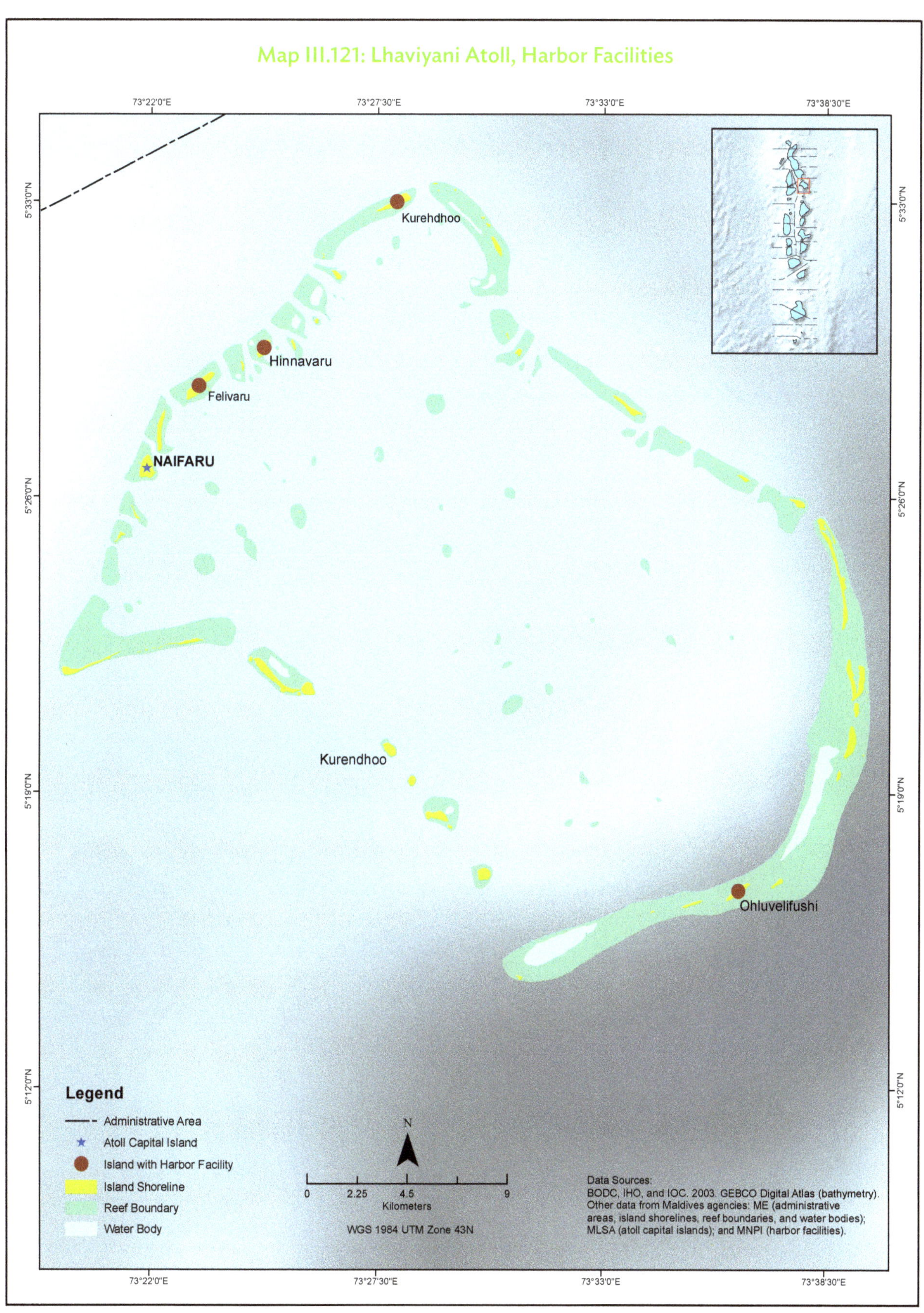

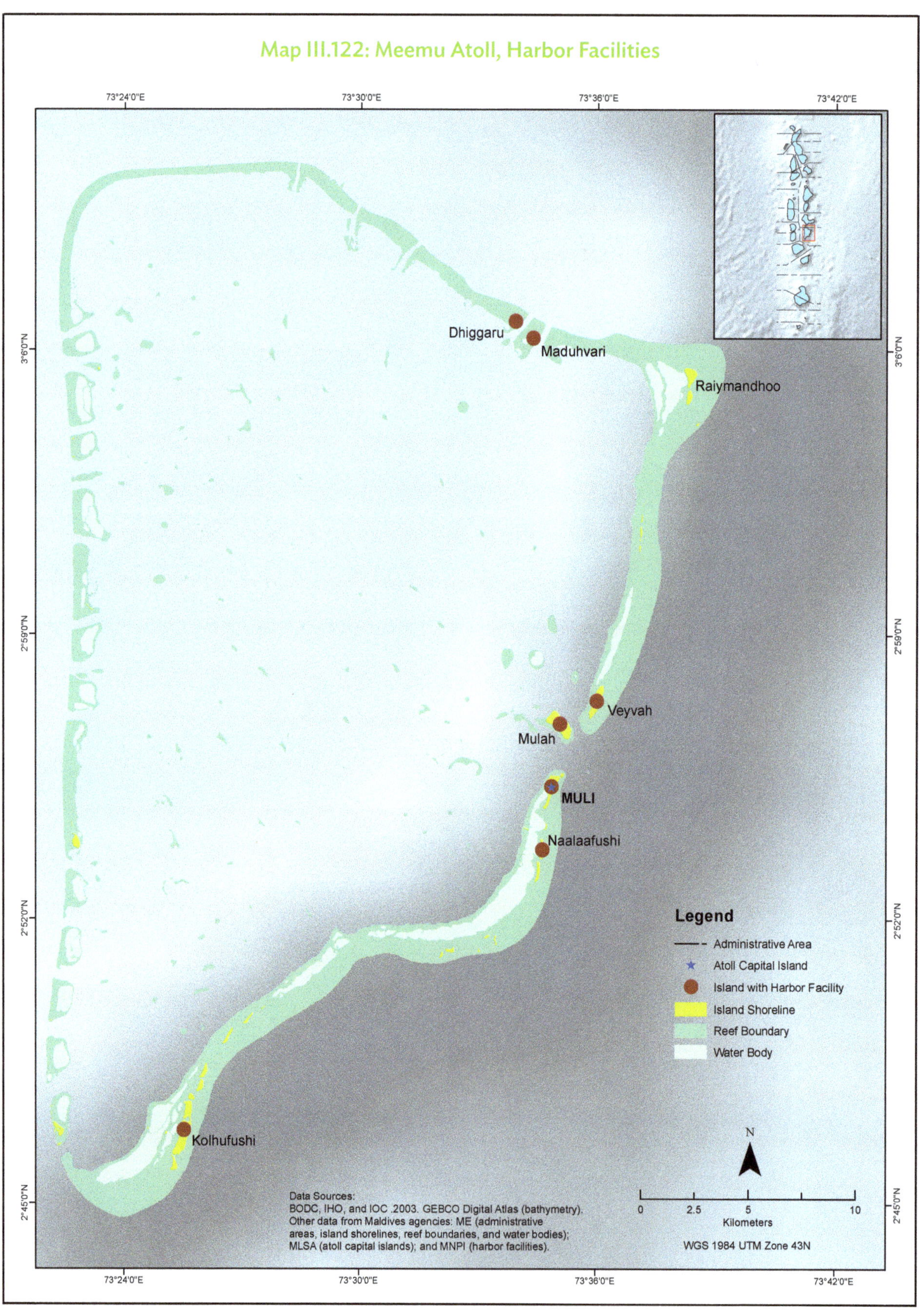
Map III.122: Meemu Atoll, Harbor Facilities
73°24'0"E
73°30'0"E
73°36'0"E
73°42'0"E
3°6'0"N
2°59'0"N
2°52'0"N
2°45'0"N
Dhiggaru
Maduhvari
Raiymandhoo
Veyvah
Mulah
MULI
Naalaafushi
Kolhufushi
Legend
Administrative Area
Atoll Capital Island
Island with Harbor Facility
Island Shoreline
Reef Boundary
Water Body
N
0 2.5 5 10
Kilometers
WGS 1984 UTM Zone 43N
Data Sources:
BODC, IHO, and IOC .2003. GEBCO Digital Atlas (bathymetry).
Other data from Maldives agencies: ME (administrative areas, island shorelines, reef boundaries, and water bodies); MLSA (atoll capital islands); and MNPI (harbor facilities).

Map III.123: Noonu Atoll, Harbor Facilities

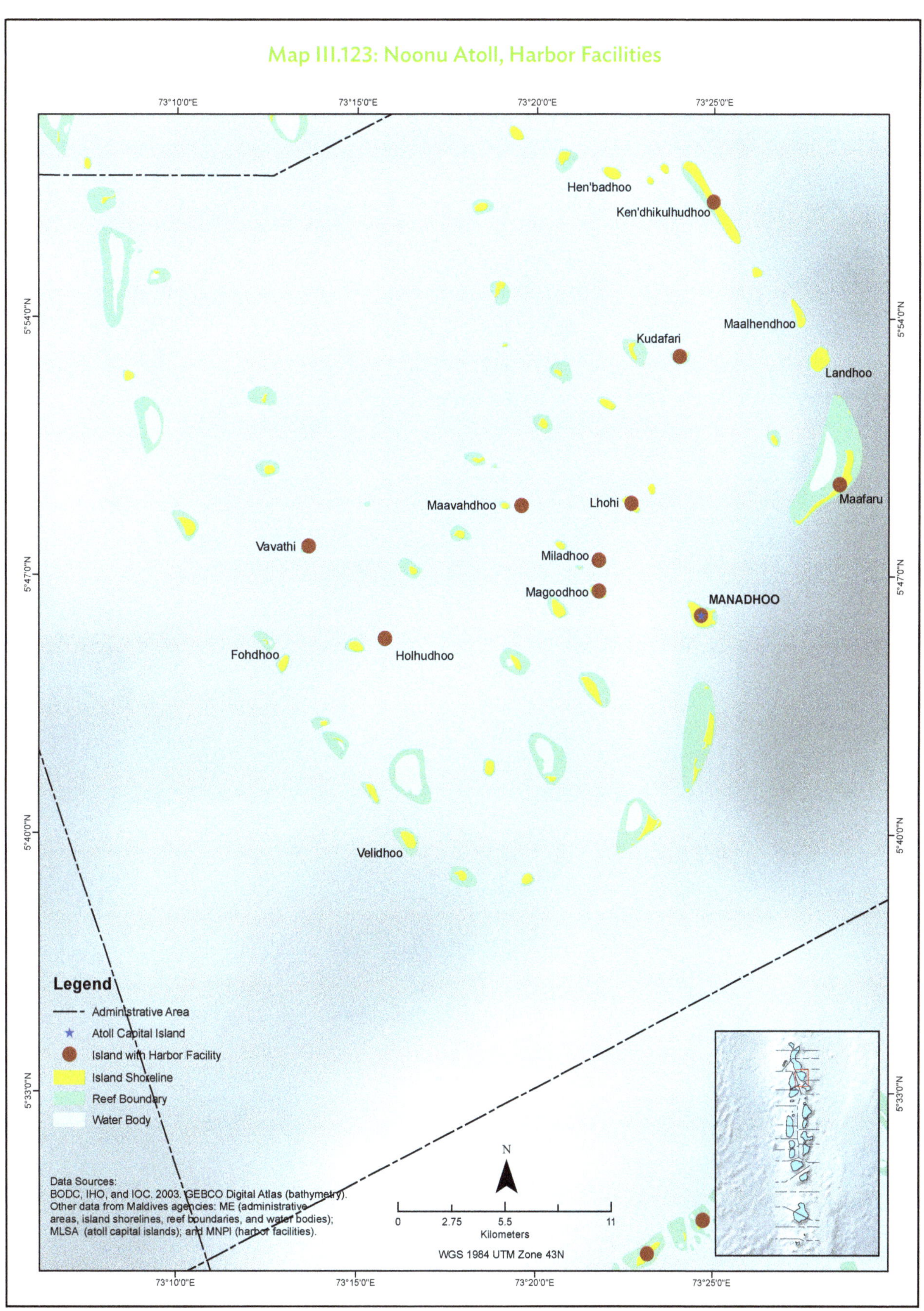

Map III.124: North Malé Atoll, Harbor Facilities

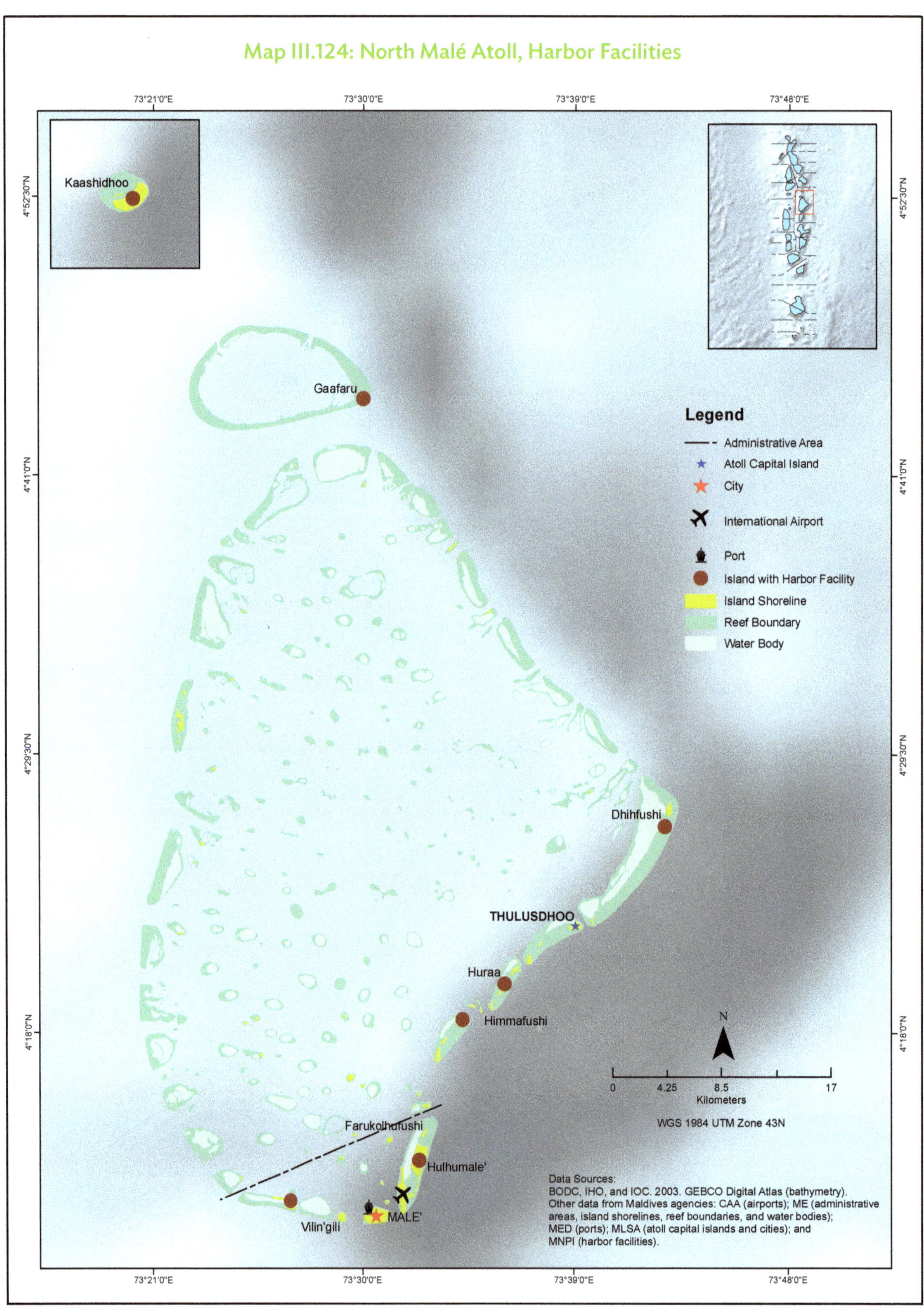

Map III.125: Raa Atoll, Harbor Facilities

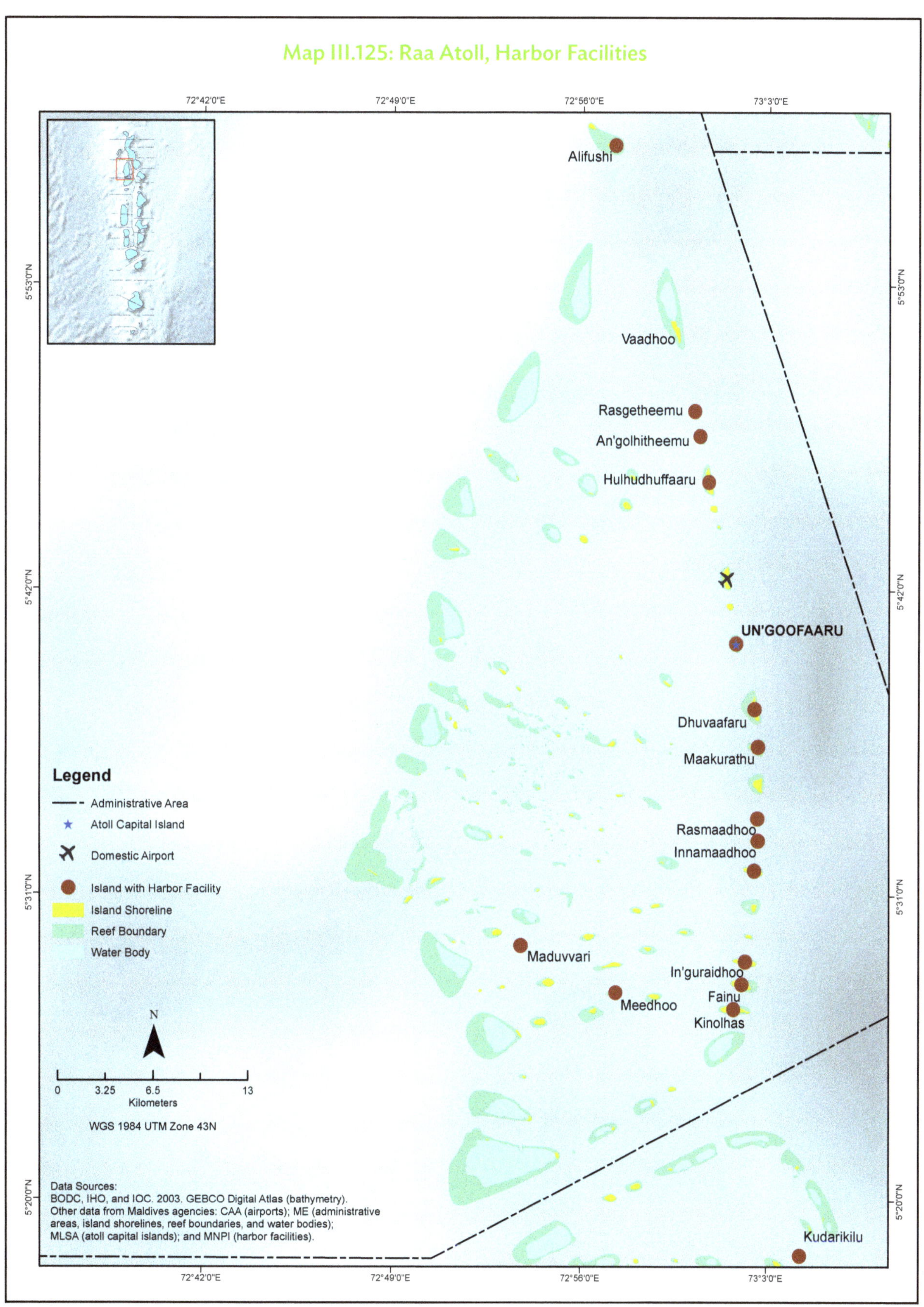

Map III.126: Shaviyani Atoll, Harbor Facilities

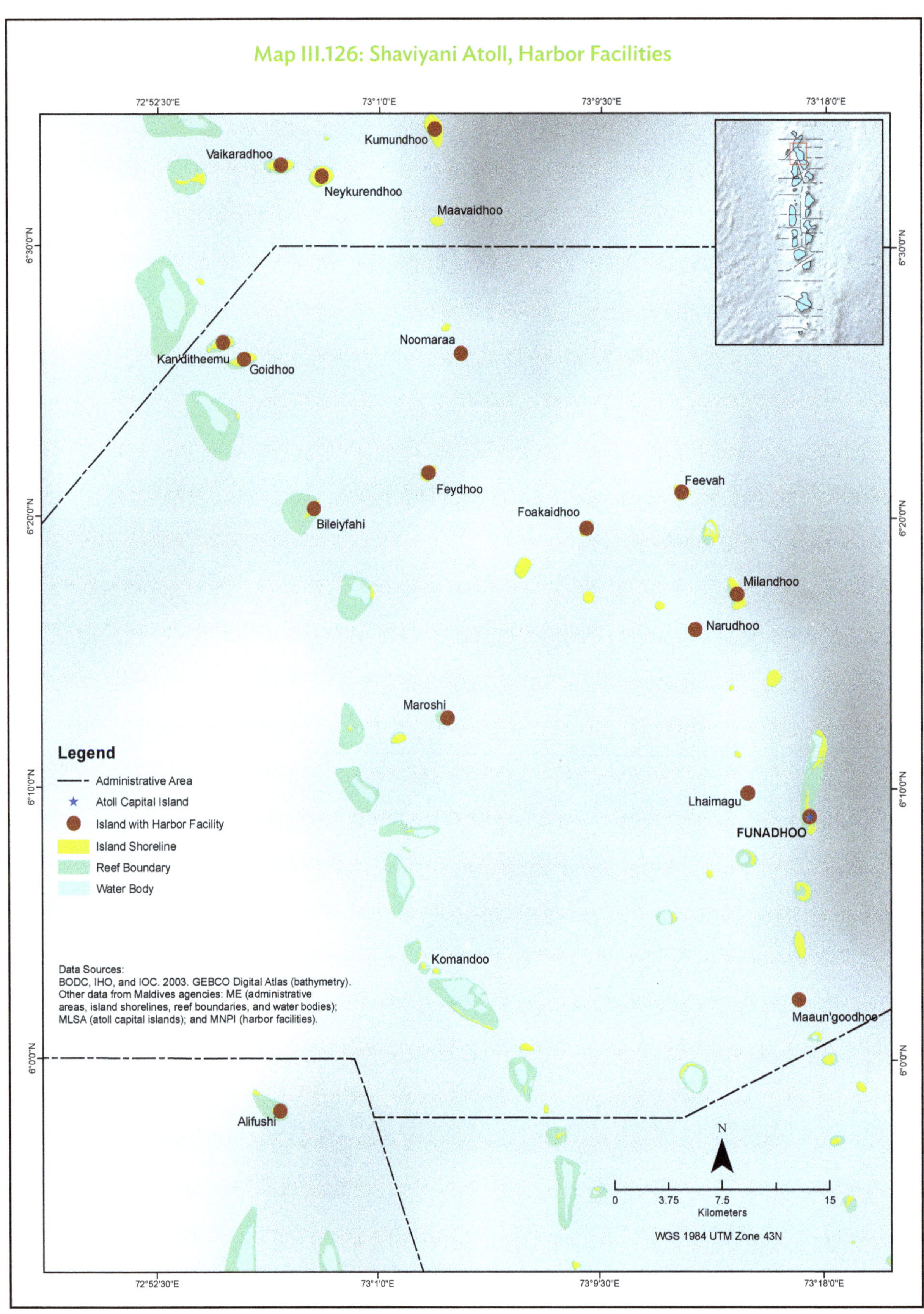

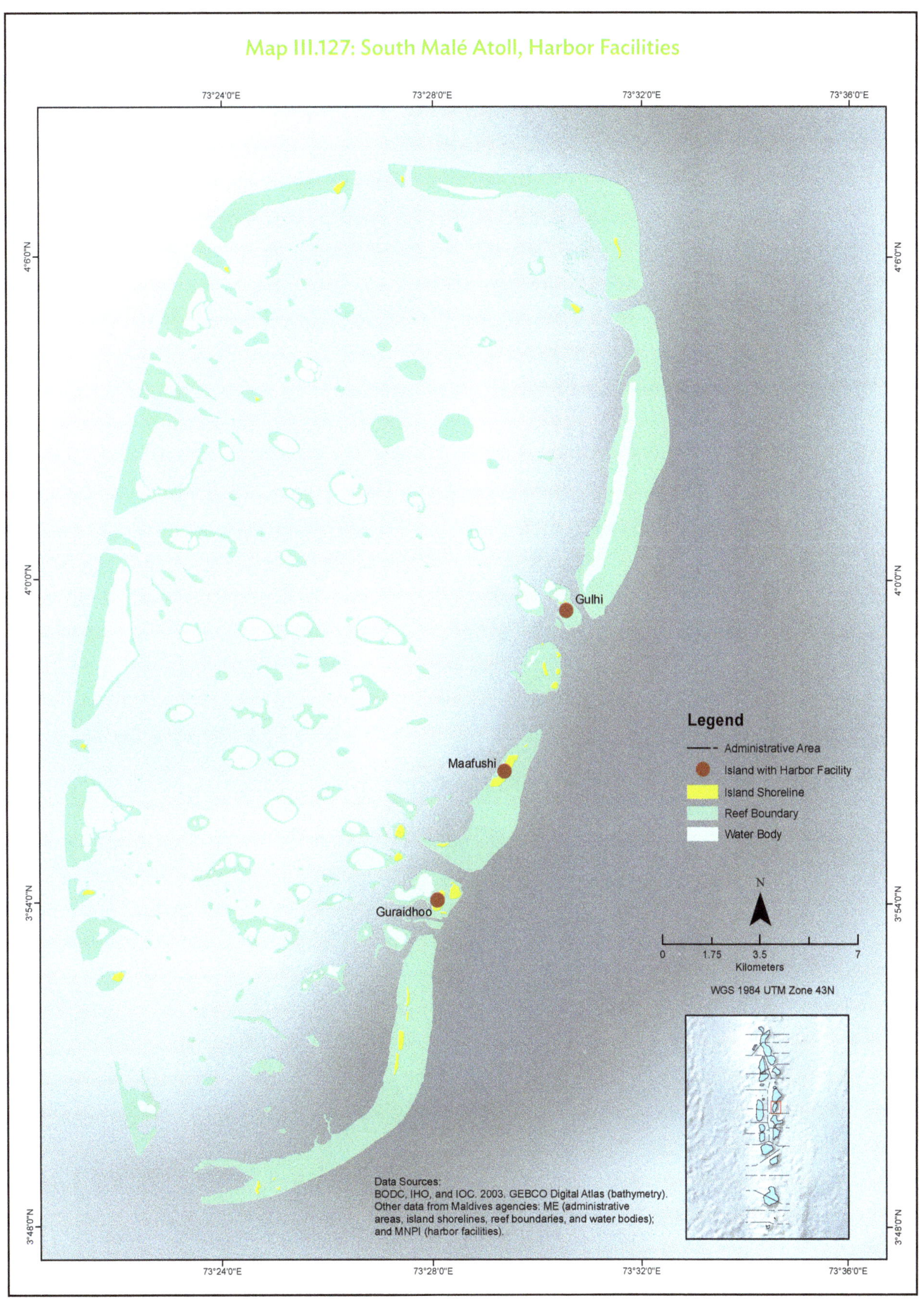
Map III.127: South Malé Atoll, Harbor Facilities
73°24'0"E
73°28'0"E
73°32'0"E
73°36'0"E
4°6'0"N
4°0'0"N
3°54'0"N
3°48'0"N
Gulhi
Maafushi
Guraidhoo
Legend
Administrative Area
Island with Harbor Facility
Island Shoreline
Reef Boundary
Water Body
N
0
1.75
3.5
7
Kilometers
WGS 1984 UTM Zone 43N
Data Sources:
BODC, IHO, and IOC. 2003. GEBCO Digital Atlas (bathymetry).
Other data from Maldives agencies: ME (administrative areas, island shorelines, reef boundaries, and water bodies); and MNPI (harbor facilities).

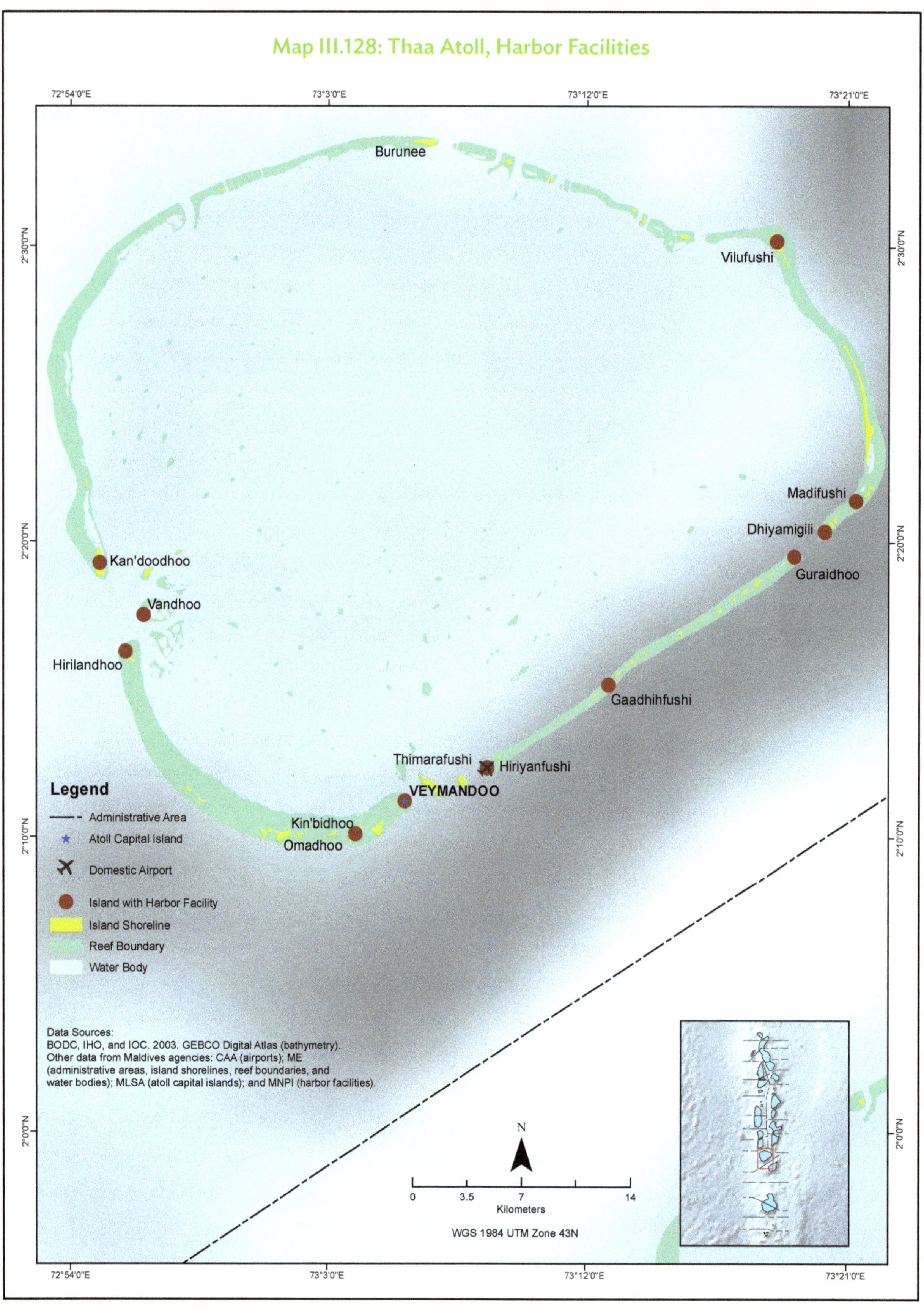
Map III.128: Thaa Atoll, Harbor Facilities
72°54'0"E
73°3'0"E
73°12'0"E
73°21'0"E
2°30'0"N
2°20'0"N
2°10'0"N
2°0'0"N
Burunee
Vilufushi
Madifushi
Dhiyamigili
Guraidhoo
Kan'doodhoo
Vandhoo
Hirilandhoo
Gaadhihfushi
Thimarafushi
Hiriyanfushi
VEYMANDOO
Kin'bidhoo
Omadhoo
Legend
Administrative Area
Atoll Capital Island
Domestic Airport
Island with Harbor Facility
Island Shoreline
Reef Boundary
Water Body
Data Sources:
BODC, IHO, and IOC. 2003. GEBCO Digital Atlas (bathymetry).
Other data from Maldives agencies: CAA (airports); ME (administrative areas, island shorelines, reef boundaries, and water bodies); MLSA (atoll capital islands); and MNPI (harbor facilities).
N
0
3.5
7
14
Kilometers
WGS 1984 UTM Zone 43N

Map III.129: Vaavu Atoll, Harbor Facilities

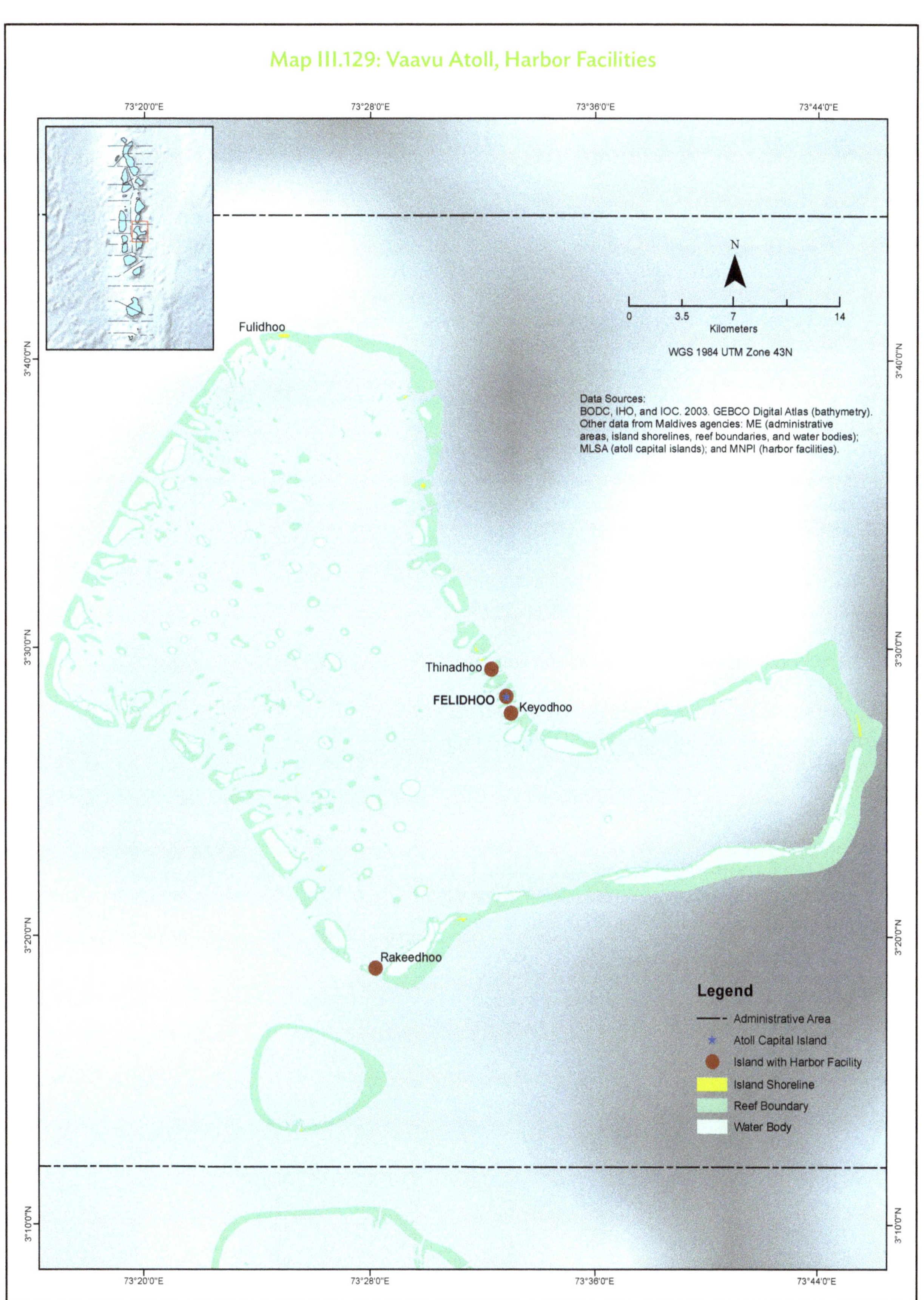

Sand Mining Applications

Sand is one of the major natural resources in Maldives. The white sand lining shores is attributed to the coralline formation of the islands.

In the country, sand mining is a common practice—mined sand is used for construction purposes as there is no other source of aggregate material. In recent years, coralline sand mining has increased to meet the demand of development (Naseer 1997). However, the high demand threatens beaches and islands. Sand mining activities could damage the integrity of shores and lead to erosion (Emerton, Baig, and Saleem 2009). It also threatens corals and other aquatic organisms (Ministry of Environment and Energy 2016). Additionally, the extraction of sand alters the natural movement of sediments and environmental processes on which organisms rely (Dhunya, Huang, and Aslam 2017).

To protect corals, sand mining is regulated through the Act on Sand Mining of 1978, which requires sand mining permits, and the Act on Coral and Sand Mining of 2000, which regulates sand mining only in designated locations for those who have permits.

There are currently 157 approved sand mining locations in Maldives, the majority of which are in Laamu (15), Baa (13), and Thaa (13) atolls (Table III.9).

Table III.9: Sand Mining in Maldives

Atoll	Number of Sand Mining Locations
Laamu	15
Baa	13
Thaa	13
Gaafu Dhaalu	12
Kaafu	11
Raa	9
Haa Alifu	9
Gaafu Alifu	9
Alifu Dhaalu	8
Dhaalu	8
Meemu	8
Haa Dhaalu	6
Alifu Alifu	6
Seenu	6
Laviyani	5
Shaviyani	5
Noonu	5
Faffu	5
Vaavu	4

Source: International Union for Conservation of Nature, Maldives, 2016.

Mineral source. Sand are mined in Maldives for construction purposes (photo by Sue Todd).

Map III.130: Maldives, Approved Sand Mining Locations

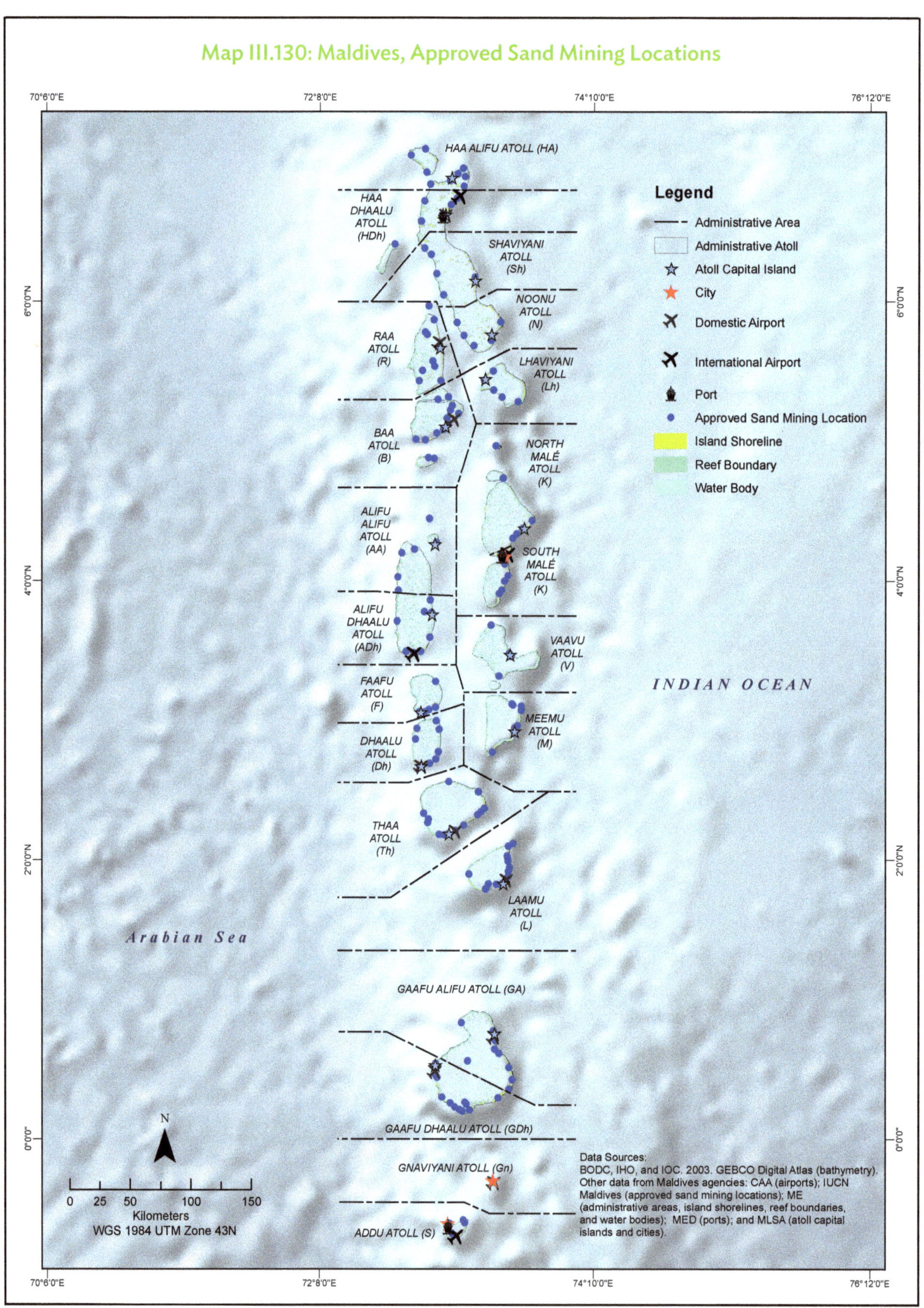

Map III.131: Addu City, Approved Sand Mining Locations

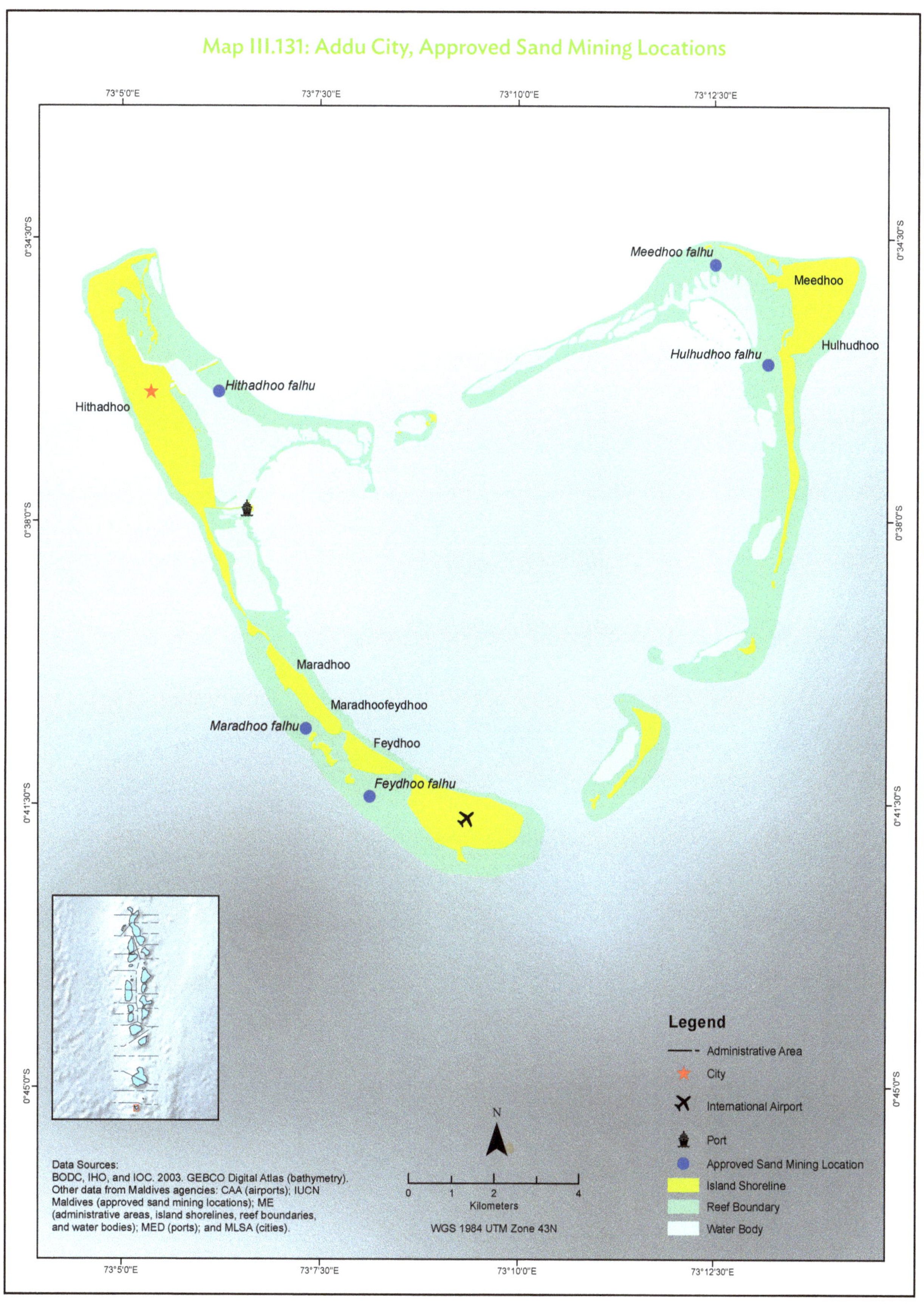

Map III.132: Alifu Alifu Atoll, Approved Sand Mining Locations

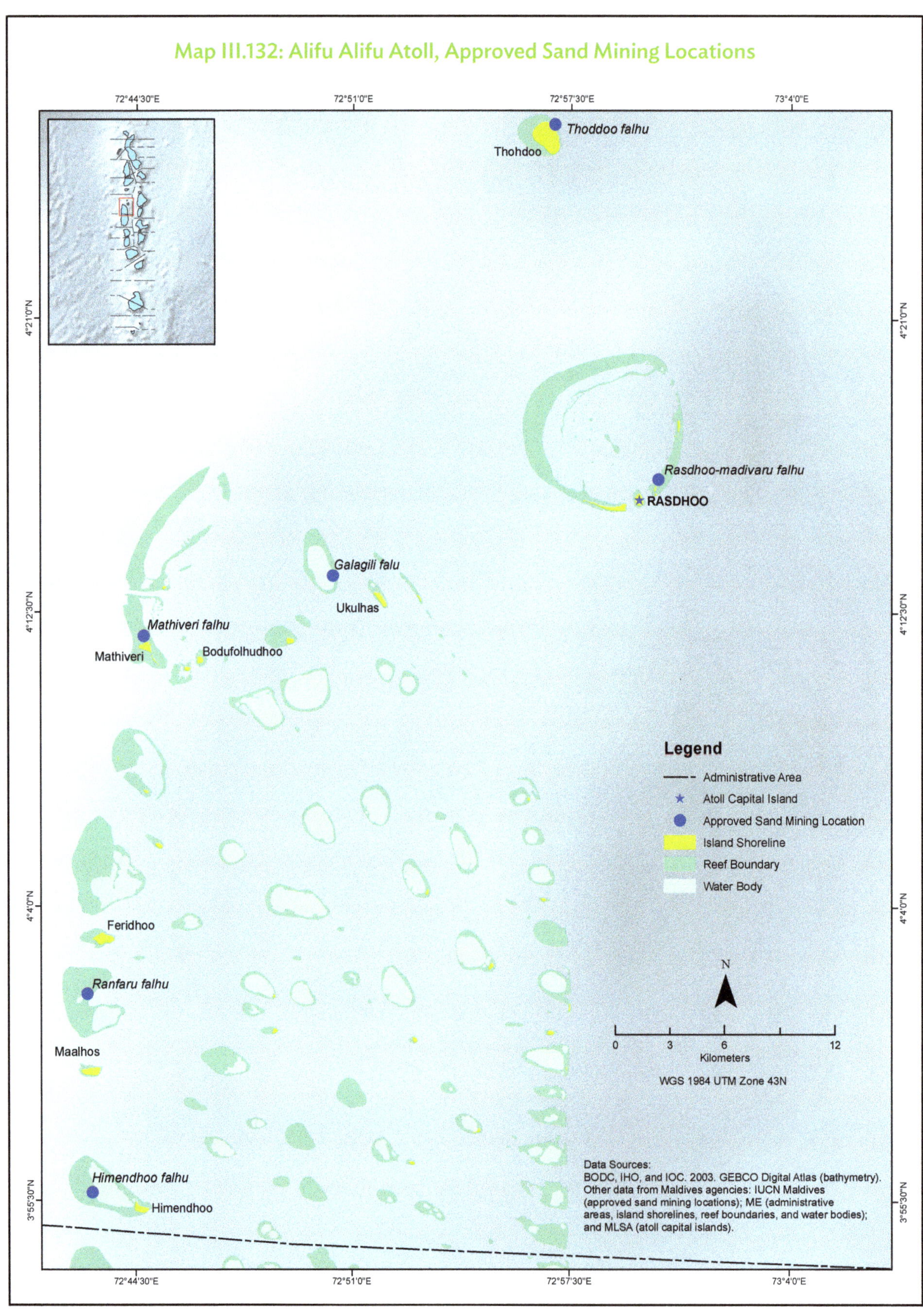

Map III.133: Alifu Dhaalu Atoll, Approved Sand Mining Locations

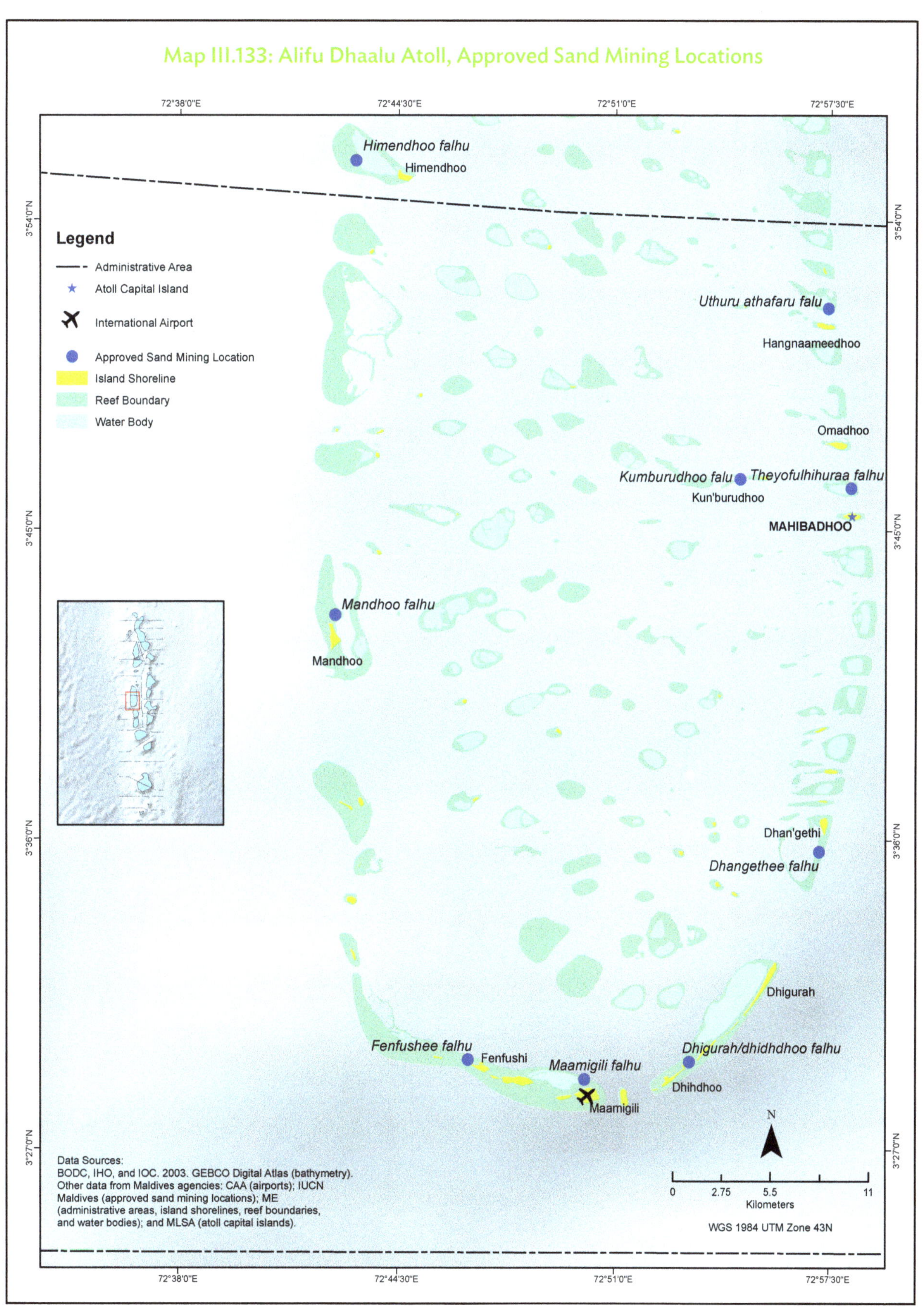

Map III.134: Baa Atoll, Approved Sand Mining Locations

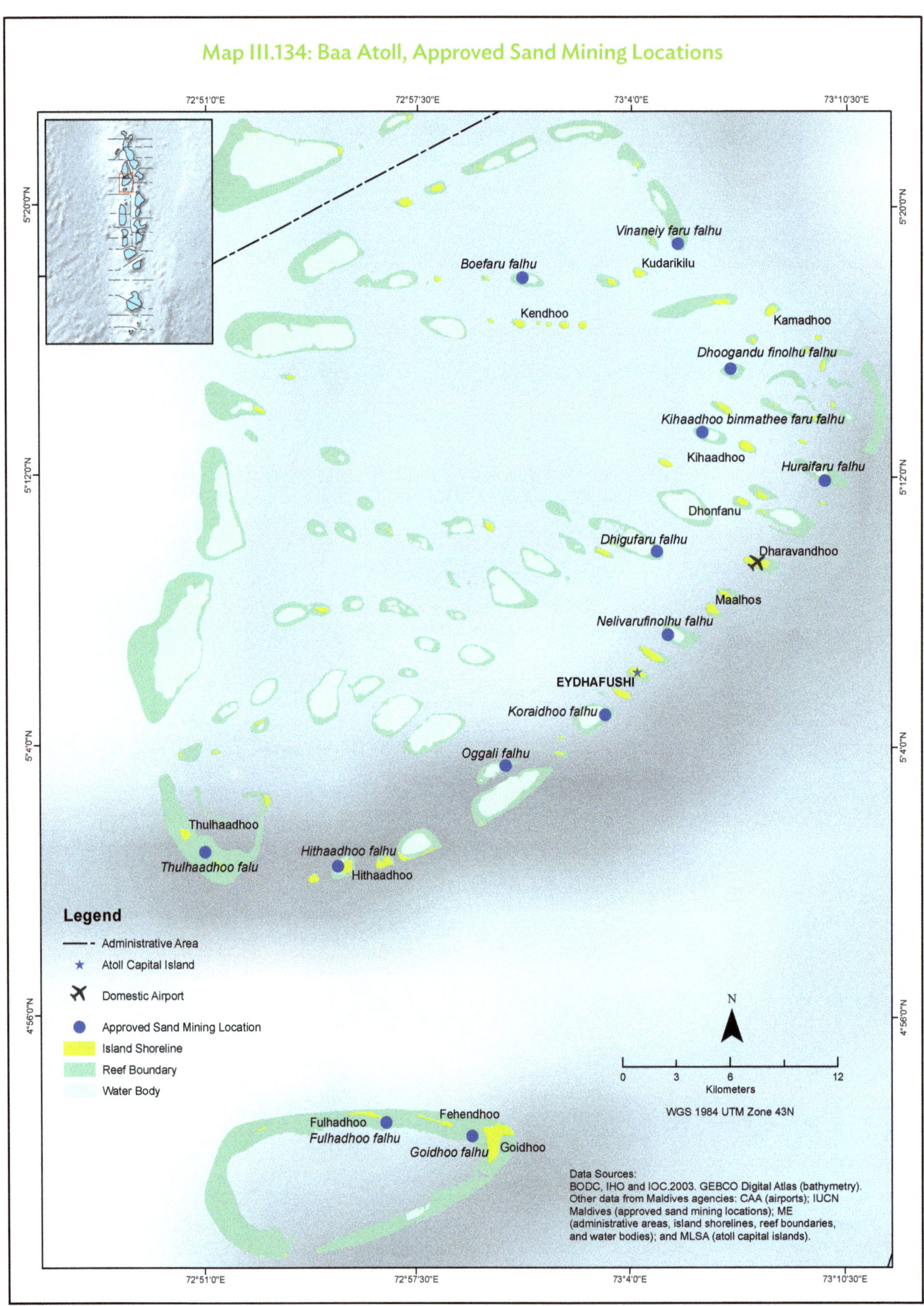

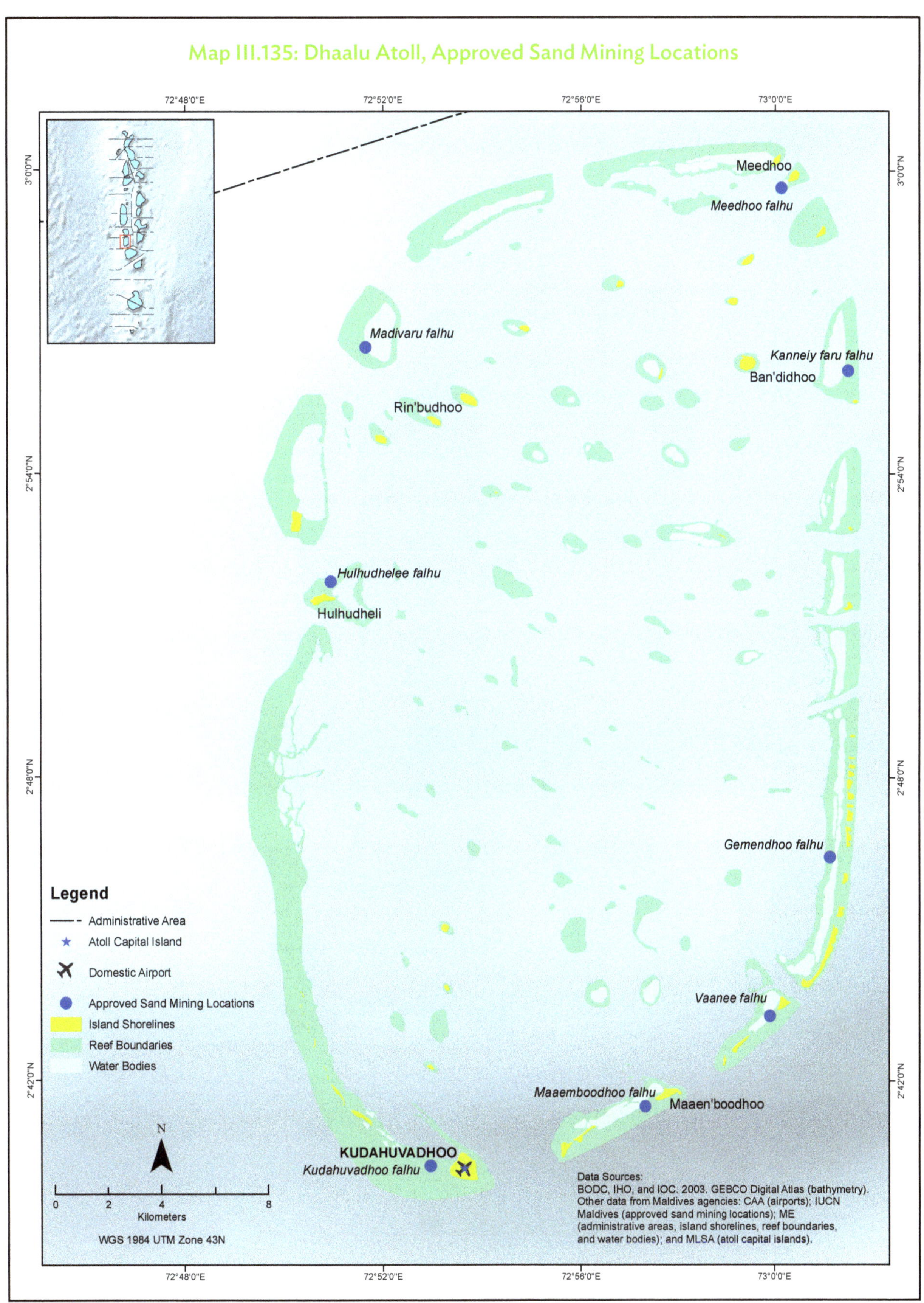
Map III.135: Dhaalu Atoll, Approved Sand Mining Locations
72°48'0"E
72°52'0"E
72°56'0"E
73°0'0"E
3°0'0"N
2°54'0"N
2°48'0"N
2°42'0"N
Meedhoo
Meedhoo falhu
Madivaru falhu
Kanneiy faru falhu
Ban'didhoo
Rin'budhoo
Hulhudhelee falhu
Hulhudheli
Gemendhoo falhu
Vaanee falhu
Maaemboodhoo falhu
Maaen'boodhoo
KUDAHUVADHOO
Kudahuvadhoo falhu
Legend
Administrative Area
Atoll Capital Island
Domestic Airport
Approved Sand Mining Locations
Island Shorelines
Reef Boundaries
Water Bodies
N
0 2 4 8
Kilometers
WGS 1984 UTM Zone 43N
Data Sources:
BODC, IHO, and IOC. 2003. GEBCO Digital Atlas (bathymetry).
Other data from Maldives agencies: CAA (airports); IUCN
Maldives (approved sand mining locations); ME
(administrative areas, island shorelines, reef boundaries,
and water bodies); and MLSA (atoll capital islands).

Map III.136: Faafu Atoll, Approved Sand Mining Locations

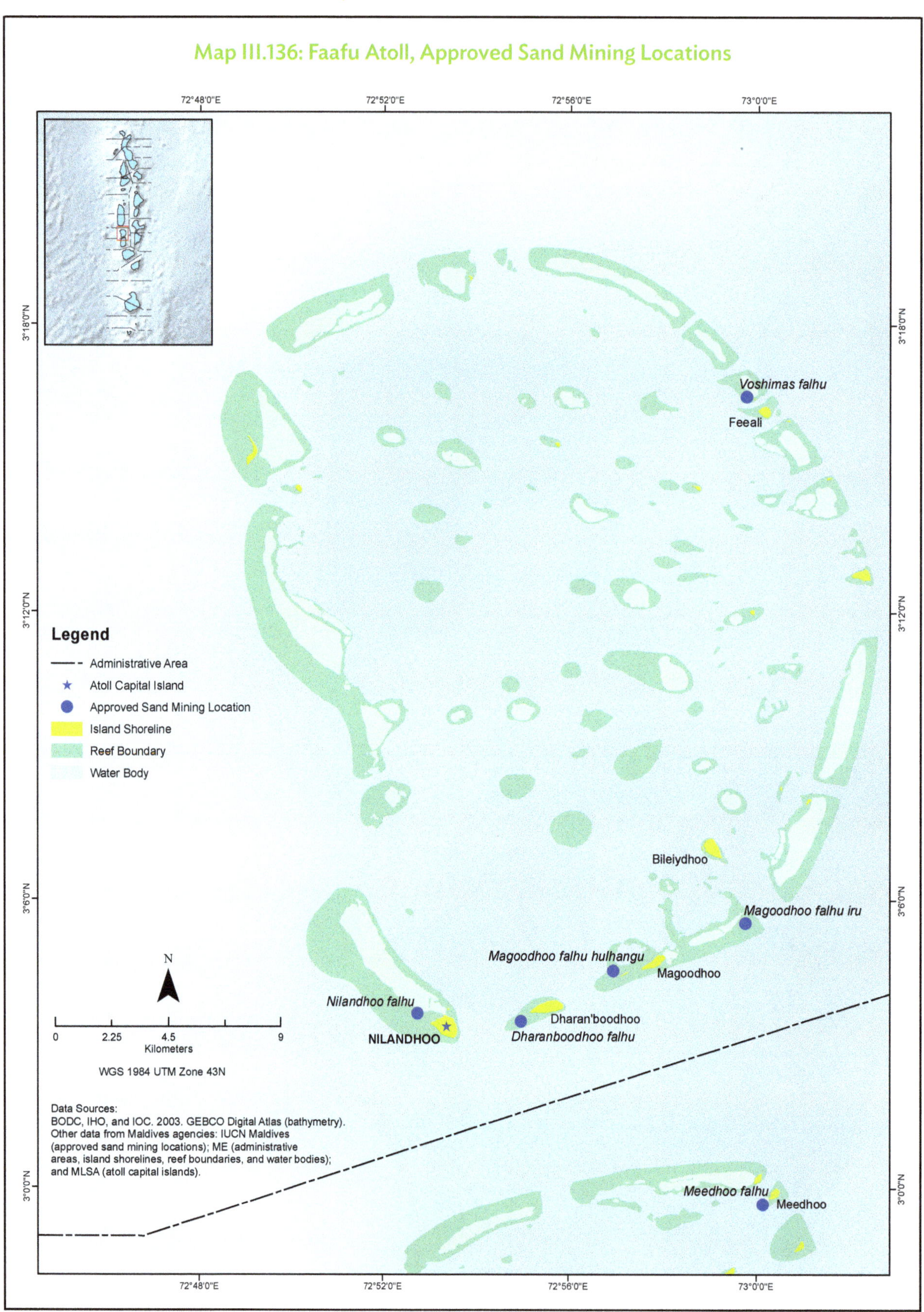

Map III.137: Gaafu Alifu Atoll, Approved Sand Mining Locations

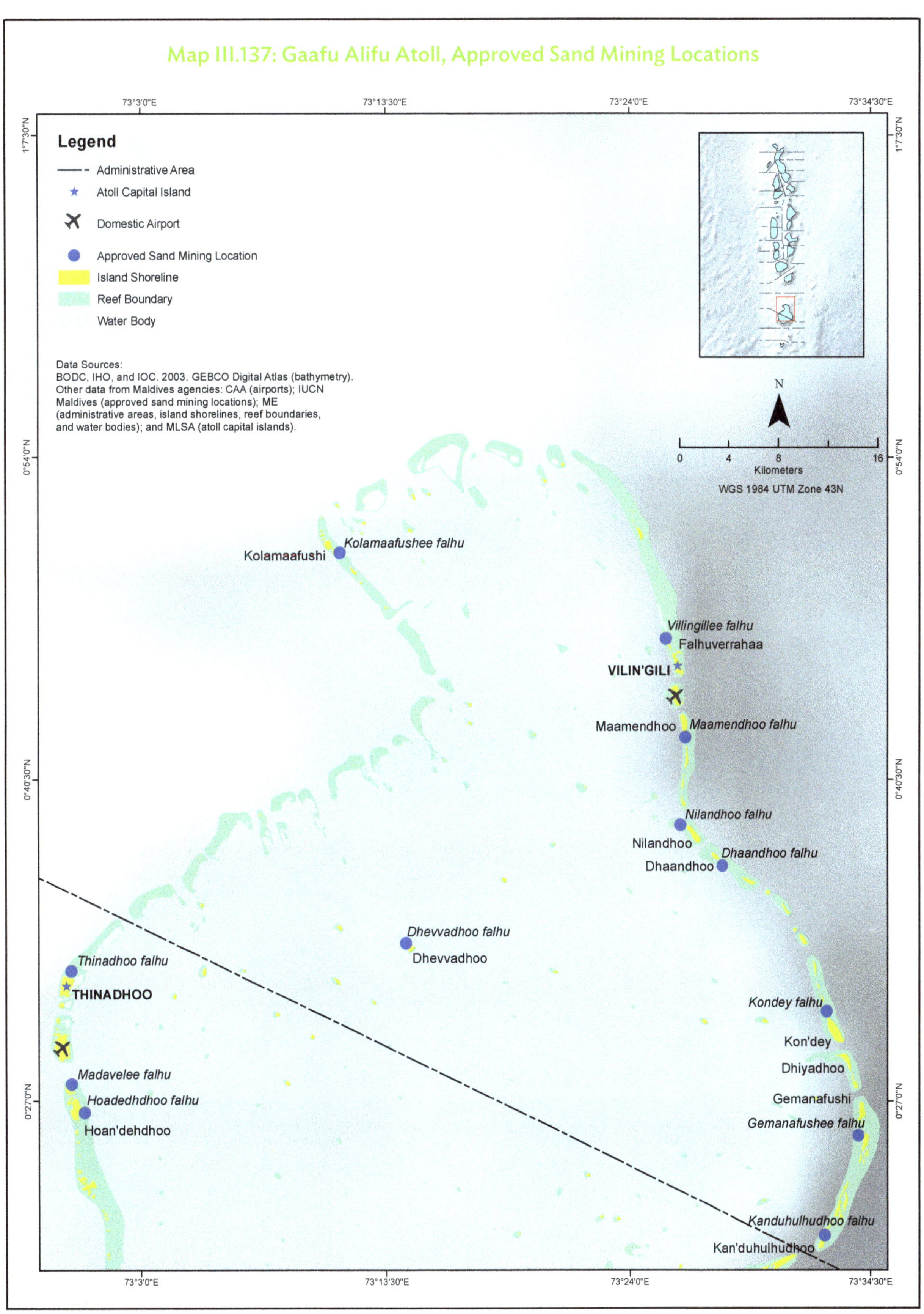

Map III.138: Gaafu Dhaalu Atoll, Approved Sand Mining Locations

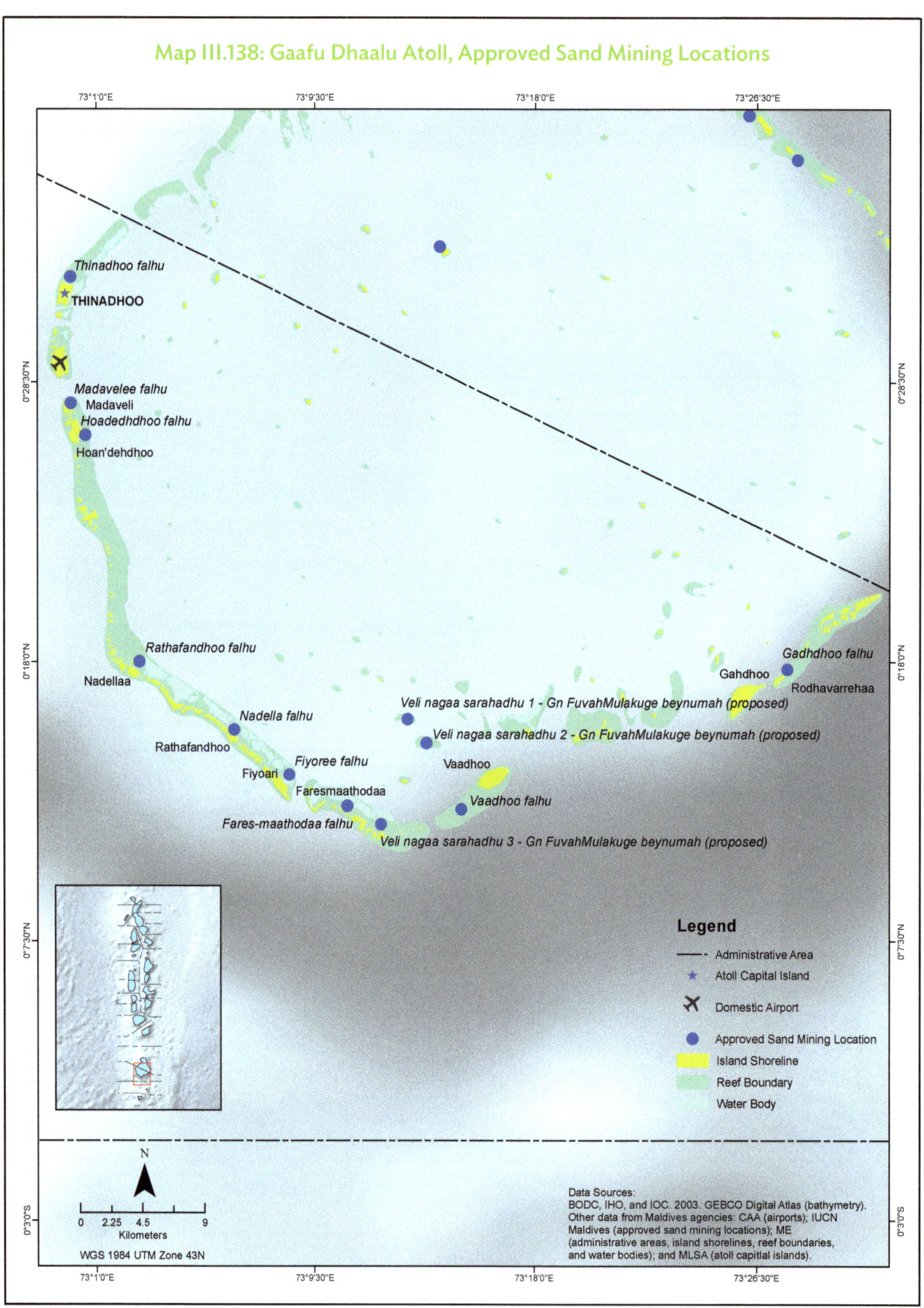

Map III.139: Haa Alifu Atoll, Approved Sand Mining Locations

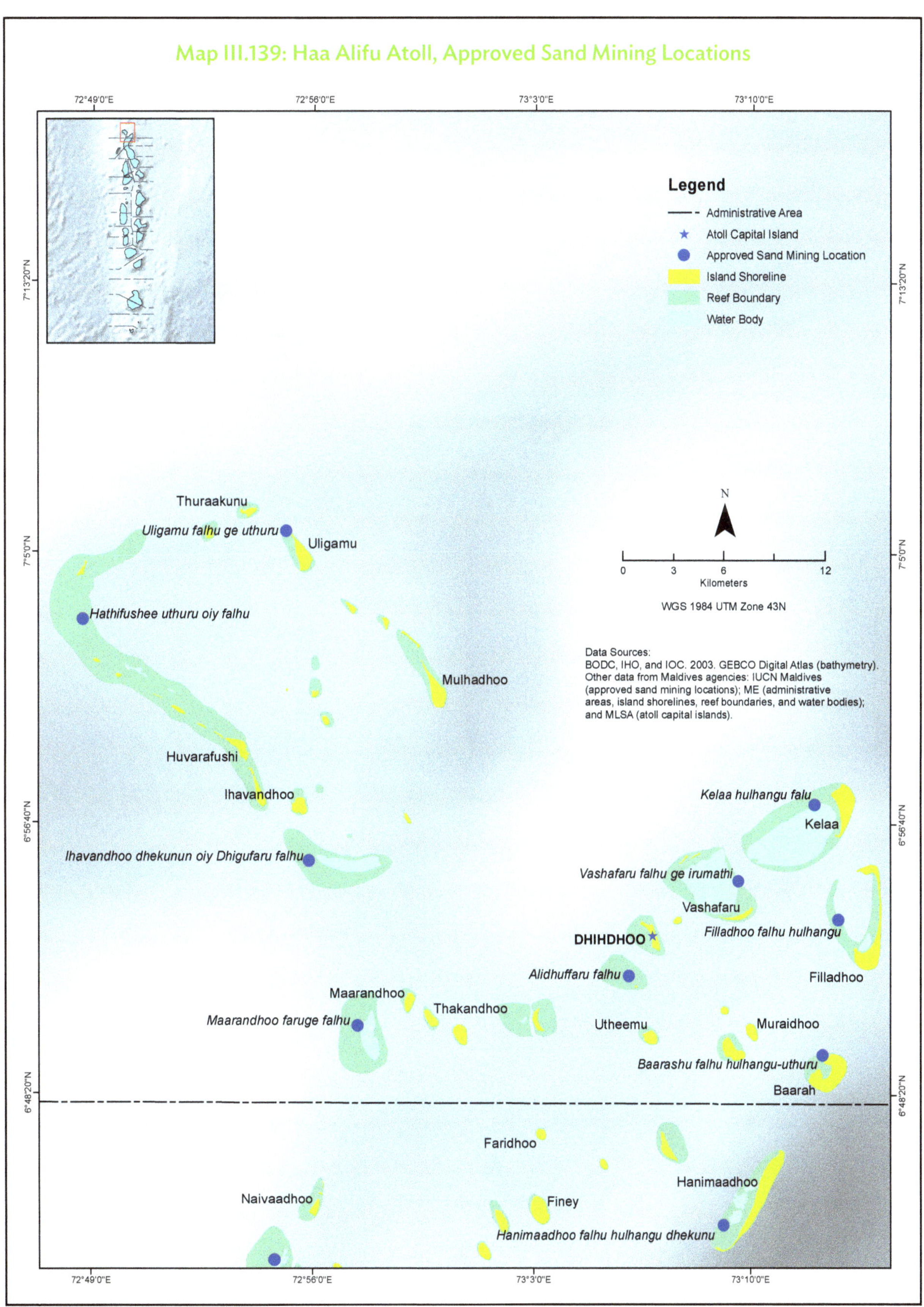

Map III.140: Haa Dhaalu Atoll, Approved Sand Mining Locations

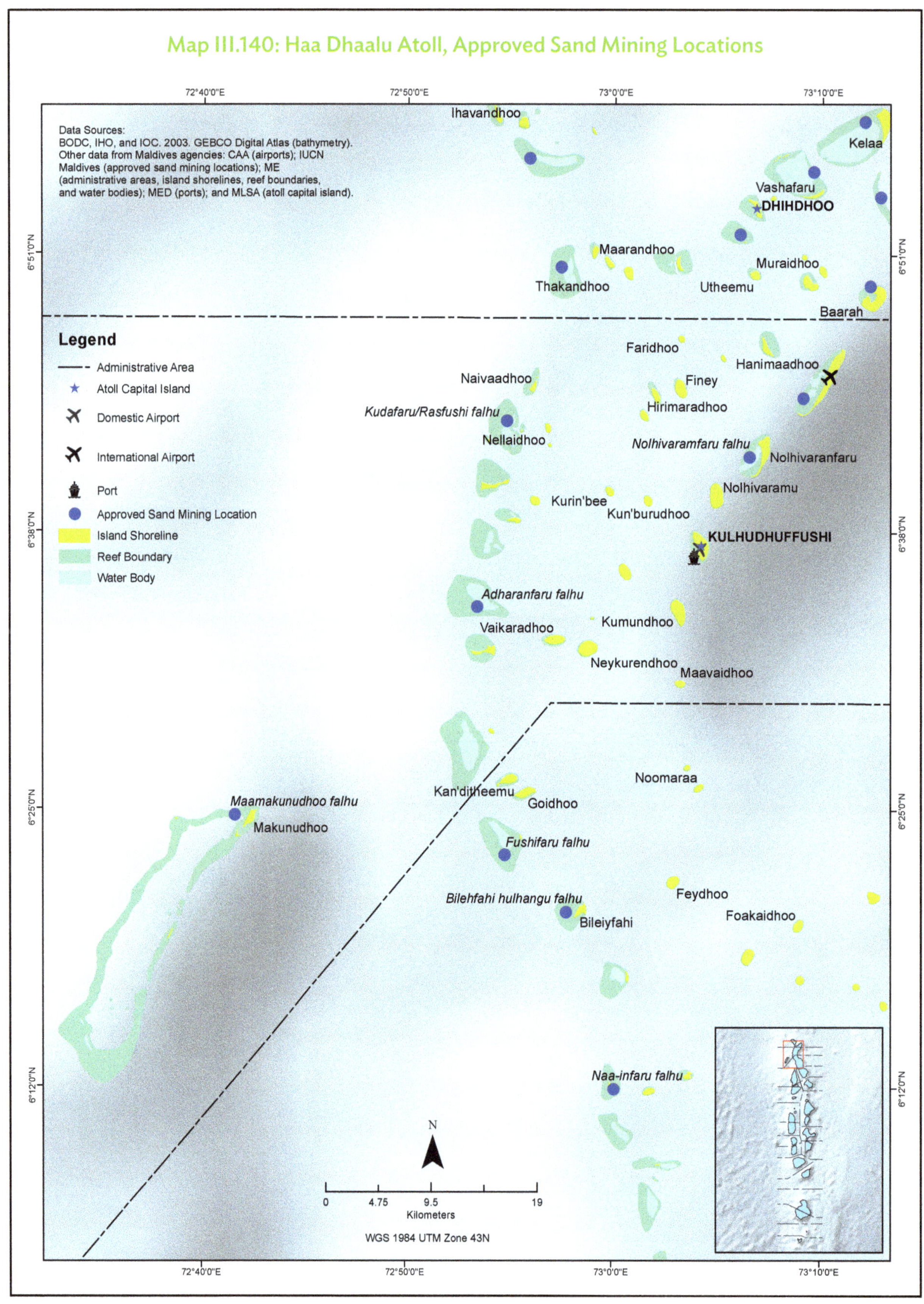

Map III.141: Laamu Atoll, Approved Sand Mining Locations

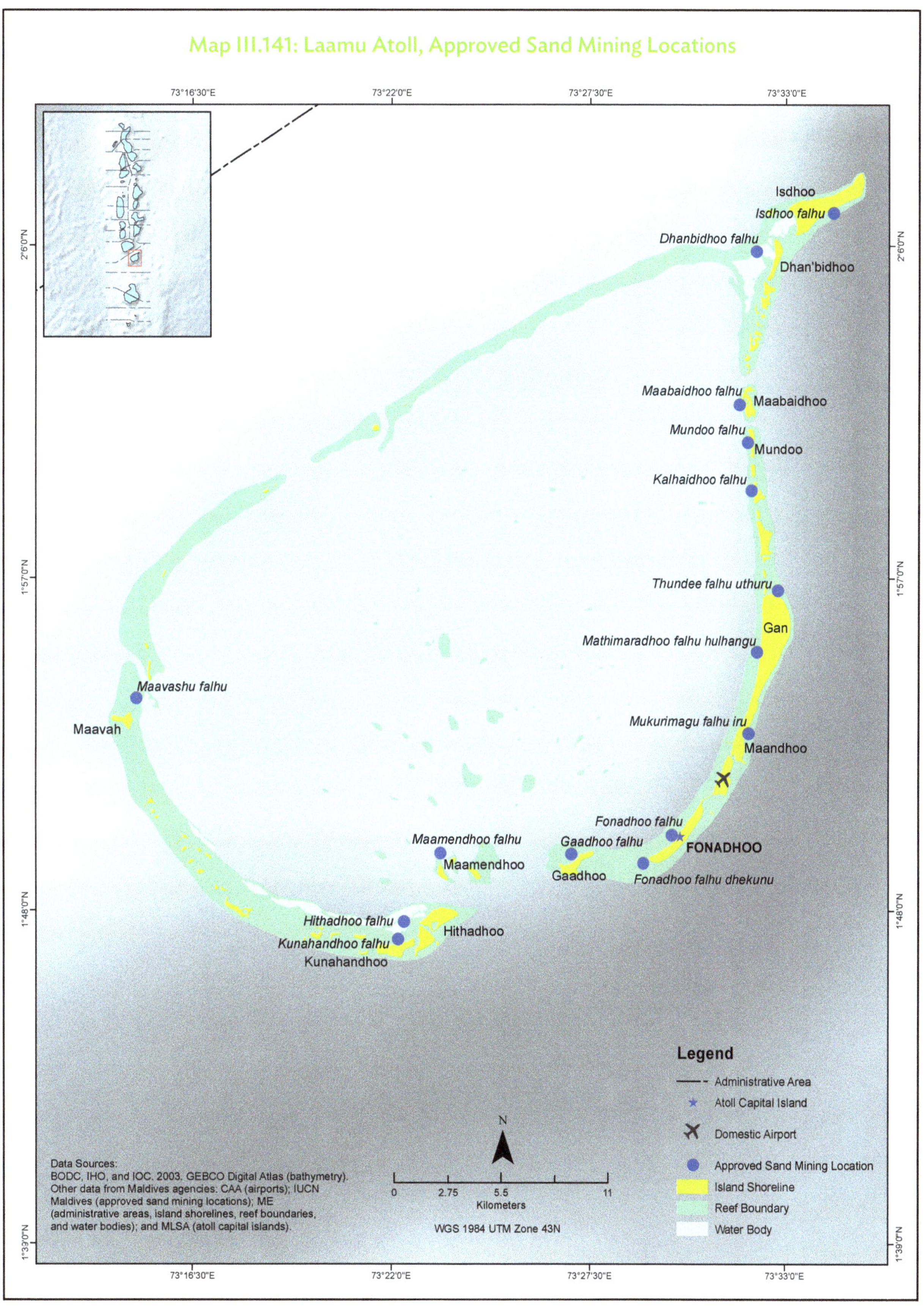

Map III.142: Lhaviyani Atoll, Approved Sand Mining Locations

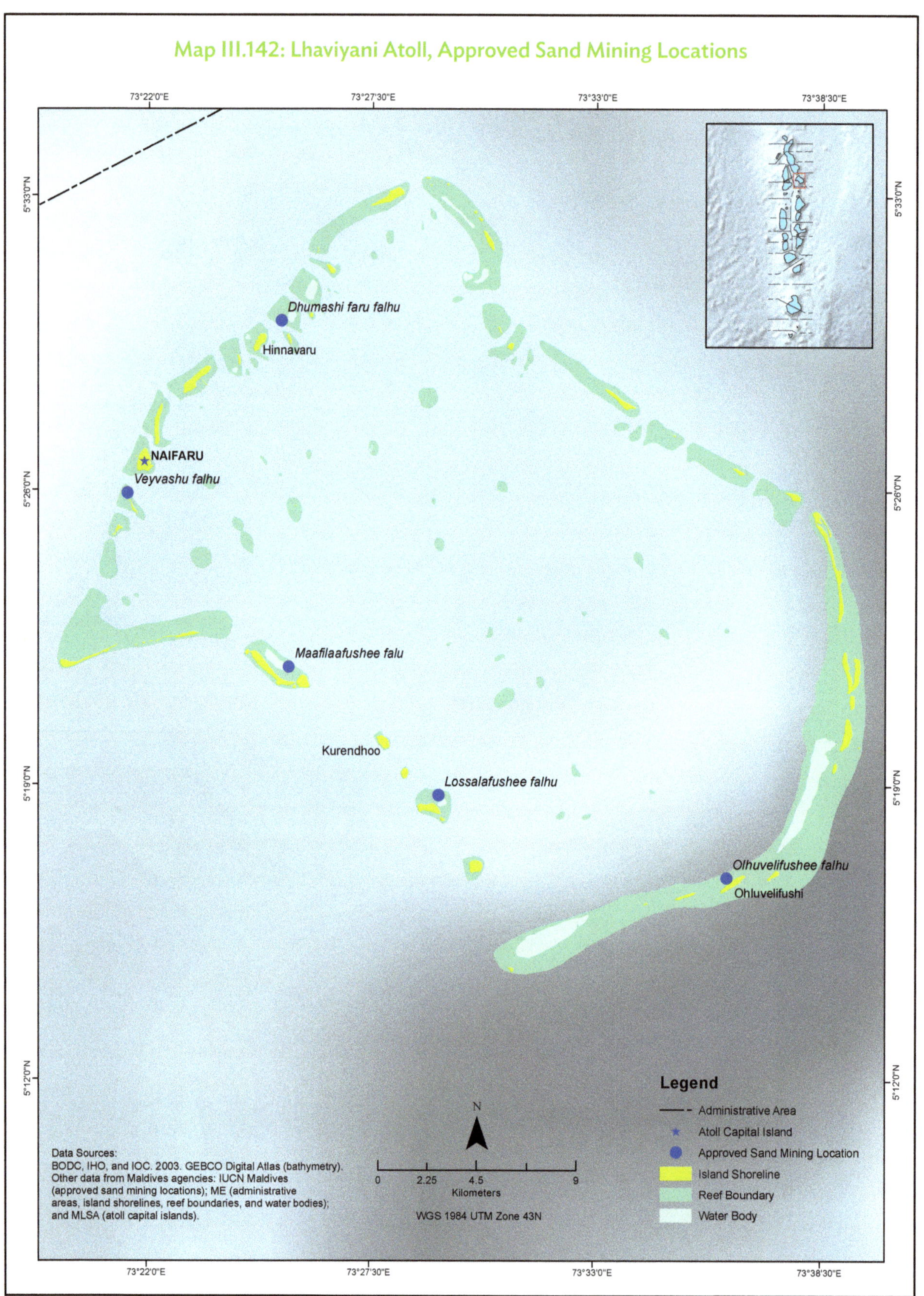

Map III.143: Meemu Atoll, Approved Sand Mining Locations

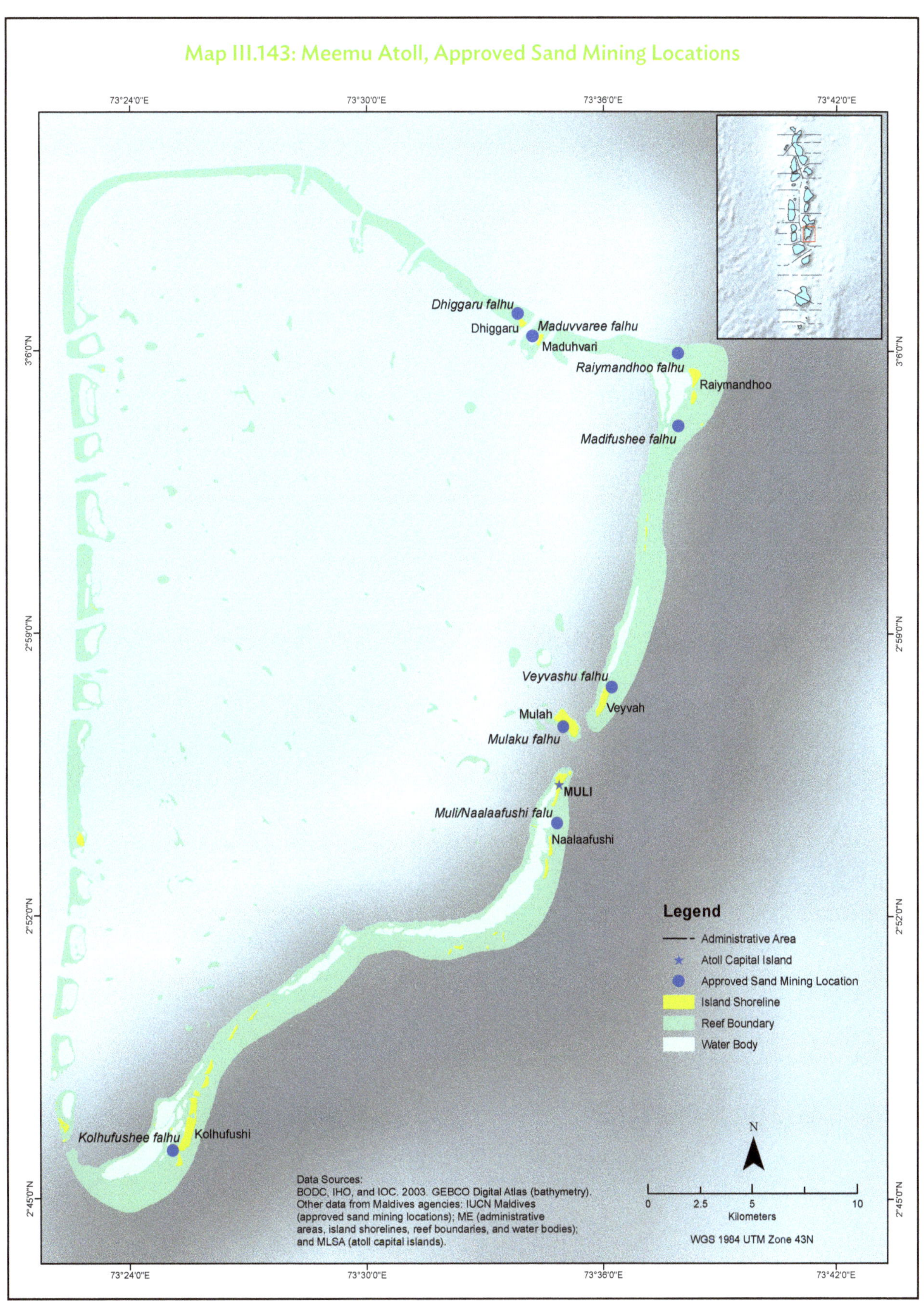

Map III.144: Noonu Atoll, Approved Sand Mining Locations

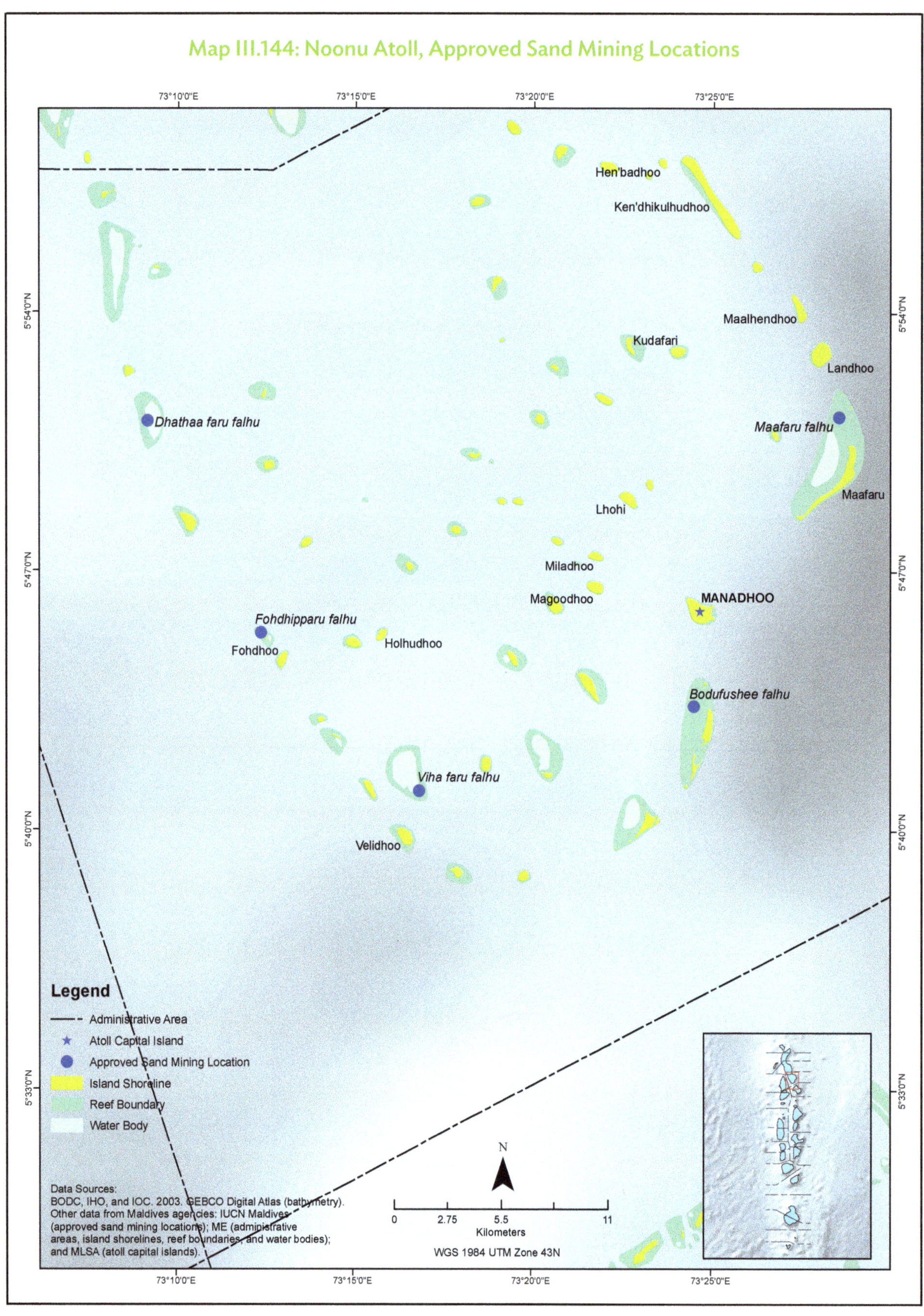

Map III.145: North Malé Atoll, Approved Sand Mining Locations

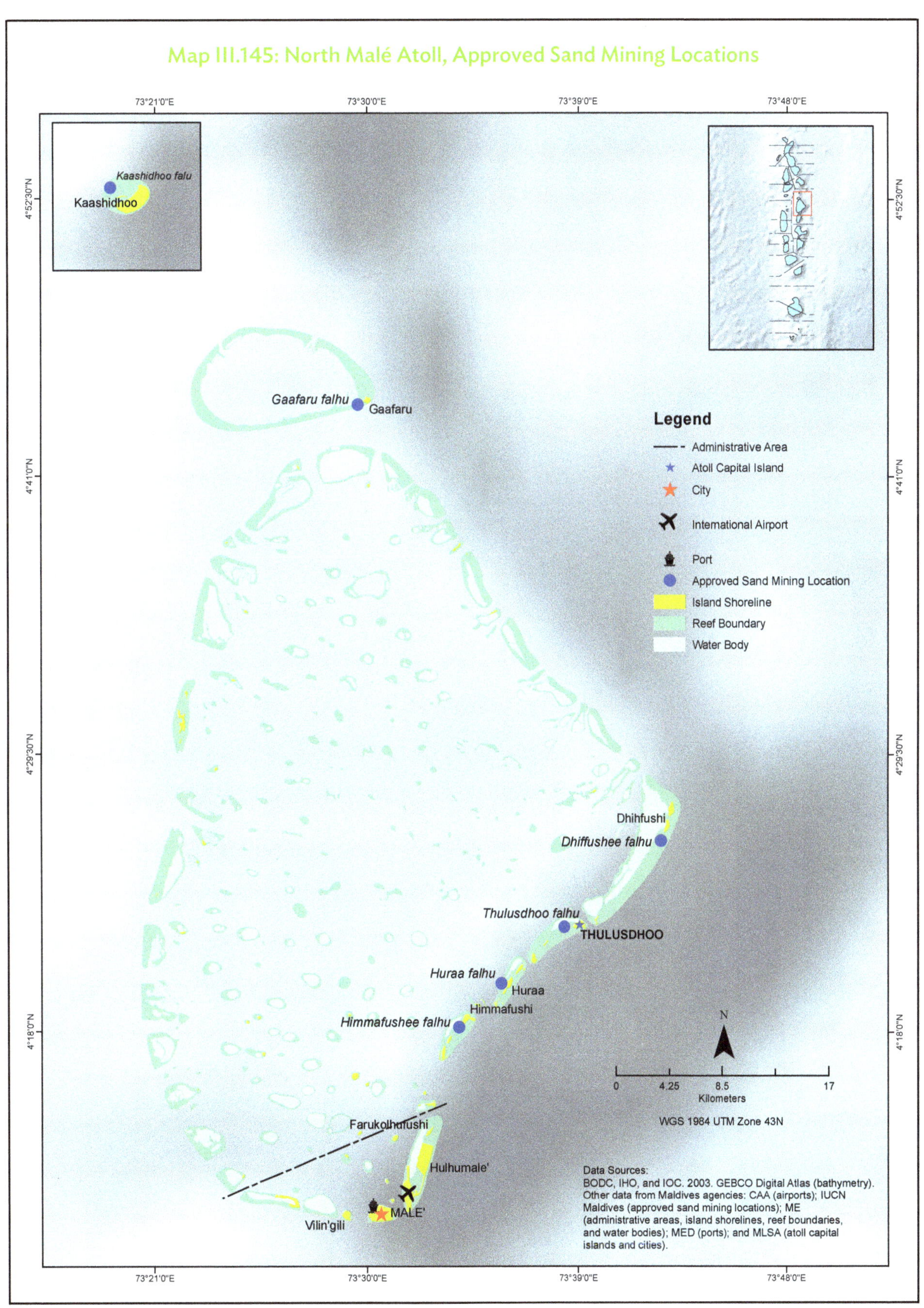

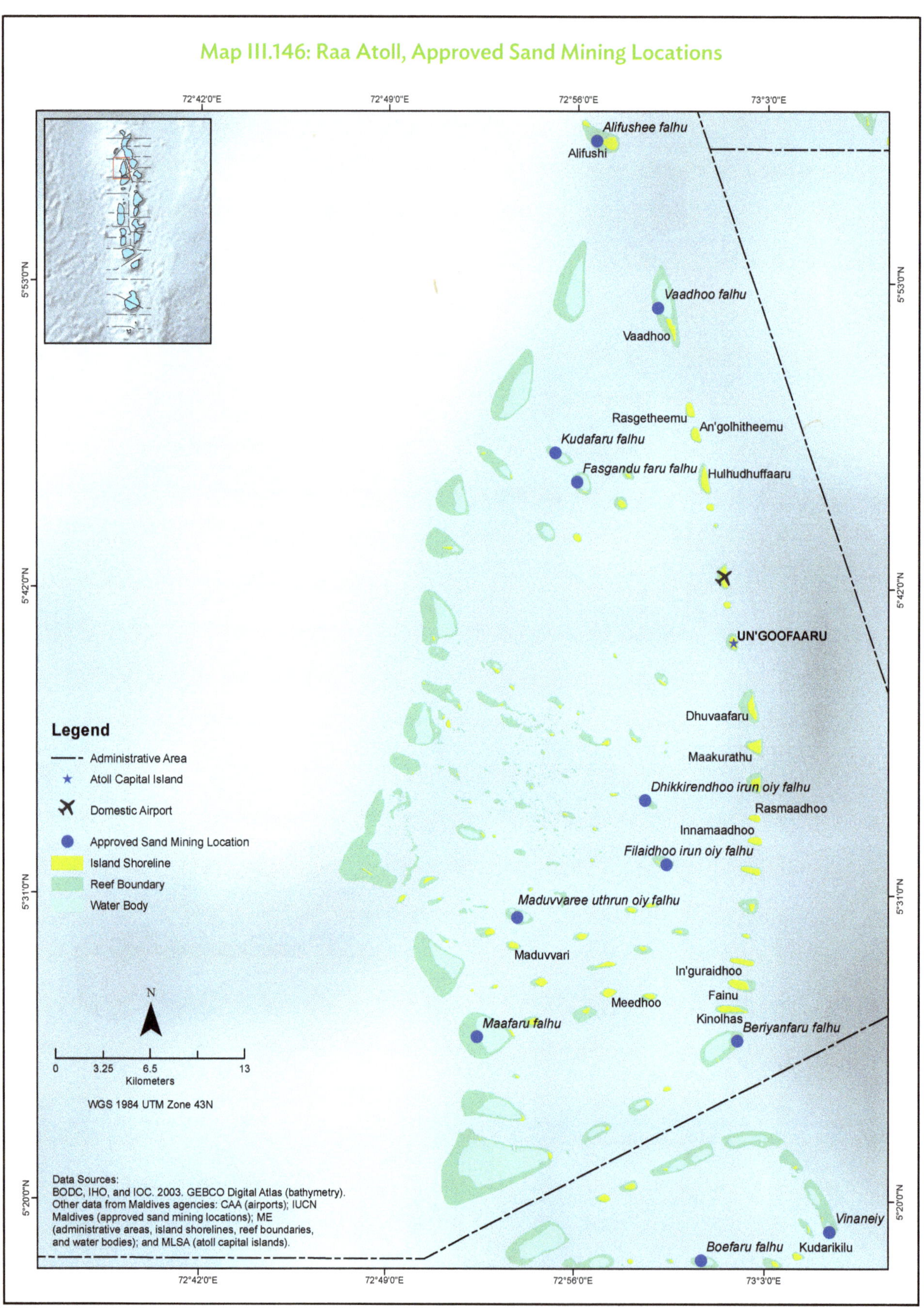

Map III.146: Raa Atoll, Approved Sand Mining Locations
72°42'0"E
72°49'0"E
72°56'0"E
73°3'0"E
5°53'0"N
5°42'0"N
5°31'0"N
5°20'0"N
Alifushee falhu
Alifushi
Vaadhoo falhu
Vaadhoo
Rasgetheemu
An'golhitheemu
Kudafaru falhu
Fasgandu faru falhu
Hulhudhuffaaru
UN'GOOFAARU
Dhuvaafaru
Maakurathu
Dhikkirendhoo irun oiy falhu
Rasmaadhoo
Innamaadhoo
Filaidhoo irun oiy falhu
Maduvvaree uthrun oiy falhu
Maduvvari
In'guraidhoo
Fainu
Meedhoo
Kinolhas
Beriyanfaru falhu
Maafaru falhu
Vinaneiy
Boefaru falhu
Kudarikilu
Legend
Administrative Area
Atoll Capital Island
Domestic Airport
Approved Sand Mining Location
Island Shoreline
Reef Boundary
Water Body
N
0 3.25 6.5 13
Kilometers
WGS 1984 UTM Zone 43N
Data Sources:
BODC, IHO, and IOC. 2003. GEBCO Digital Atlas (bathymetry).
Other data from Maldives agencies: CAA (airports); IUCN Maldives (approved sand mining locations); ME (administrative areas, island shorelines, reef boundaries, and water bodies); and MLSA (atoll capital islands).

Map III.147: Shaviyani Atoll, Approved Sand Mining Locations

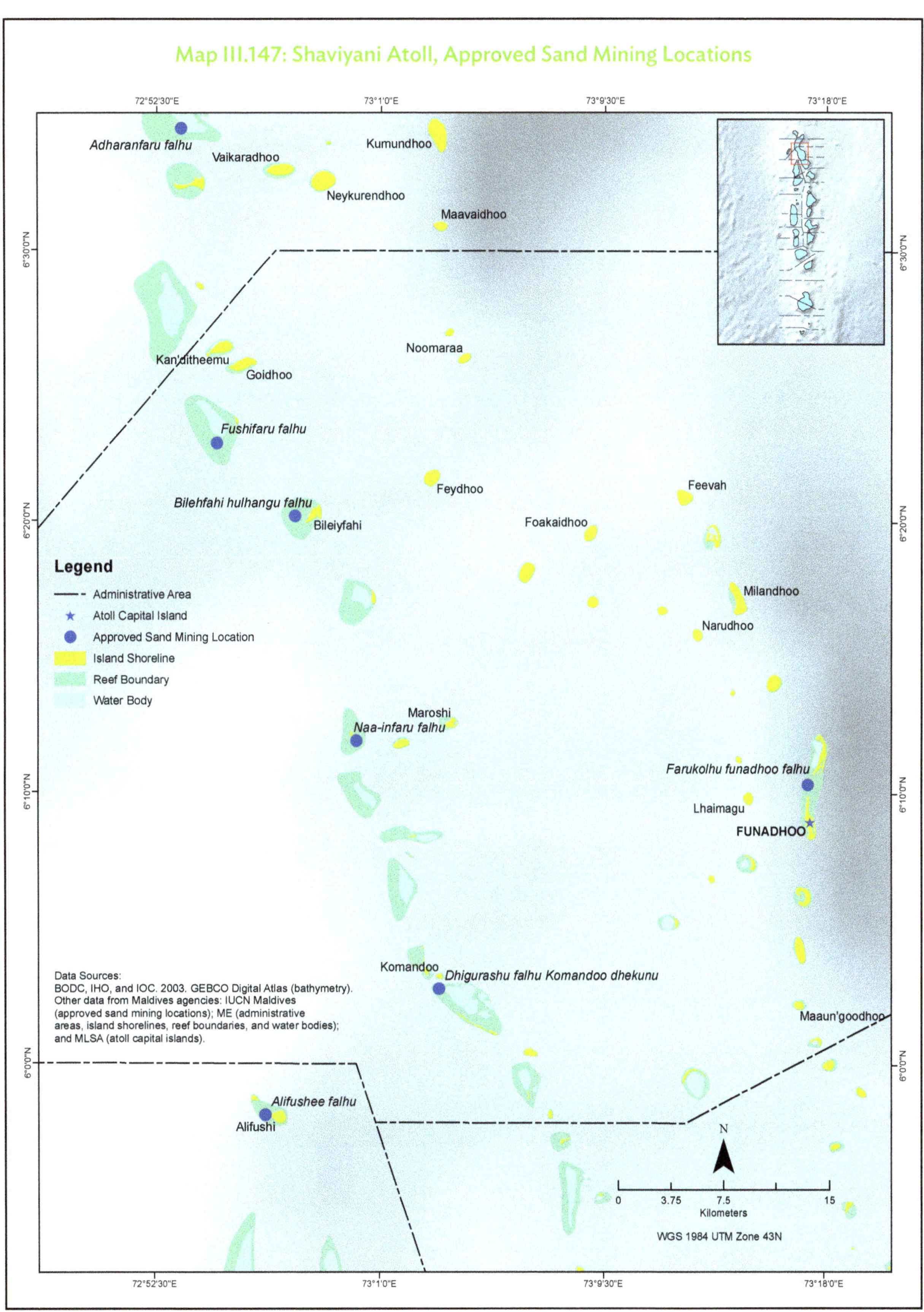

Map III.148: South Malé Atoll, Approved Sand Mining Locations

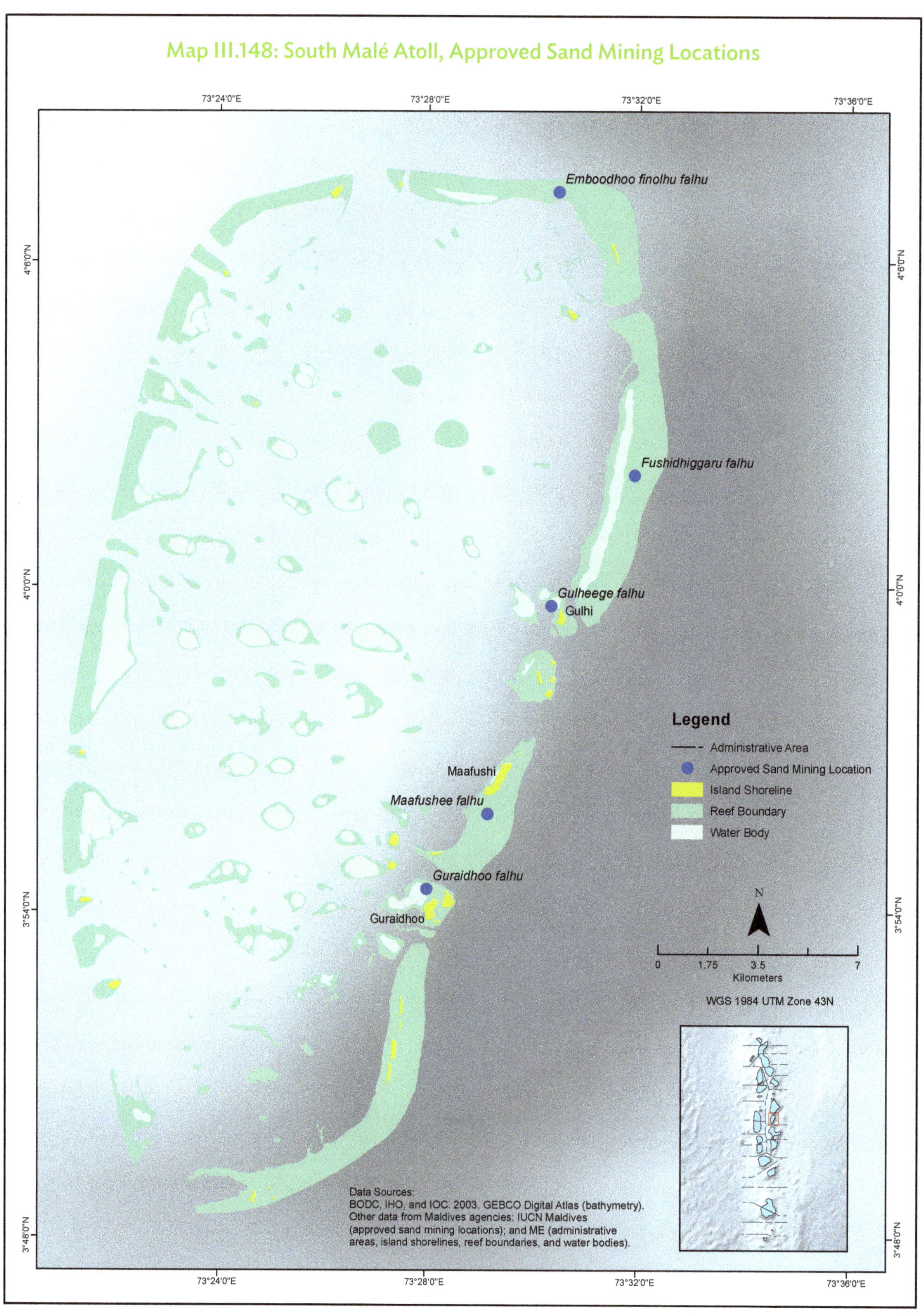

Map III.149: Thaa Atoll, Approved Sand Mining Locations

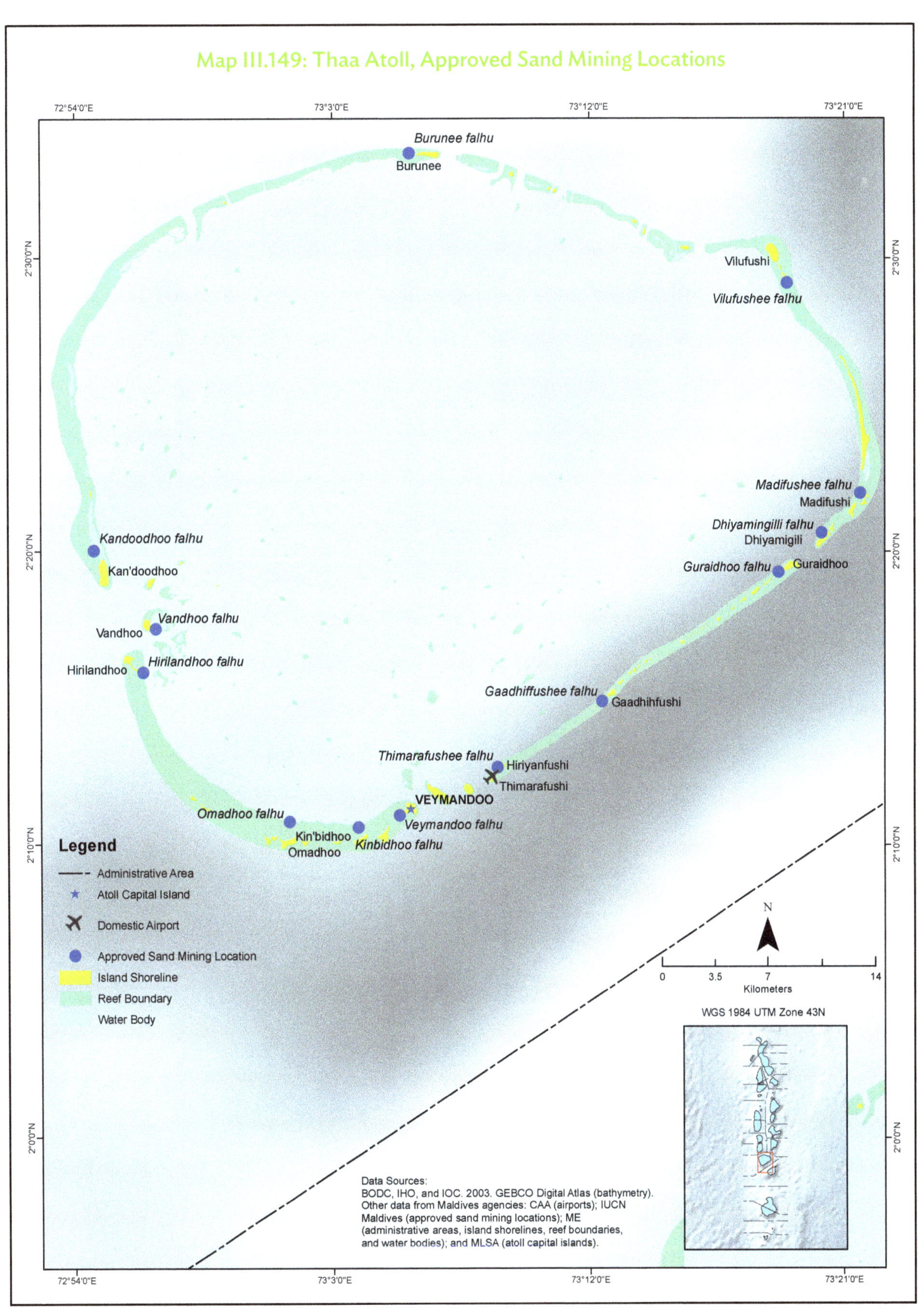

Map III.150: Vaavu Atoll, Approved Sand Mining Locations

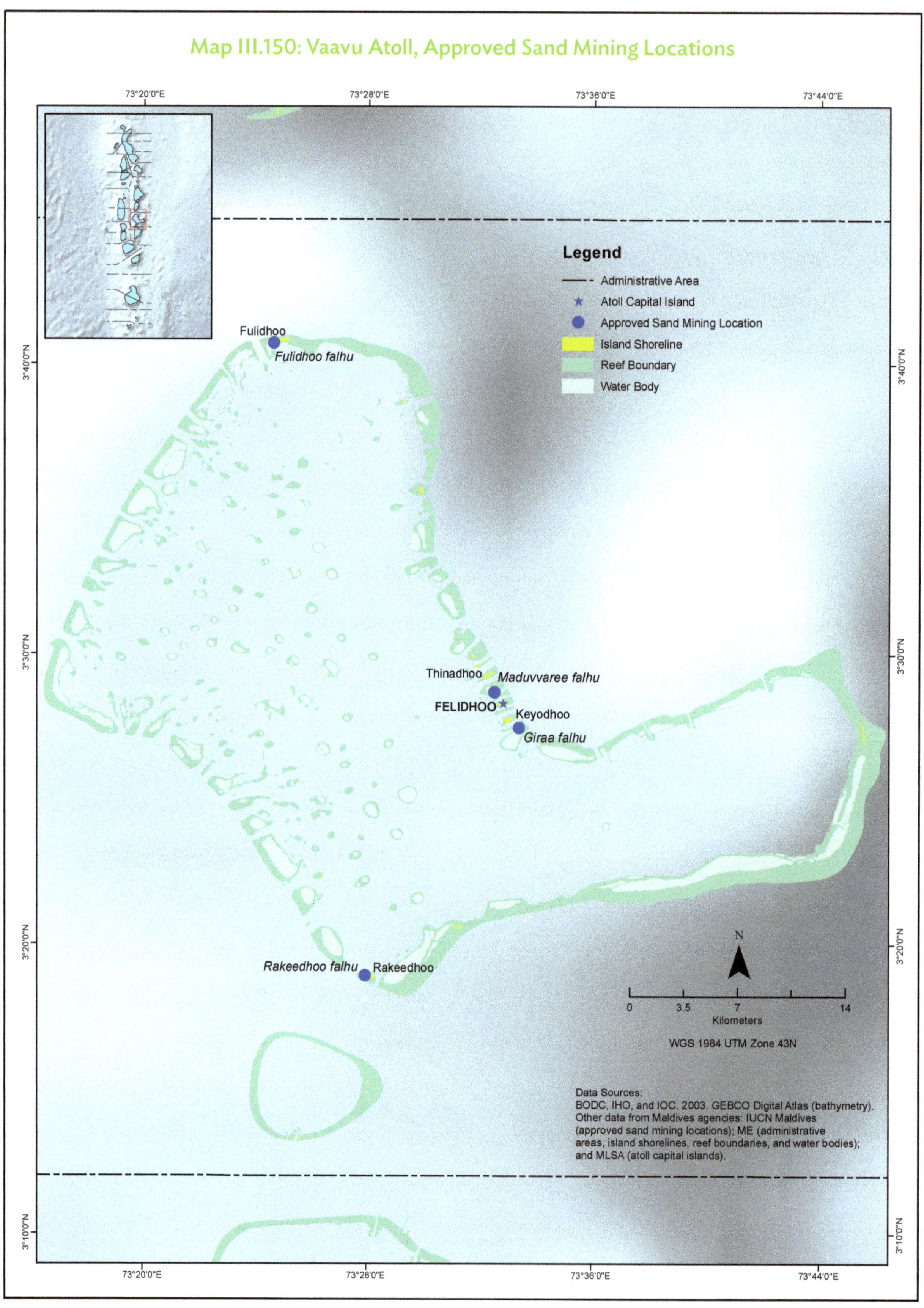

Power Stations

All the inhabited islands of Maldives have access to continuous electric supply. Two major electric supply companies—Fenaka Corporation Limited and State Electric Company Limited—provide electricity to the islands. The largest electric demand comes from Malé City.

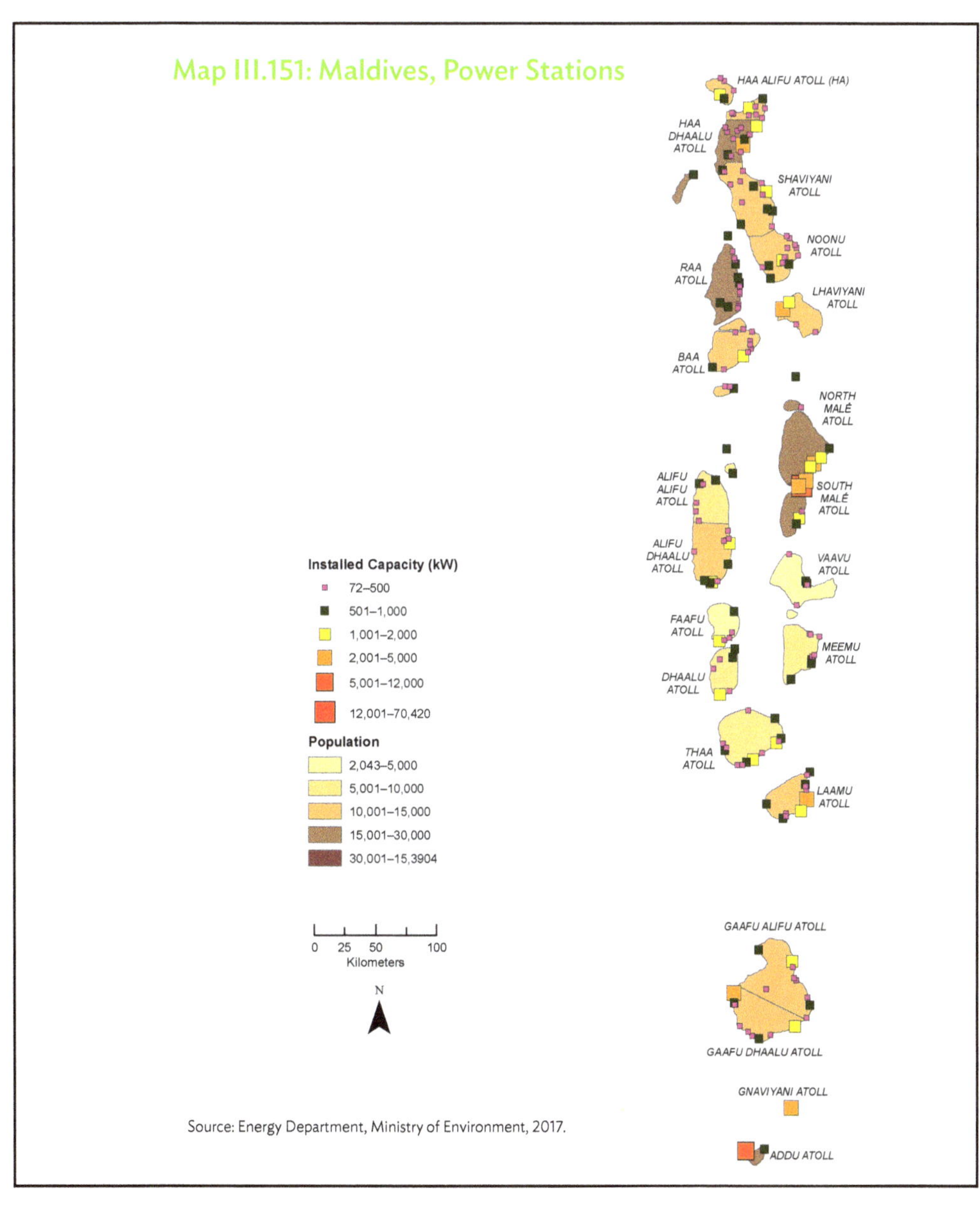

Map III.151: Maldives, Power Stations

Source: Energy Department, Ministry of Environment, 2017.

Map III.152: Addu City, Power Stations

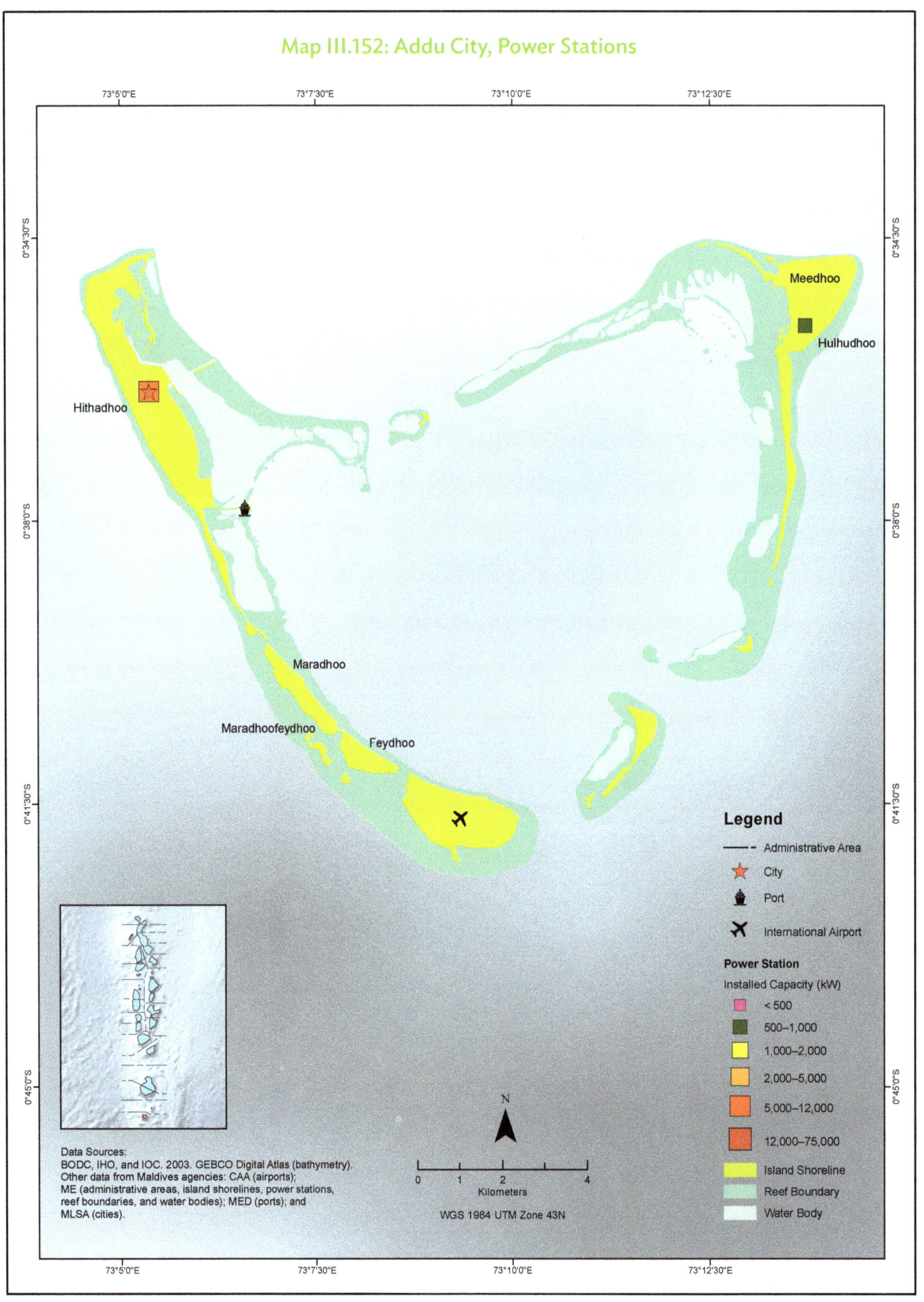

Map III.153: Alifu Alifu Atoll, Power Stations

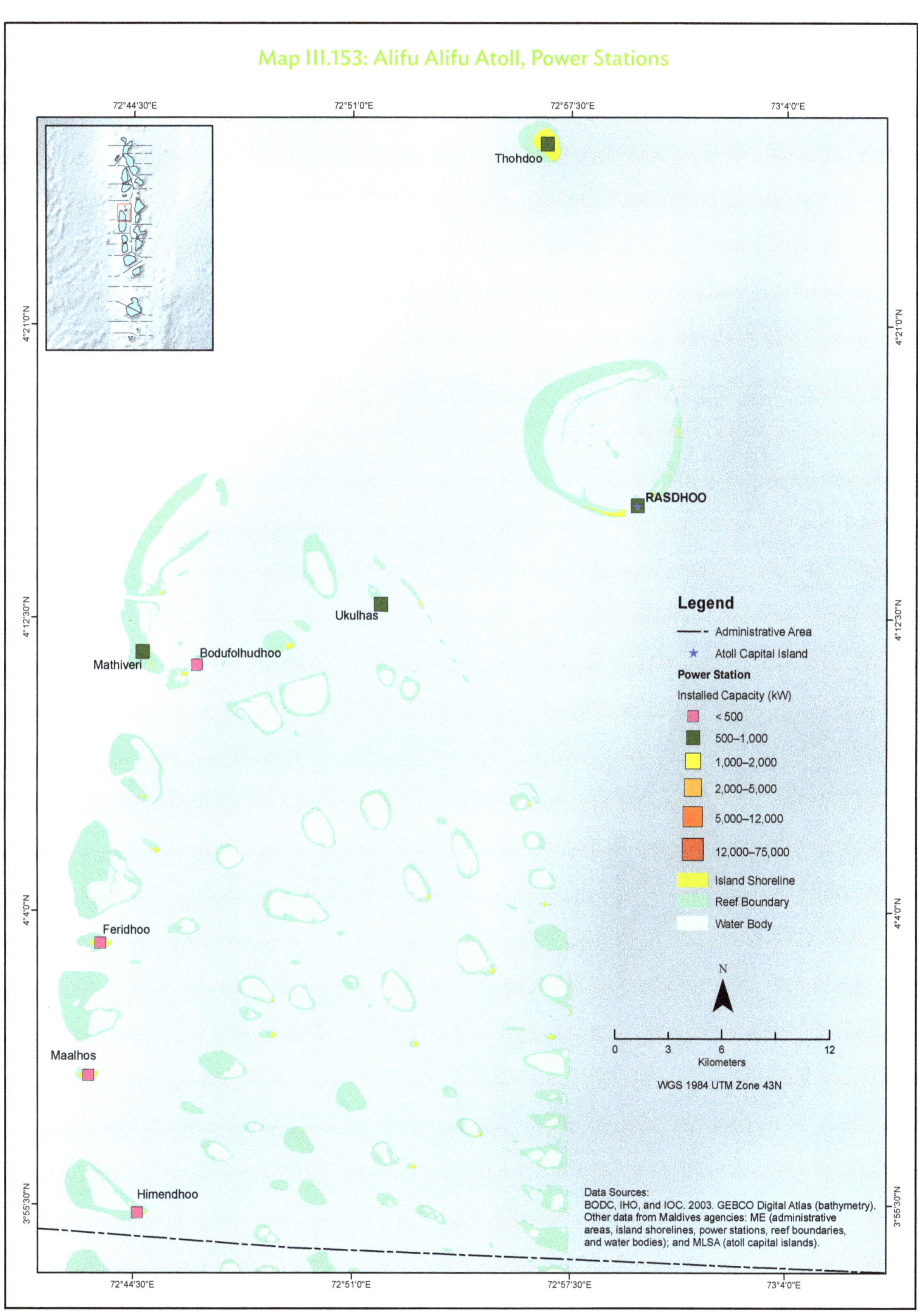

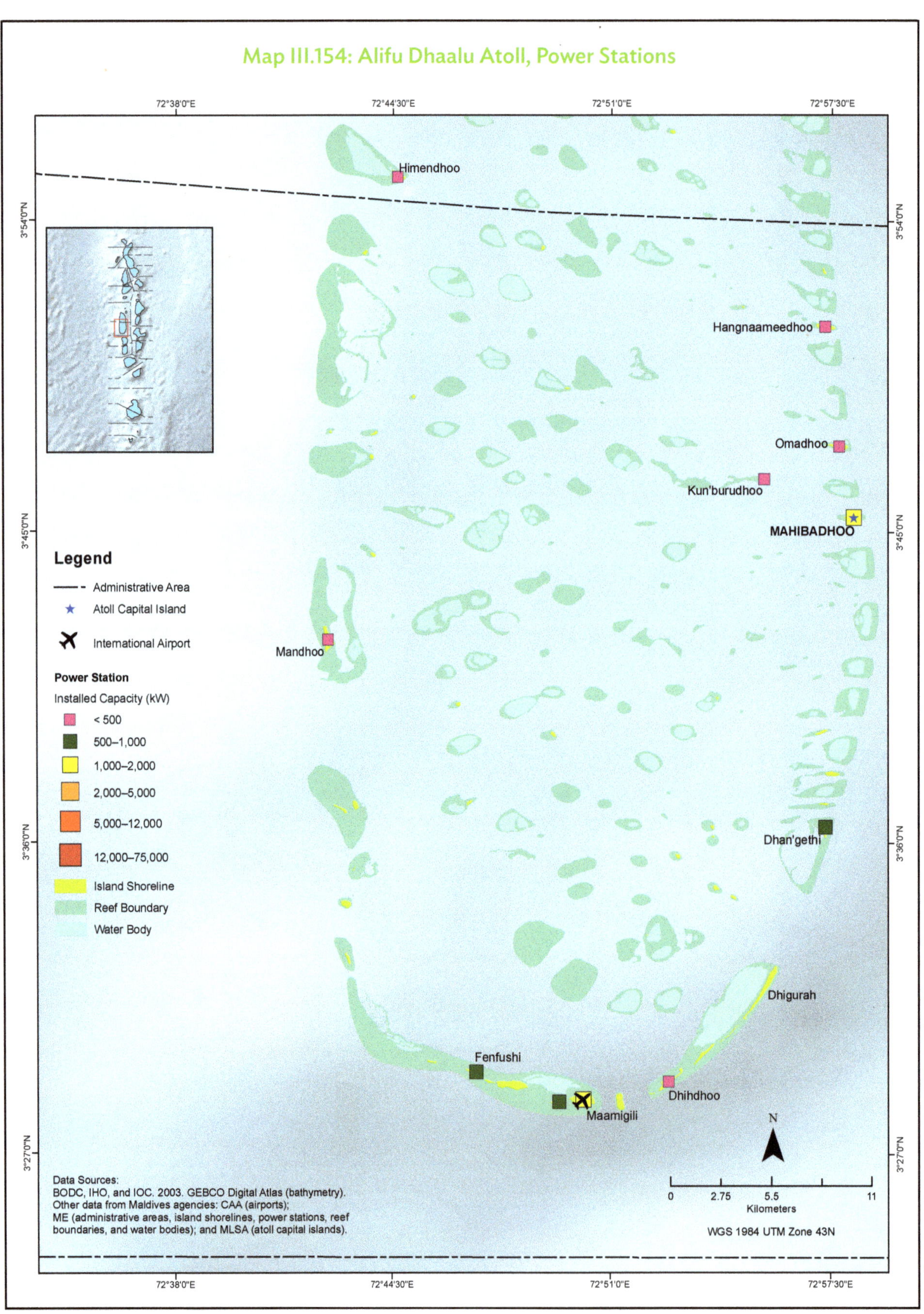
Map III.154: Alifu Dhaalu Atoll, Power Stations
72°38'0"E
72°44'30"E
72°51'0"E
72°57'30"E
3°54'0"N
3°45'0"N
3°36'0"N
3°27'0"N
Himendhoo
Hangnaameedhoo
Omadhoo
Kun'burudhoo
MAHIBADHOO
Mandhoo
Dhan'gethi
Dhigurah
Fenfushi
Dhihdhoo
Maamigili
Legend
Administrative Area
Atoll Capital Island
International Airport
Power Station
Installed Capacity (kW)
< 500
500–1,000
1,000–2,000
2,000–5,000
5,000–12,000
12,000–75,000
Island Shoreline
Reef Boundary
Water Body
N
0 2.75 5.5 11
Kilometers
WGS 1984 UTM Zone 43N
Data Sources:
BODC, IHO, and IOC. 2003. GEBCO Digital Atlas (bathymetry).
Other data from Maldives agencies: CAA (airports);
ME (administrative areas, island shorelines, power stations, reef boundaries, and water bodies); and MLSA (atoll capital islands).

Map III.155: Baa Atoll, Power Stations

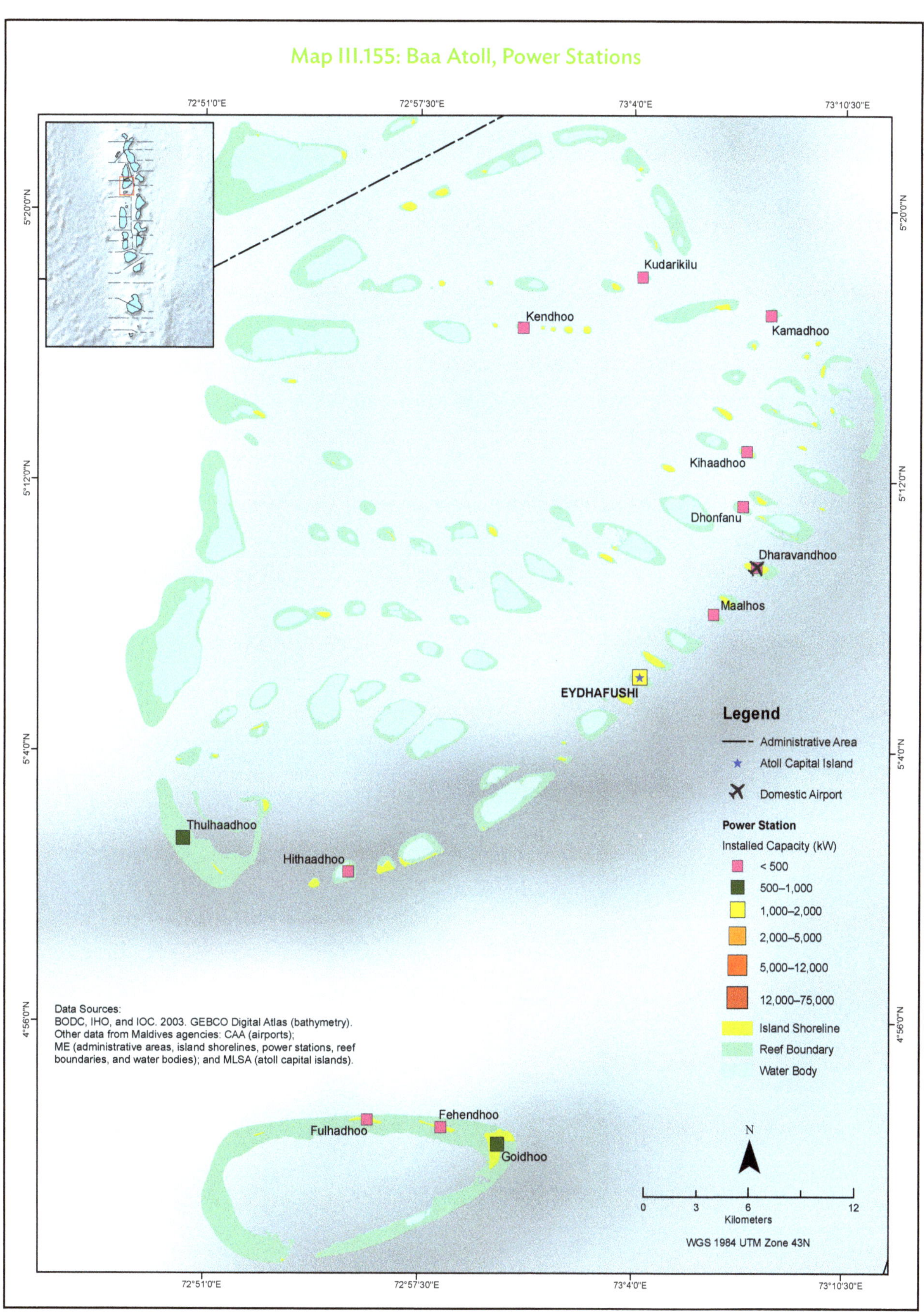

Map III.156: Dhaalu Atoll, Power Stations

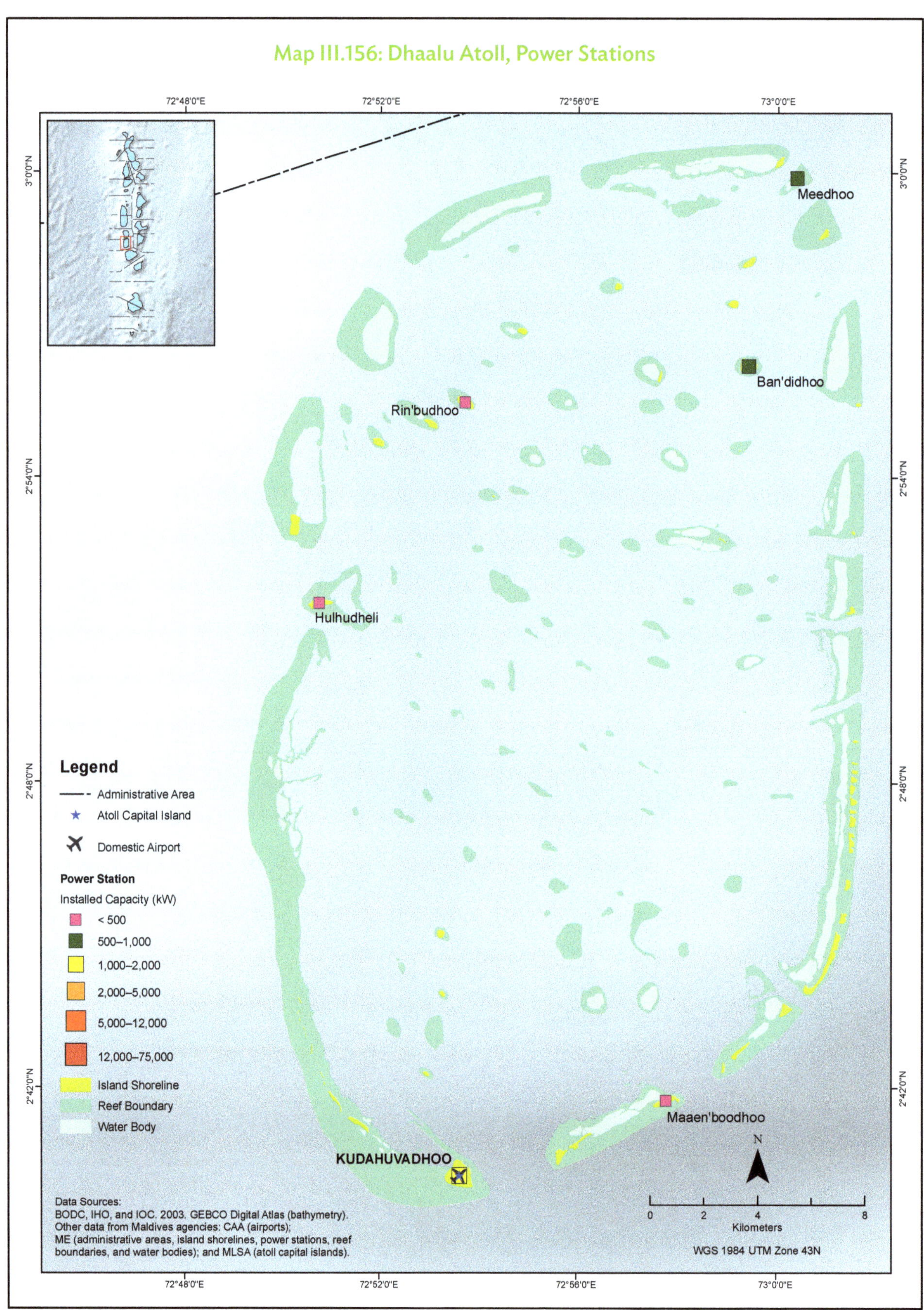

Map III.157: Faafu Atoll, Power Stations

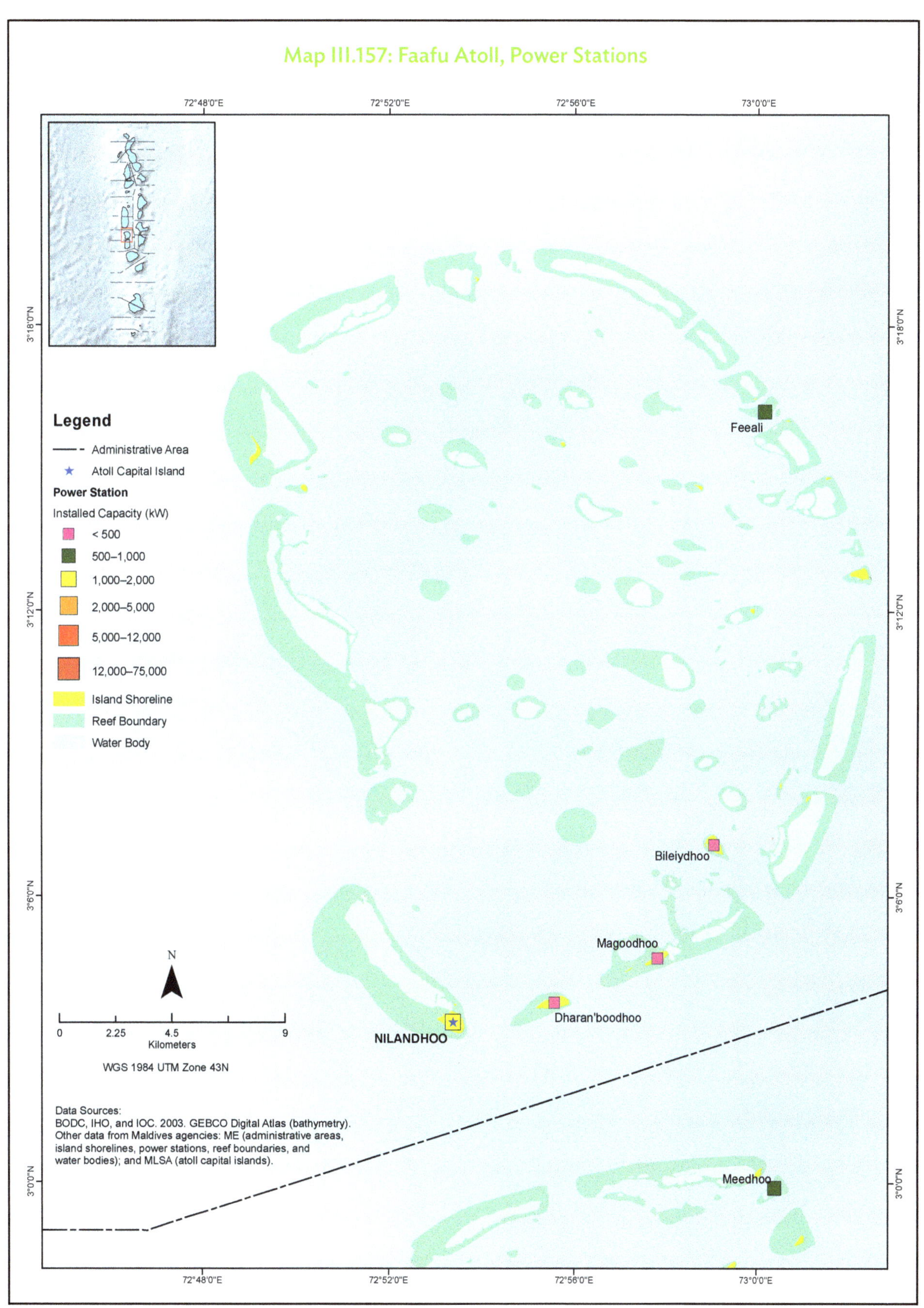

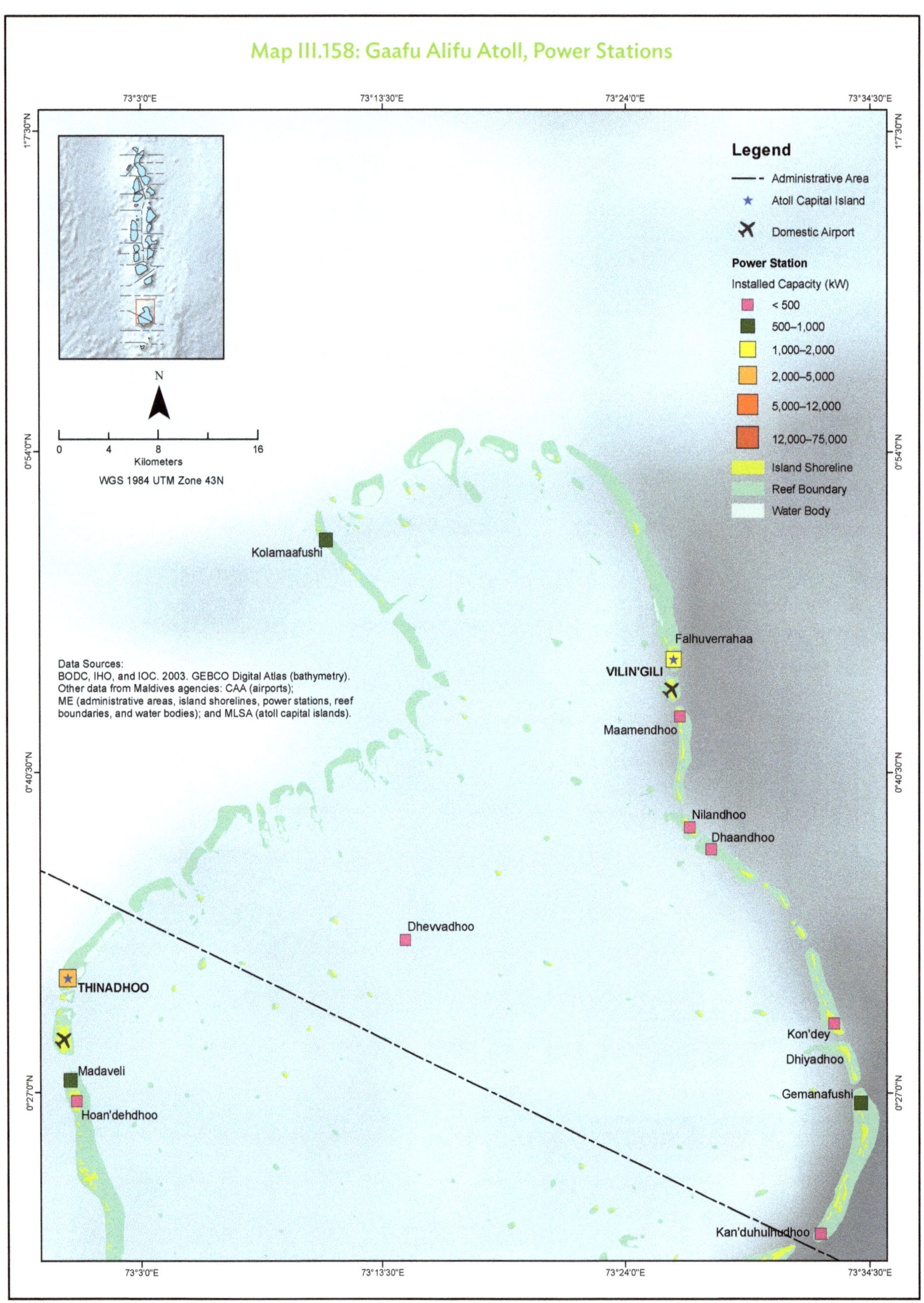
Map III.158: Gaafu Alifu Atoll, Power Stations
73°3'0"E
73°13'30"E
73°24'0"E
73°34'30"E
1°7'30"N
0°54'0"N
0°40'30"N
0°27'0"N
Legend
Administrative Area
Atoll Capital Island
Domestic Airport
Power Station
Installed Capacity (kW)
< 500
500–1,000
1,000–2,000
2,000–5,000
5,000–12,000
12,000–75,000
Island Shoreline
Reef Boundary
Water Body
N
0 4 8 16
Kilometers
WGS 1984 UTM Zone 43N
Data Sources:
BODC, IHO, and IOC. 2003. GEBCO Digital Atlas (bathymetry).
Other data from Maldives agencies: CAA (airports);
ME (administrative areas, island shorelines, power stations, reef boundaries, and water bodies); and MLSA (atoll capital islands).
Kolamaafushi
Falhuverrahaa
VILIN'GILI
Maamendhoo
Nilandhoo
Dhaandhoo
Dhevvadhoo
THINADHOO
Madaveli
Hoan'dehdhoo
Kon'dey
Dhiyadhoo
Gemanafushi
Kan'duhulhudhoo

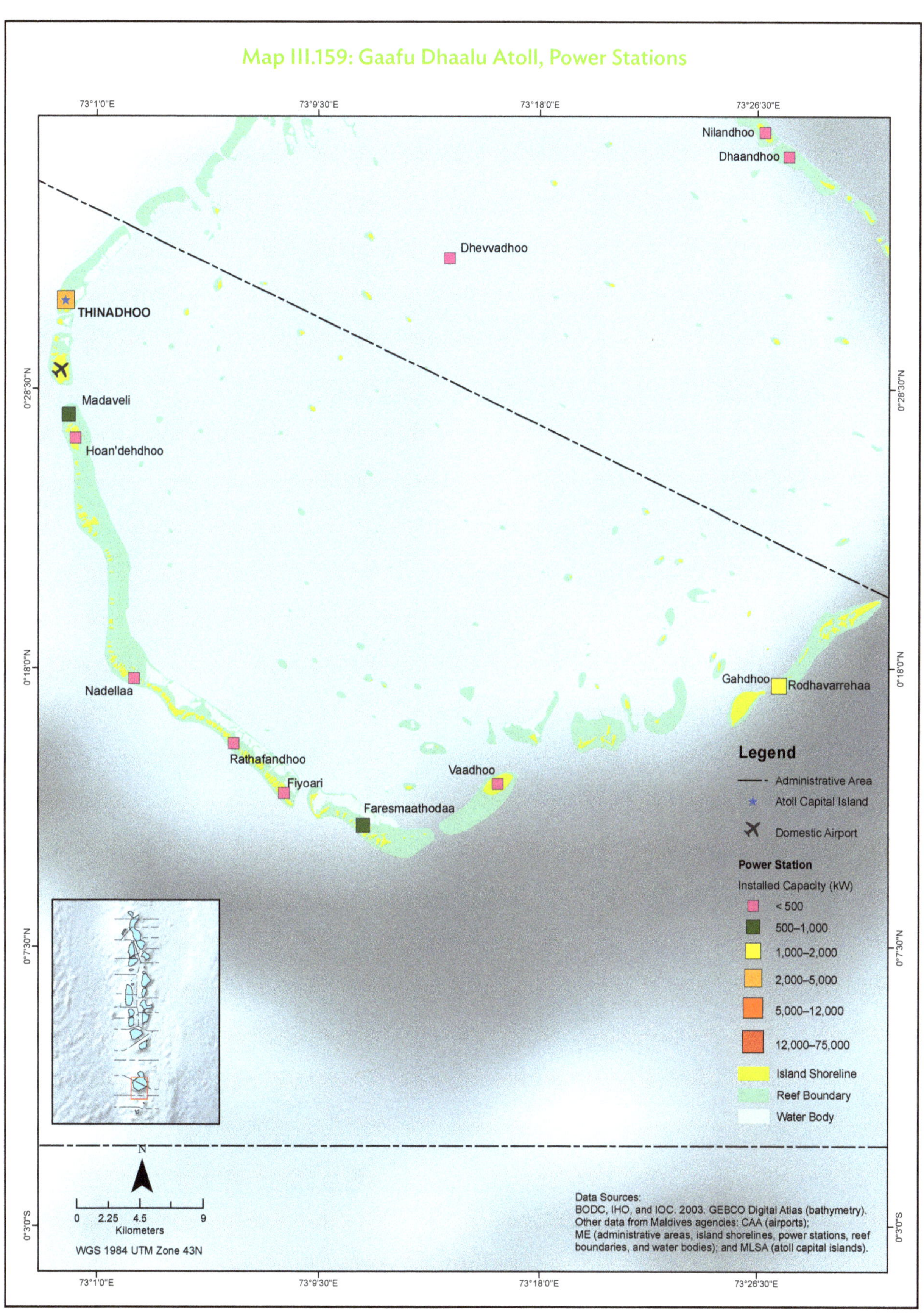
Map III.159: Gaafu Dhaalu Atoll, Power Stations
73°1'0"E
73°9'30"E
73°18'0"E
73°26'30"E
0°28'30"N
0°18'0"N
0°7'30"N
0°3'0"S
Nilandhoo
Dhaandhoo
Dhevvadhoo
THINADHOO
Madaveli
Hoan'dehdhoo
Nadellaa
Rathafandhoo
Fiyoari
Faresmaathodaa
Vaadhoo
Gahdhoo
Rodhavarrehaa
Legend
Administrative Area
Atoll Capital Island
Domestic Airport
Power Station
Installed Capacity (kW)
< 500
500–1,000
1,000–2,000
2,000–5,000
5,000–12,000
12,000–75,000
Island Shoreline
Reef Boundary
Water Body
N
0 2.25 4.5 9
Kilometers
WGS 1984 UTM Zone 43N
Data Sources:
BODC, IHO, and IOC. 2003. GEBCO Digital Atlas (bathymetry).
Other data from Maldives agencies: CAA (airports);
ME (administrative areas, island shorelines, power stations, reef boundaries, and water bodies); and MLSA (atoll capital islands).

Map III.160: Gnaviyani Atoll, Power Stations

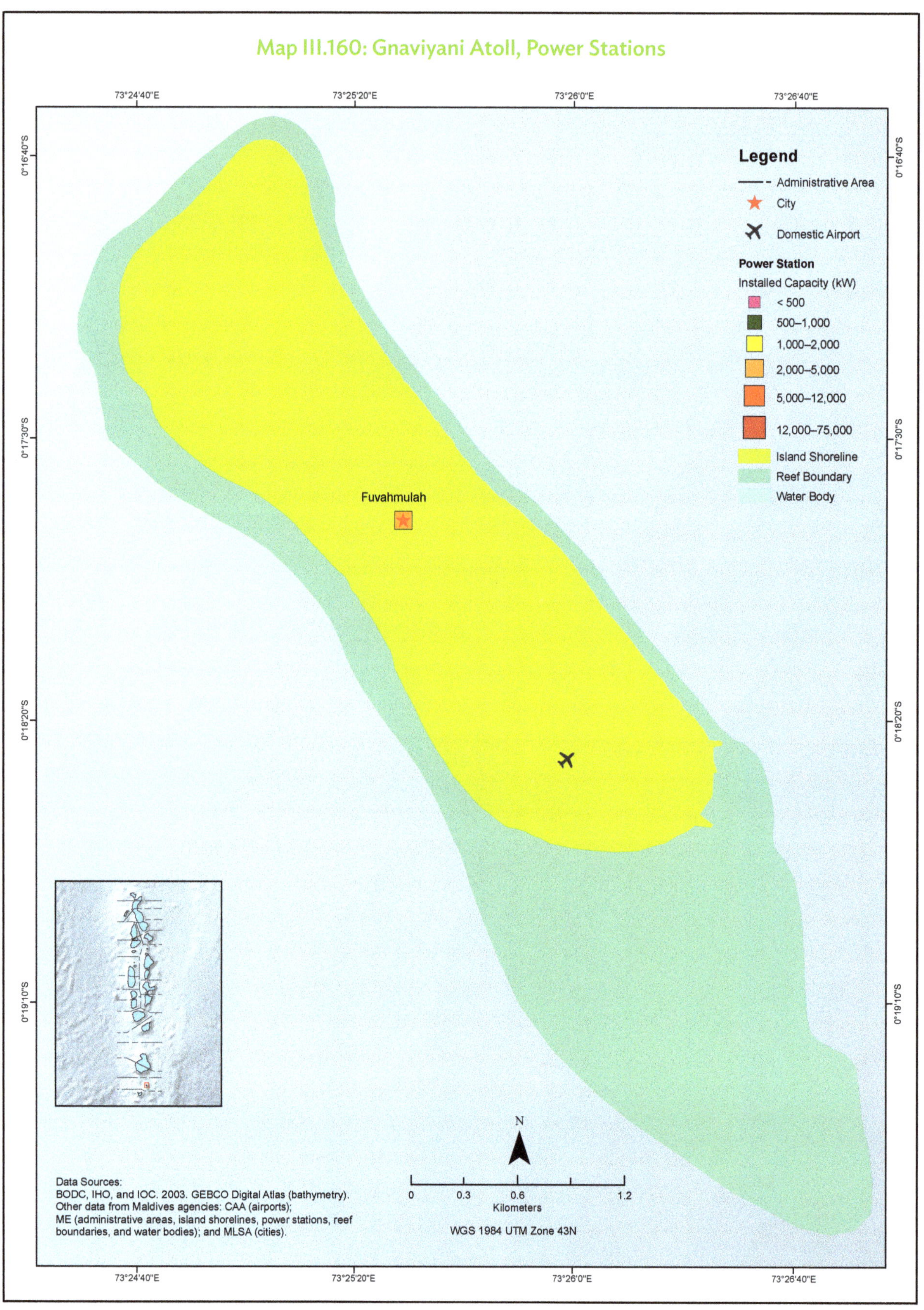

Map III.161: Haa Alifu Atoll, Power Stations

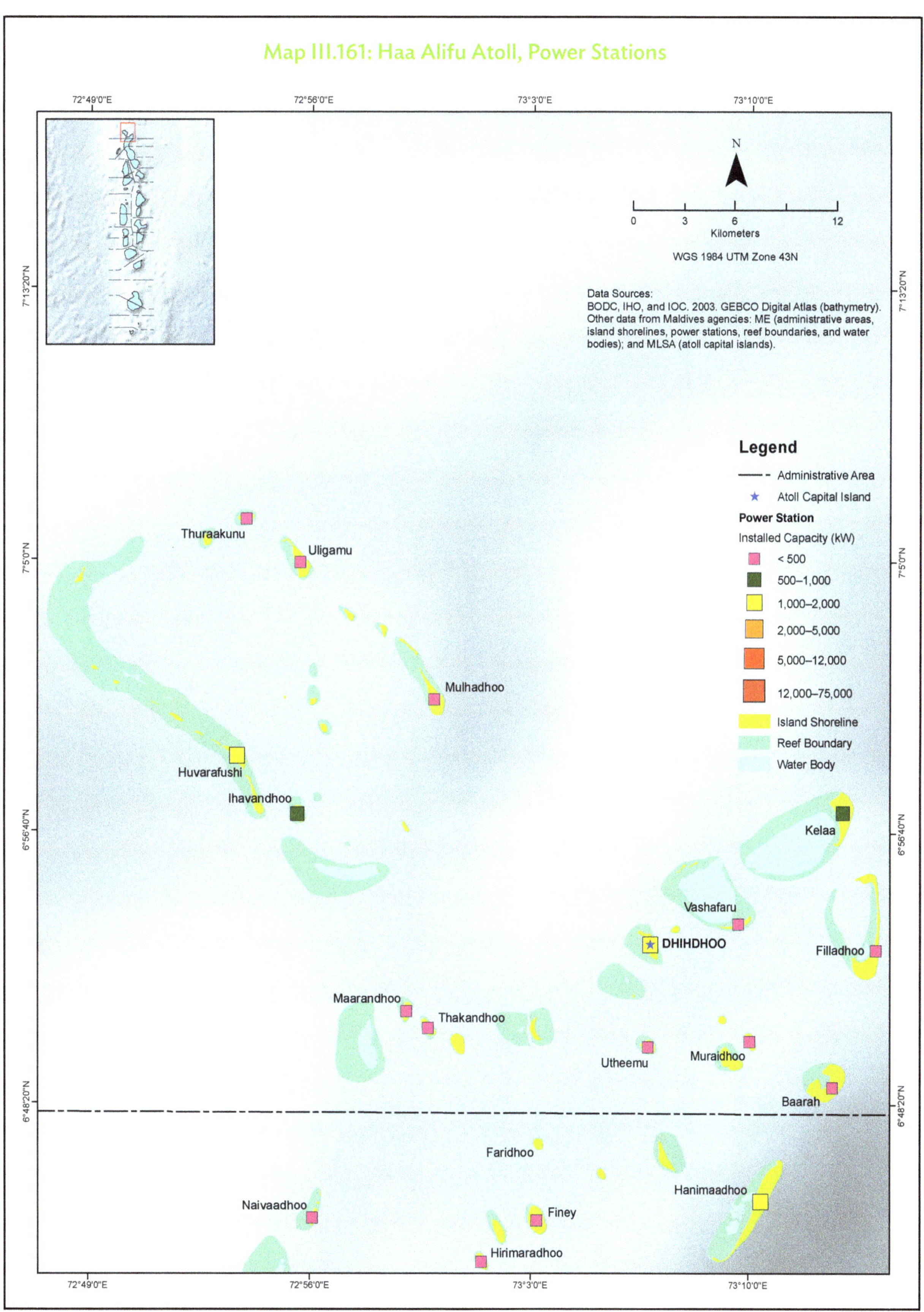

Map III.162: Haa Dhaalu Atoll, Power Stations

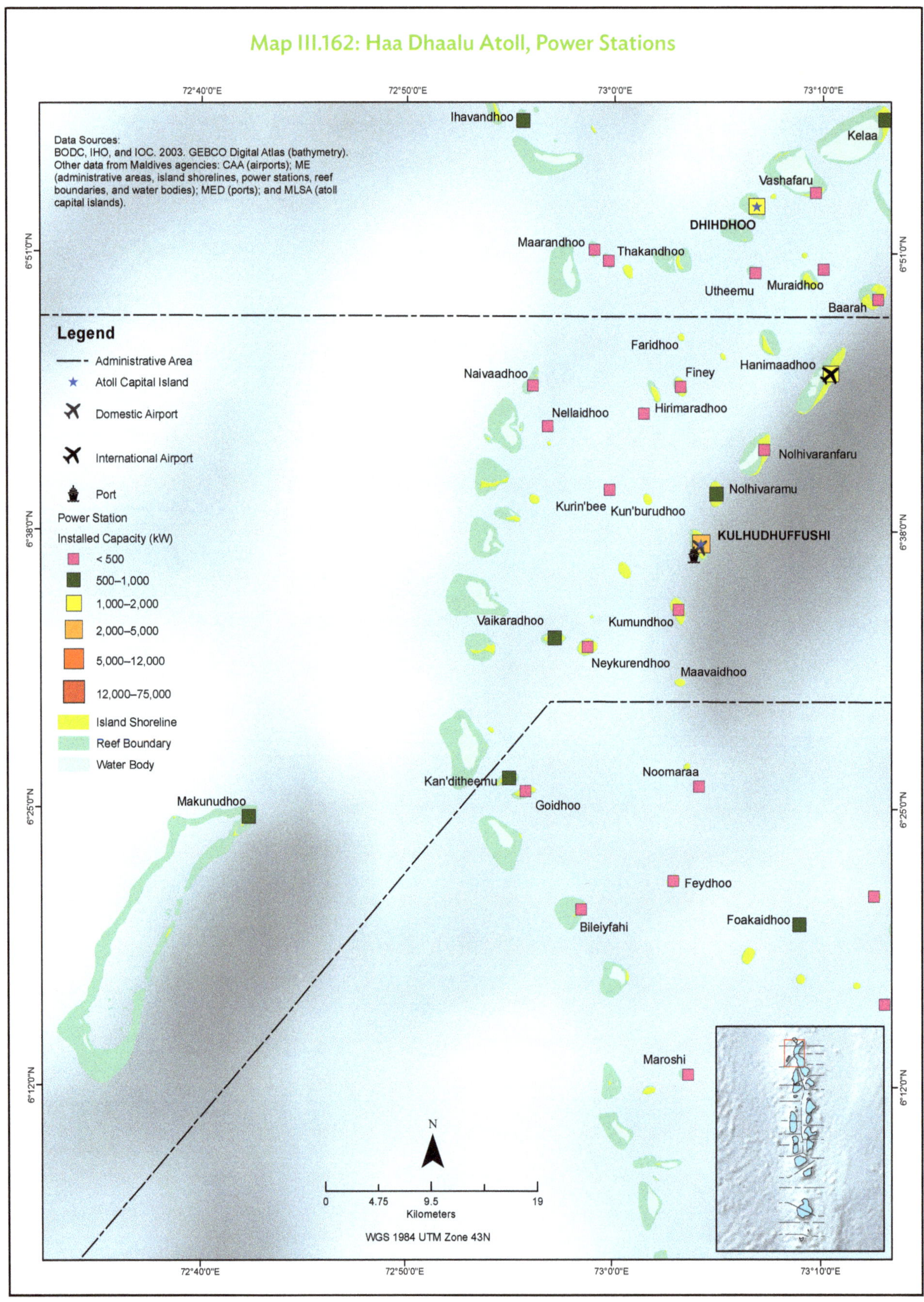

Map III.163: Laamu Atoll, Power Stations

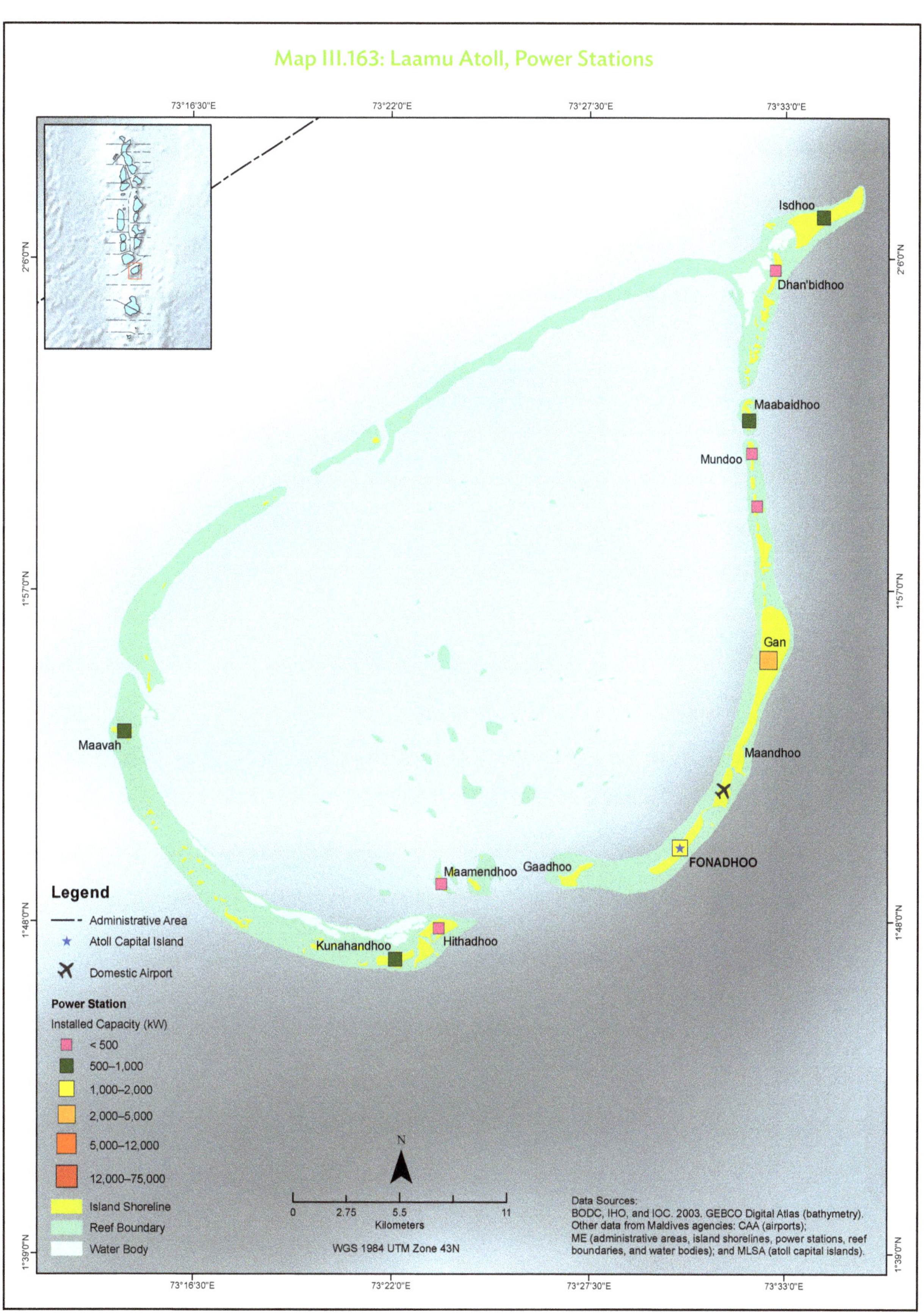

Map III.164: Lhaviyani Atoll, Power Stations

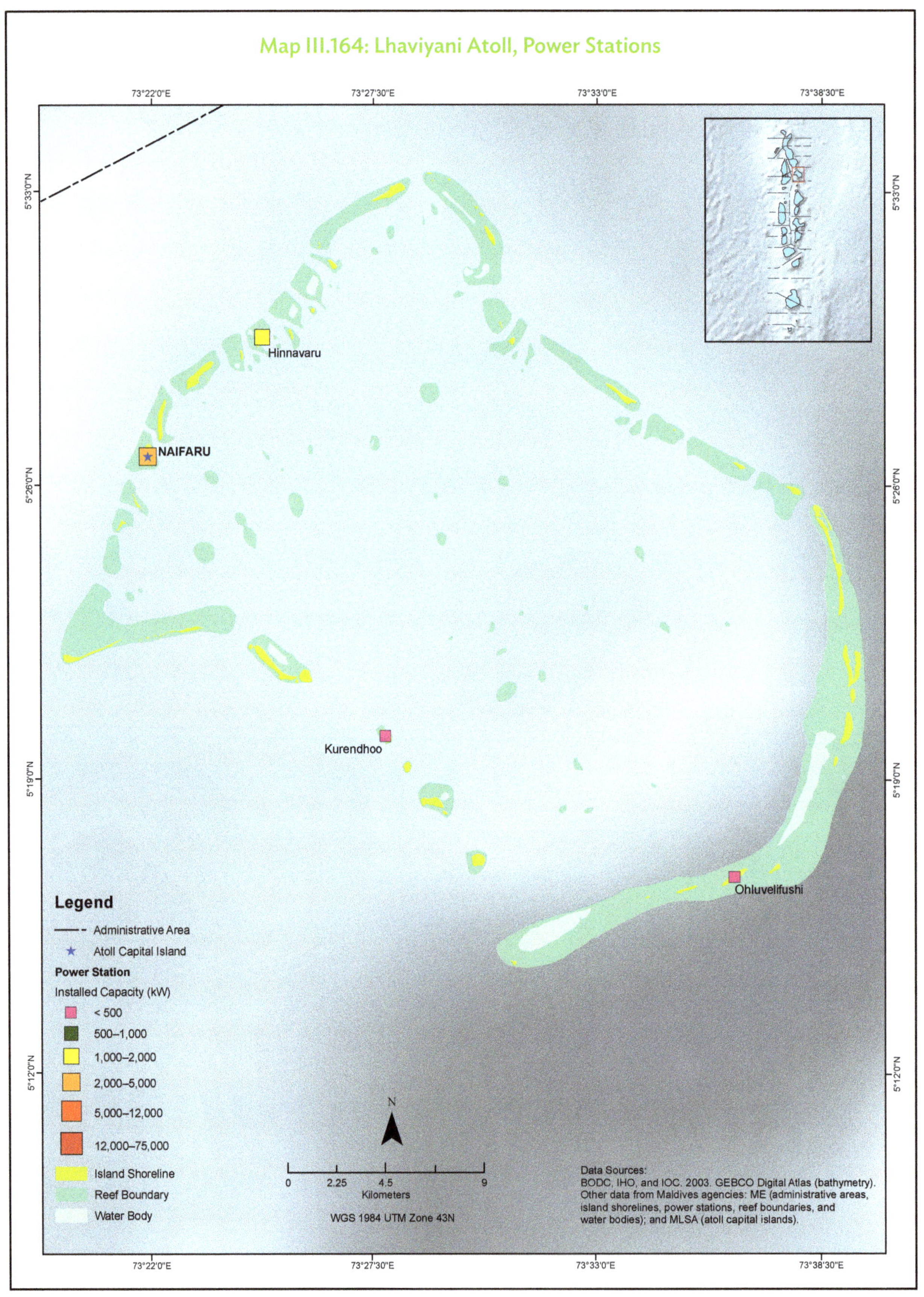

Map III.165: Meemu Atoll, Power Stations

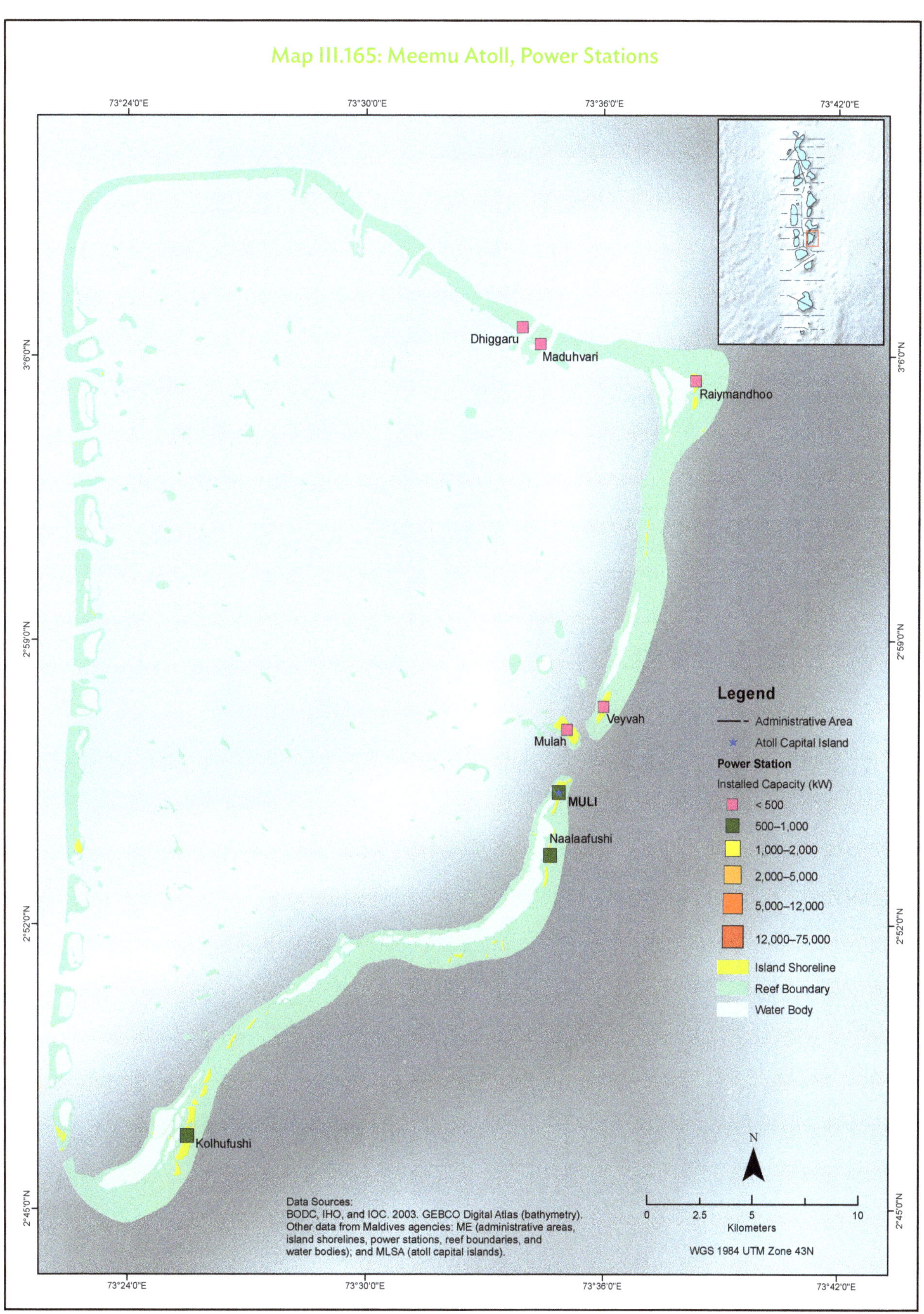

Map III.166: Noonu Atoll, Power Stations

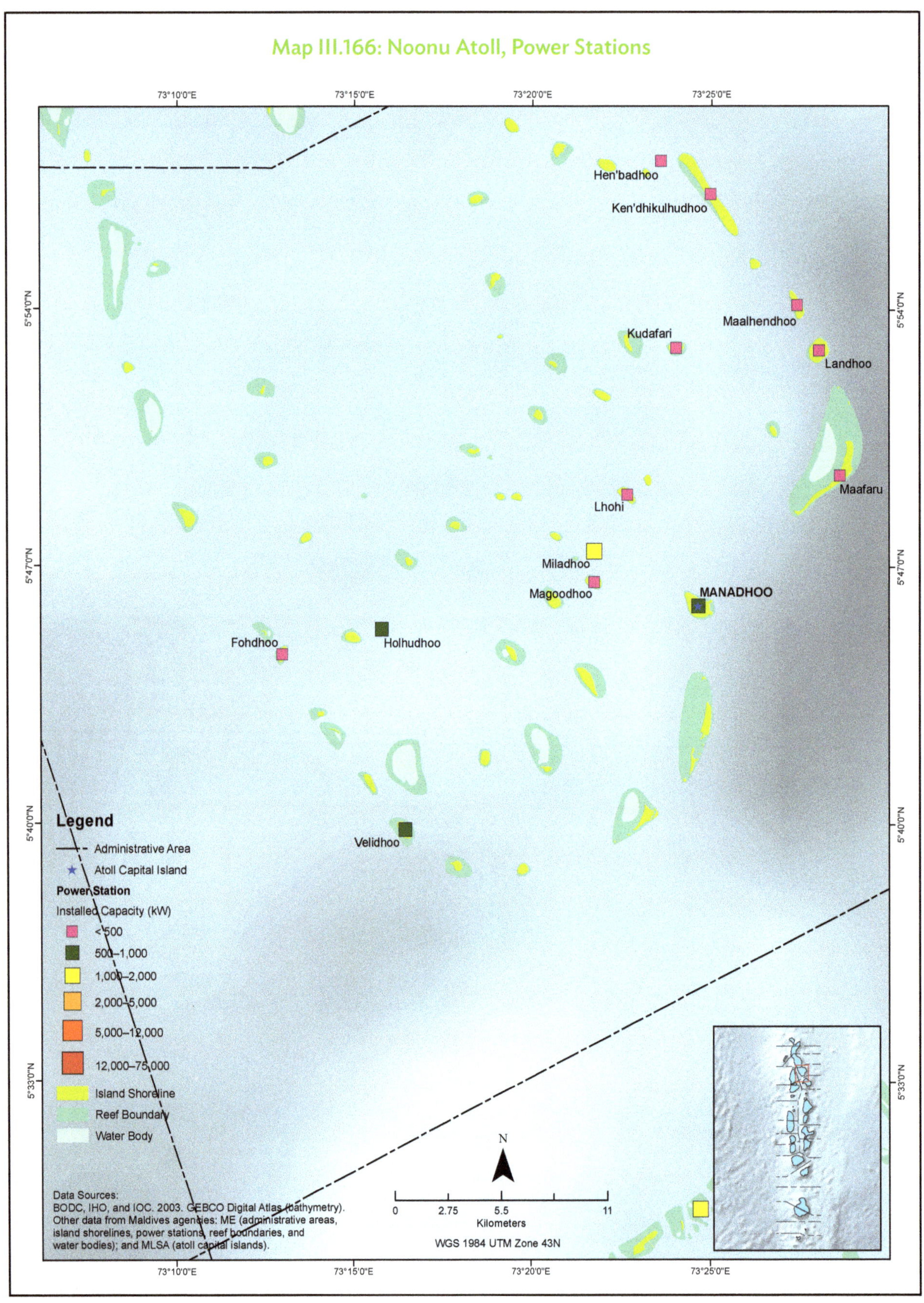

Map III.167: North Malé Atoll, Power Stations

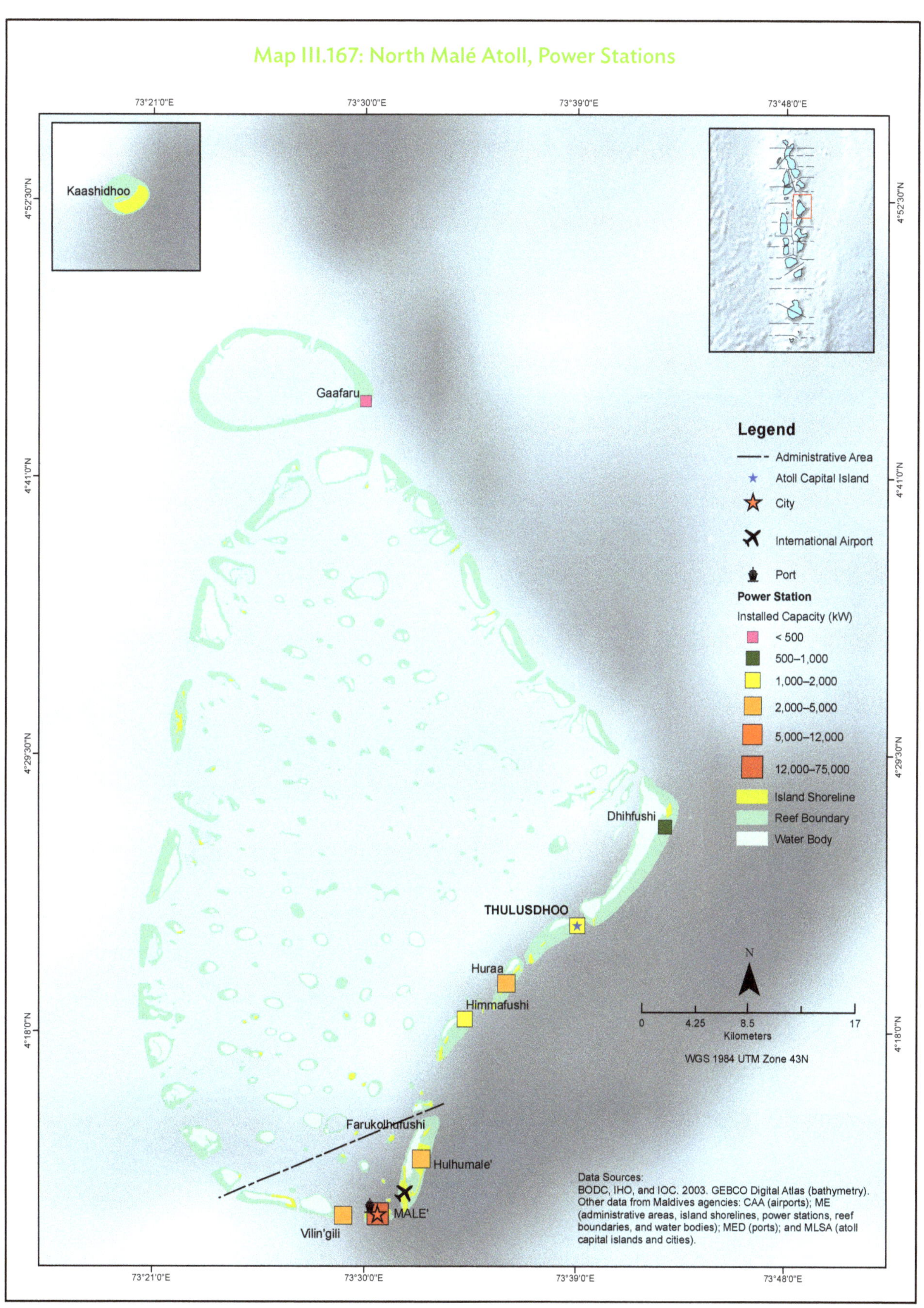

Map III.168: Raa Atoll, Power Stations

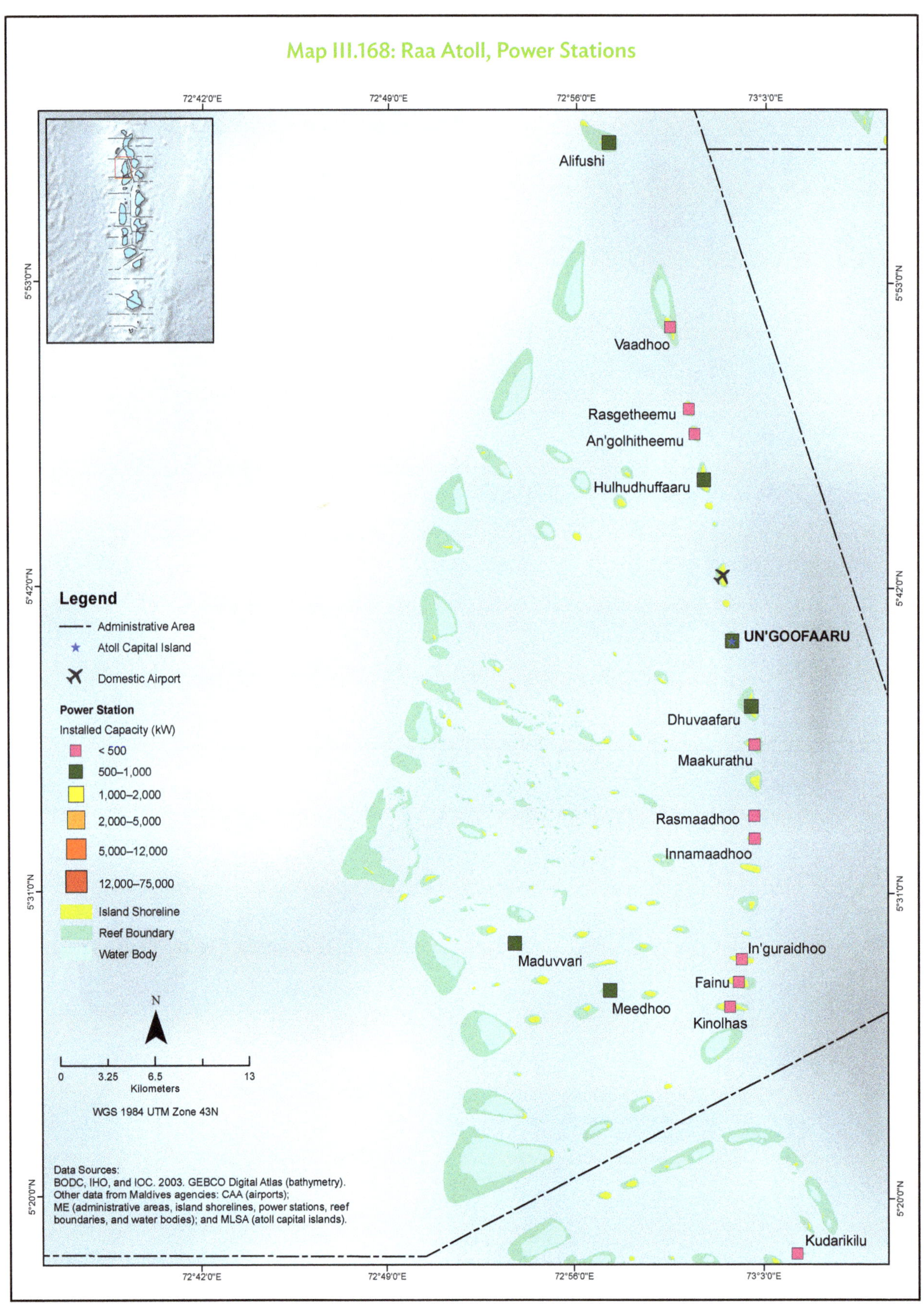

Map III.169: Shaviyani Atoll, Power Stations

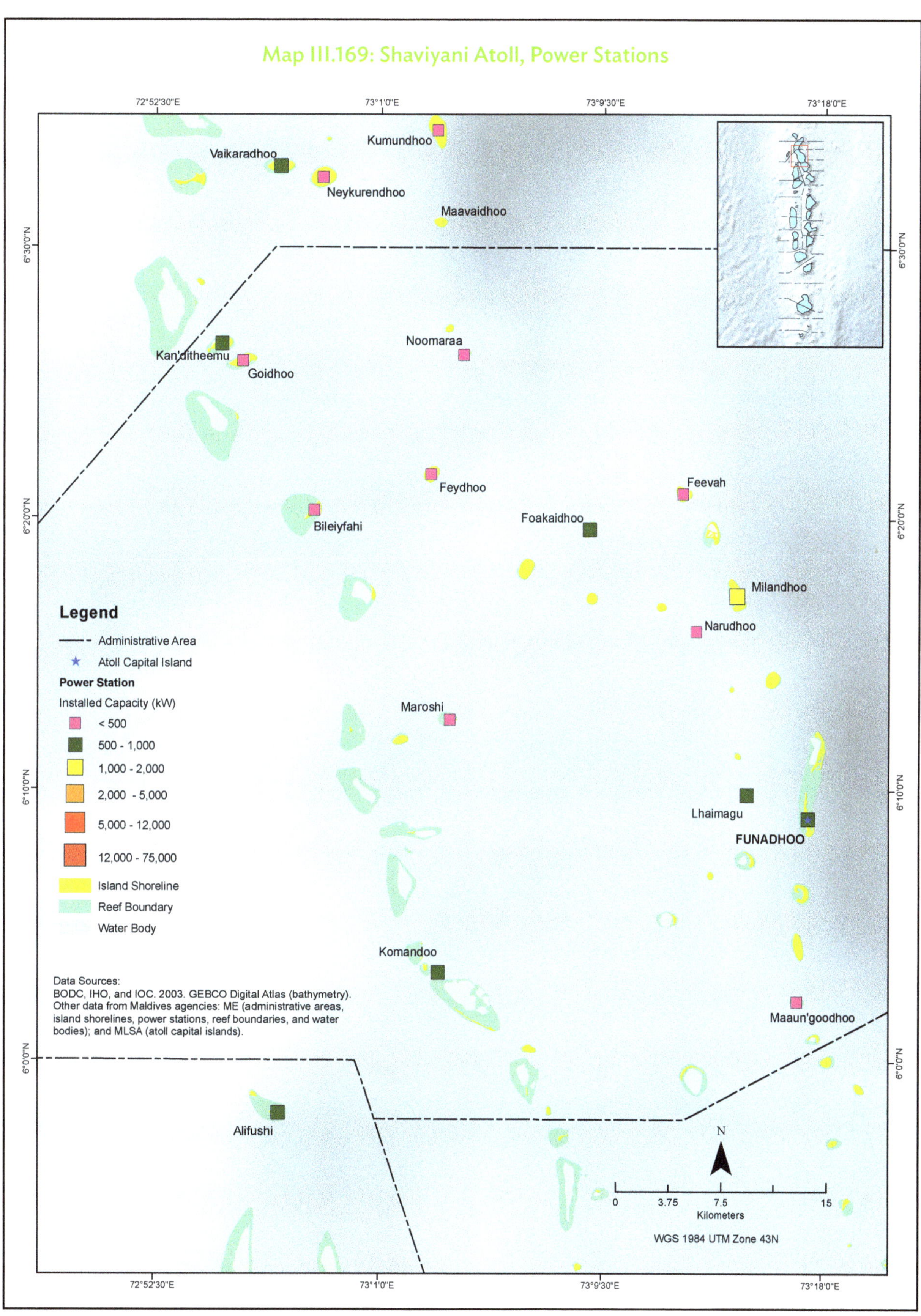

Map III.170: South Malé Atoll, Power Stations

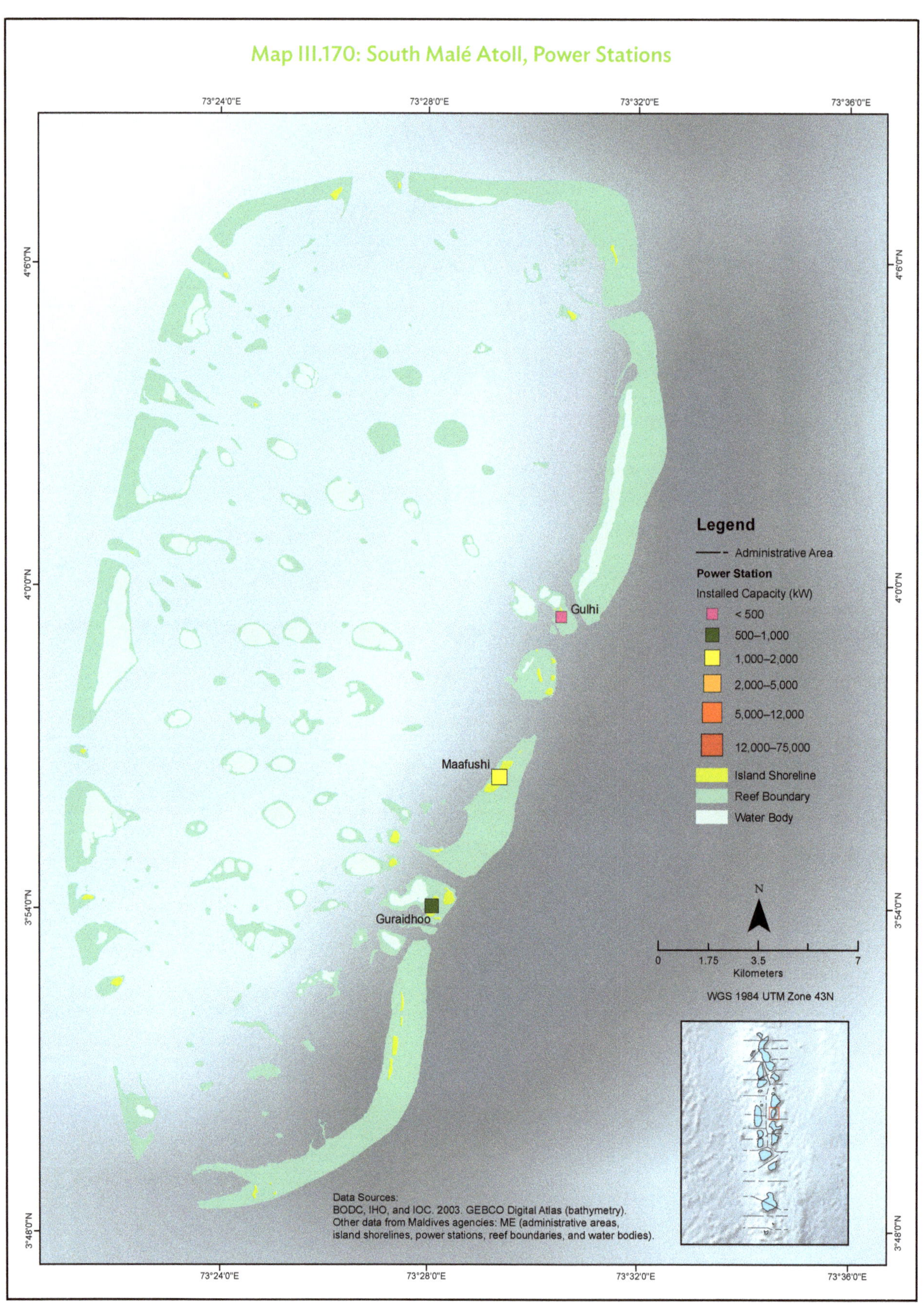

Map III.171: Thaa Atoll, Power Stations

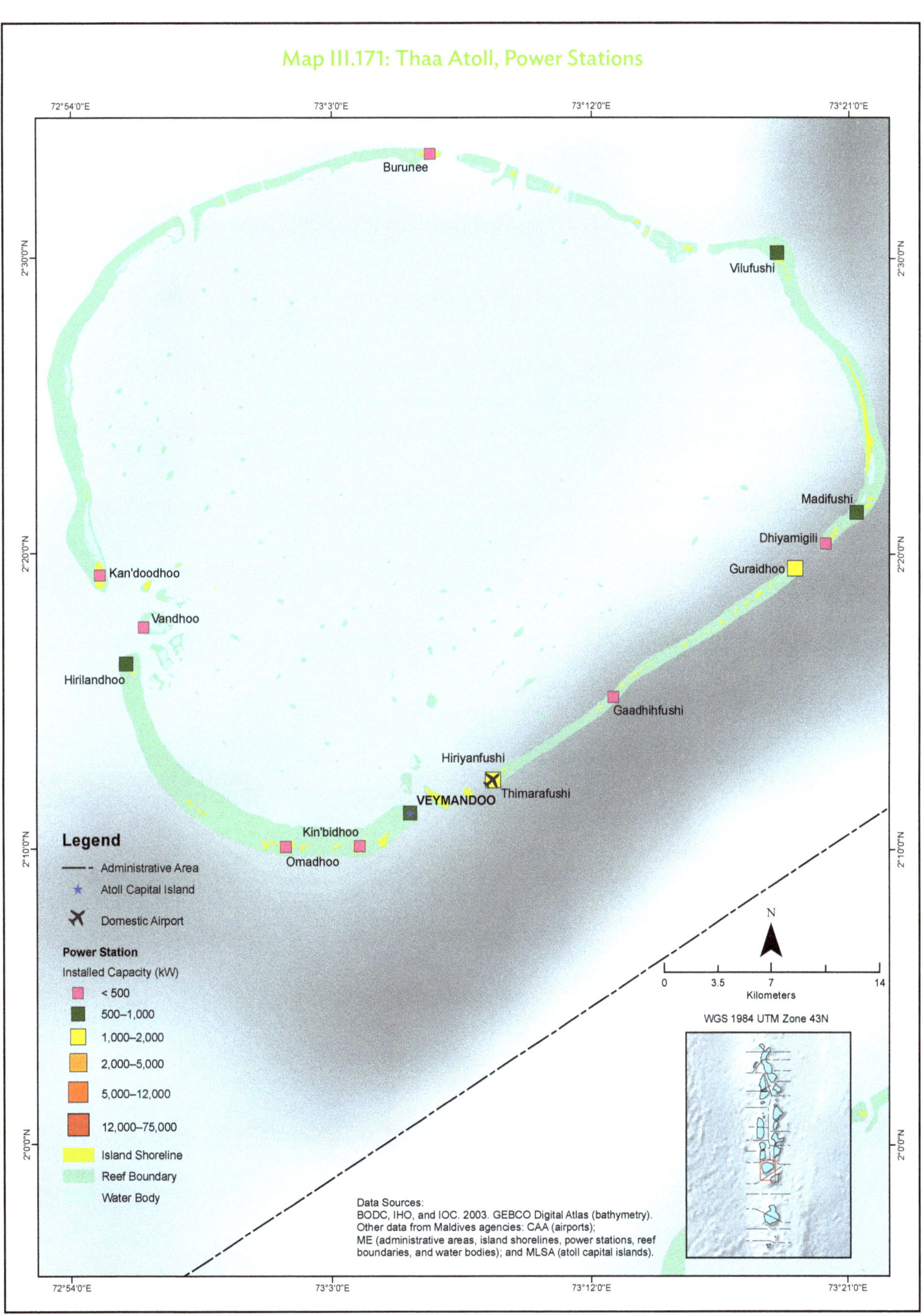

Map III.172: Vaavu Atoll, Power Stations

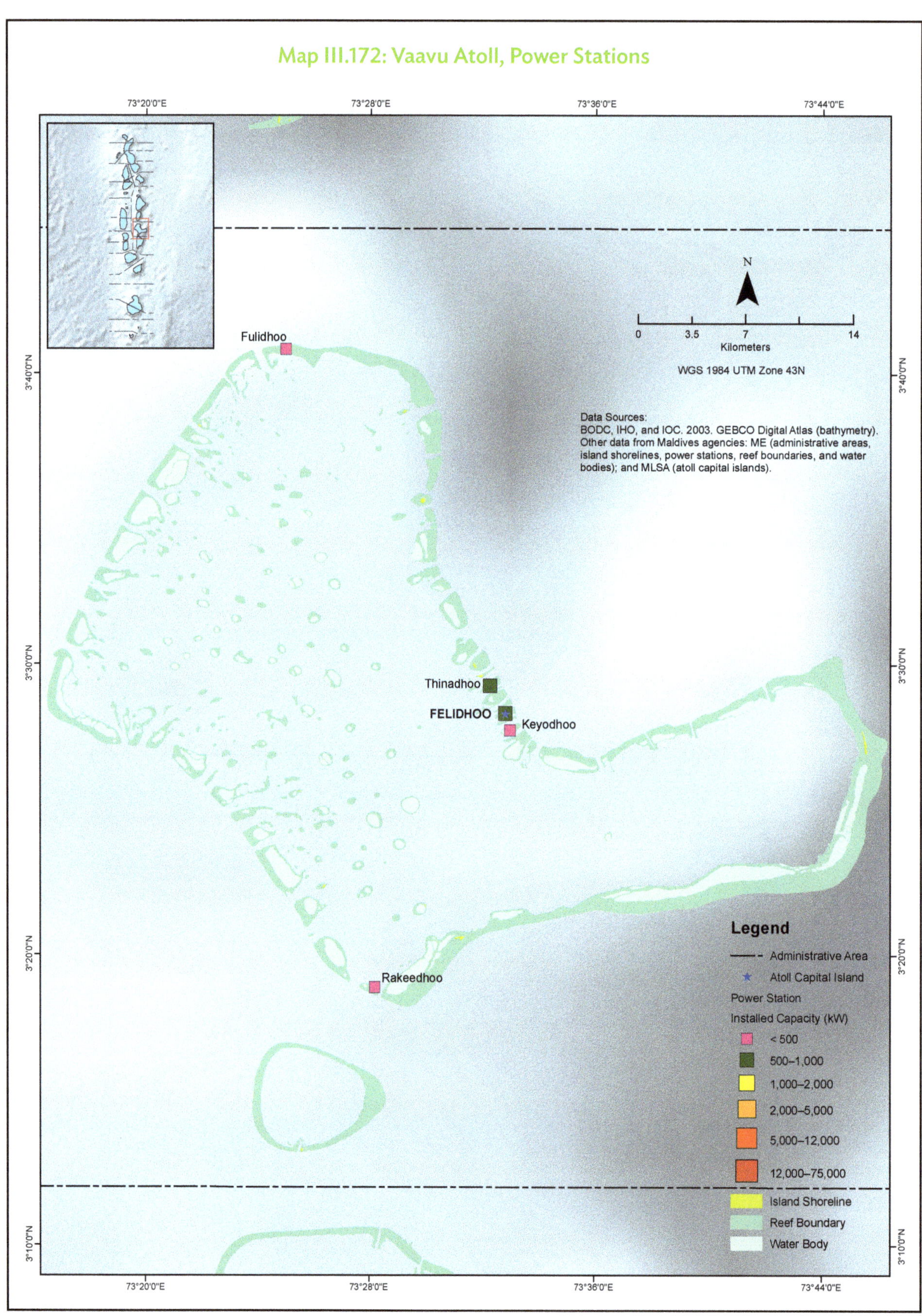

Human Development Index

The Human Development Index is a summary measure developed by the United Nations Development Programme to assess national development based on three basic dimensions of human development: a long and healthy life, access to knowledge, and a decent standard of living (UNDP 2018). In Maldives, ratings are high for Faafu, Dhaalu, and Meemu atolls as well as Malé City.

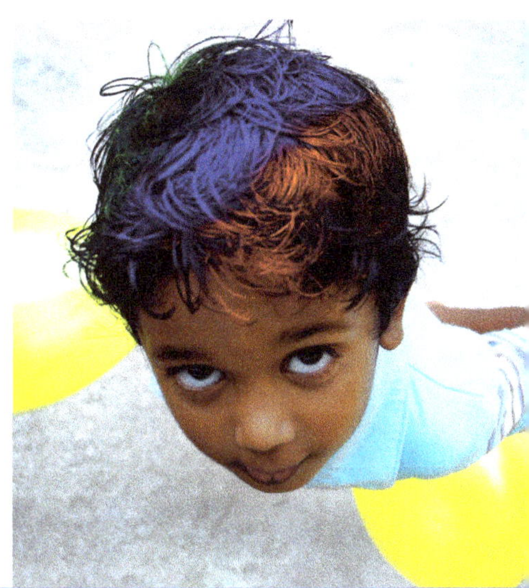

Maldivians on a sunny day at the beach. Human development include activities and interactions with other people and relationship with the environment (photos by Sue Todd).

Map III.173: Maldives, Human Development Index

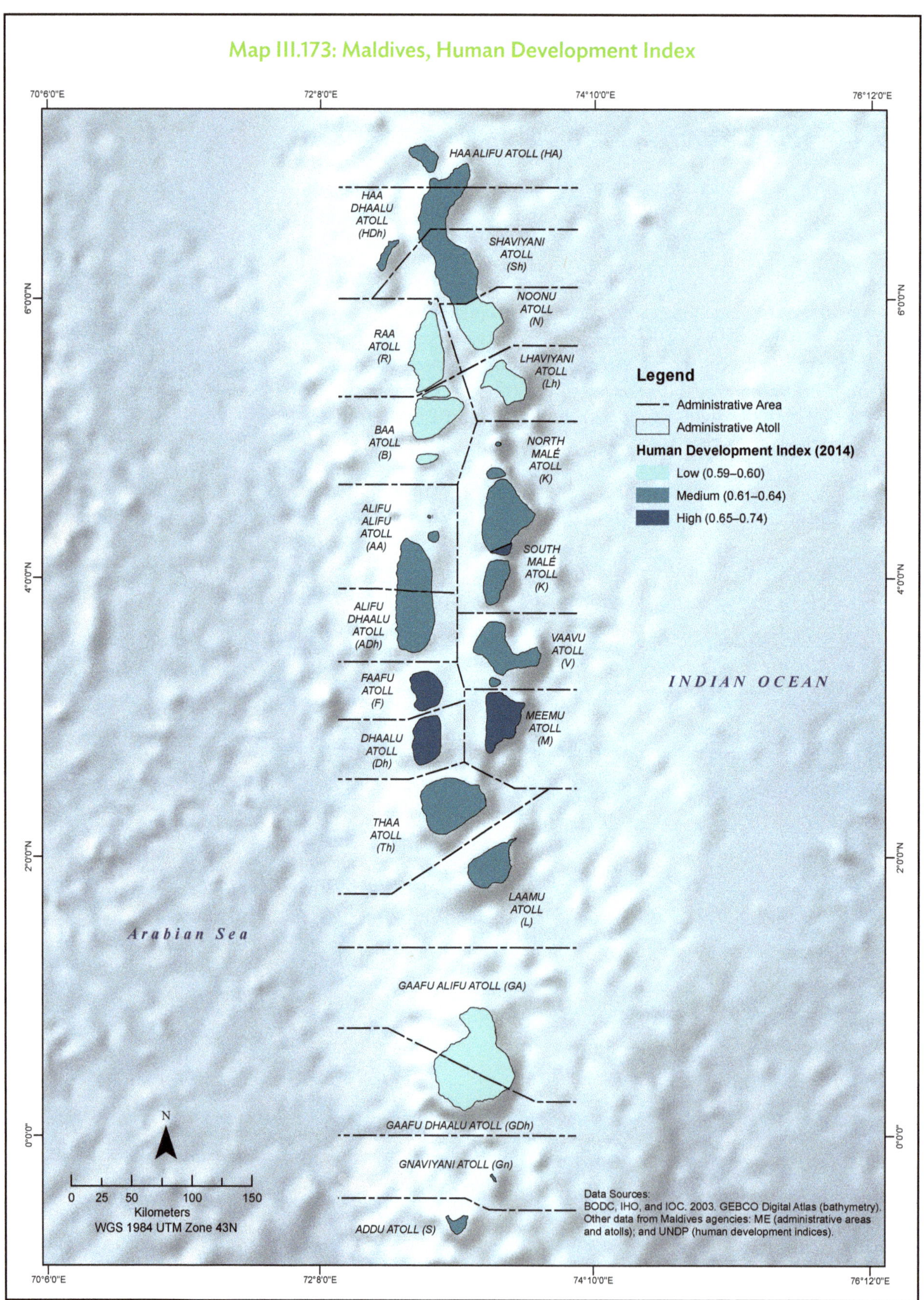

Map Data Sources

Government Ministries, Departments, and Agencies in Maldives
- Civil Aviation Authority
  - Airports
  - Floatplane platform
- Land and Survey Authority
  - Atoll capital islands
  - Cities
- Ministry of Economic Development
  - Ports
- Ministry of Education
  - Education centers
- Ministry of Environment
  - Administrative areas
  - Administrative atolls
  - Island shorelines
  - Power stations
  - Reef boundaries
  - Water bodies
- Ministry of Health
  - Healthcare facilities
- Ministry of National Planning and Infrastructure
  - Harbor facilities
- Ministry of Tourism
  - Resort islands
- National Bureau of Statistics
  - Population

International Institutions
- United Nations Development Programme
  - Human development index 2014

International Institution in Maldives
- International Union for Conservation of Nature, Maldives
  - Approved sand mining locations

References

Ahmed, M., and S. Suphachalasai. 2014. *Assessing the Costs of Climate Change and Adaptation in South Asia.* Manila: Asian Development Bank, UK Aid.

British Oceanographic Data Centre (BODC), International Hydrographic Organisation (IHO) and the Intergovernmental Oceanographic Commission (IOC) of the United Nations Educational, Scientific and Cultural Organization. 2003. *General Bathymetric Chart of the Oceans (GEBCO) Digital Atlas.* UK: British Oceanographic Data Centre.

Dhunya, A., Q. Huang, and A. Aslam. 2017. Coastal Habitats of Maldives: Status, Trends, Threats, and Potential Conservation Strategies. *International Journal of Scientific and Engineering Research.* 8 (3). pp. 47–49.

Emerton, L., S. Baig, and M. Saleem. 2009. *Valuing Biodiversity: The Economic Case for Biodiversity Conservation in the Maldives.* Homagama: Ecosystems and Livelihoods Group Asia; International Union for the Conservation of Nature for the Atoll Ecosystem Conservation Project; Ministry of Housing, Transport, and Environment, Government of Maldives.

Hosterman, H., and J. Smith. 2014. *Economic Costs and Benefits of Climate Change Impacts and Adaptation to the Maldives Tourism Industry.* Malé: Ministry of Tourism.

Intergovernmental Panel on Climate Change. 2001. *Climate Change 2001: The Scientific Basis. Contribution of Working Group I to the Third Assessment Report of the Intergovernmental Panel on Climate Change.* Cambridge, United Kingdom, and New York, USA: Cambridge University Press.

Ministry of Environment and Energy. 2016. *State of the Environment.* Malé: Ministry of Environment and Energy.

Naseer, A. 1997. *Paper 5: Status of Coral Mining in the Maldives: Impacts and Management Options. Food and Agriculture Organization of the United Nations.* http://www.fao.org/docrep/X5623E/x5623e0o.htm.

National Bureau of Statistics. 2014. *Maldives Population and Housing Census: Statistical Release 1 - Population and Households.* Malé.

United Nations Development Programme. 2018. *Briefing Note for Countries on the 2018 Statistical Update: Maldives.* http://hdr.undp.org/sites/all/themes/hdr_theme/country-notes/MDV.pdf.

United Nations Educational, Scientific and Cultural Organization's Institute for Statistics. 2014. *Maldives.* https://en.unesco.org/countries/maldives.

Maps were prepared by the Country Consultant Team and the Manila Observatory on behalf of the Asian Development Bank.

www.ingramcontent.com/pod-product-compliance
Lightning Source LLC
LaVergne TN
LVHW060617110826
845147LV00019B/1041
* 9 7 8 9 2 9 2 6 2 0 4 8 6 *